T0092018

Spatiotemporal Analysis of Air Pollution and Its Application in Public Health

Spatiotemporal Analysis of Air Pollution and Its Application in Public Health

Edited by

Lixin Li

Xiaolu Zhou

Weitian Tong

ELSEVIER

Elsevier
Radarweg 29, PO Box 211, 1000 AE Amsterdam, Netherlands
The Boulevard, Langford Lane, Kidlington, Oxford OX5 1GB, United Kingdom
50 Hampshire Street, 5th Floor, Cambridge, MA 02139, United States

Notices
Knowledge and best practice in this field are constantly changing. As new research and experience
broaden our understanding, changes in research methods, professional practices, or medical treatment
may become necessary.

Practitioners and researchers must always rely on their own experience and knowledge in evaluating
and using any information, methods, compounds, or experiments described herein. In using such
information or methods they should be mindful of their own safety and the safety of others, including
parties for whom they have a professional responsibility.

To the fullest extent of the law, neither the Publisher nor the authors, contributors, or editors, assume
any liability for any injury and/or damage to persons or property as a matter of products liability,
negligence or otherwise, or from any use or operation of any methods, products, instructions, or ideas
contained in the material herein.

Library of Congress Cataloging-in-Publication Data
A catalog record for this book is available from the Library of Congress

British Library Cataloguing-in-Publication Data
A catalogue record for this book is available from the British Library

ISBN: 978-0-12-815822-7

For information on all Elsevier publications
visit our website at https://www.elsevier.com/books-and-journals

Publisher: Candice Janco
Acquisition Editor: Laura S Kelleher
Editorial Project Manager: Vincent Gabrielle
Production Project Manager: Debasish Ghosh
Cover Designer: Miles Hitchen

Typeset by SPi Global, India

Contents

Contributors

Atin Adhikari
Department of Biostatistics, Epidemiology and Environmental Health Sciences, Jiann-Ping Hsu College of Public Health, Georgia Southern University, Statesboro, GA, United States

Anzhelika (Angela) Antipova
Department of Earth Sciences, University of Memphis, Memphis, TN, United States

Kerry Ard
School of Environment & Natural Resources, The Ohio State University, Columbus, OH, United States

Clair Bullock
School of Environment & Natural Resources, The Ohio State University, Columbus, OH, United States

Guofeng Cao
Department of Geosciences; Center for Geospatial Technology, Texas Tech University, Lubbock, TX, United States

Zheng Cao
School of Geographical Sciences, Guangzhou University, Guangzhou, China

Yanwei Chai
College of Urban and Environmental Sciences, Peking University, Beijing, China

Lingling Chen
School of Geographical Sciences, Guangzhou University, Guangzhou, China

Wenbo Guo
School of Geography and the Environment, University of Oxford, Oxford, United Kingdom; College of Urban and Environmental Sciences, Peking University, Beijing, China

Yongping Hao
Department of Housing and Urban Development, Washington, DC, United States

Marc Kalo
Department of Computer Science, Georgia Southern University, Statesboro, GA, United States

Jing Kersey
School of Mathematics and Natural Sciences, East Georgia State College; Department of Biostatistics, Epidemiology, and Environmental Health Sciences, Jiann-Ping Hsu College of Public Health, Georgia Southern University, Statesboro, GA, United States

Mei-Po Kwan
Department of Geography and Geographic Information Science, University of Illinois at Urbana-Champaign, Urbana, IL, United States

Hope Landrine
Center for Health Disparities Research, East Carolina University, Greenville, NC, United States

Lixin Li
Department of Computer Science, Georgia Southern University, Statesboro, GA, United States

Ying Liu
Department of Geosciences; Center for Geospatial Technology, Texas Tech University, Lubbock, TX, United States

Reinhard Piltner
Department of Mathematical Sciences, Georgia Southern University, Statesboro, GA, United States

Michele Ver Ploeg
USDA Economic Research Service, Washington, DC, United States

Weitian Tong
Department of Computer Science, Eastern Michigan University, Ypsilanti, MI, United States

Wei Tu
Department of Geology and Geography, Georgia Southern University, Statesboro, GA, United States

Zhifeng Wu
School of Geographical Sciences, Guangzhou University, Guangzhou, China

Jingjing Yin
Department of Biostatistics, Epidemiology, and Environmental Health Sciences, Jiann-Ping Hsu College of Public Health, Georgia Southern University, Statesboro, GA, United States

Hao Zhang
Department of Information Technology, College of Engineering and Computing, Georgia Southern University, Statesboro, GA, United States

Xingyou Zhang
USDA Economic Research Service, Washington, DC, United States

Naizhuo Zhao
Center for Geospatial Technology, Texas Tech University, Lubbock, TX, United States

Xiaolu Zhou
Department of Geography, Texas Christian University, Fort Worth, TX, United States

Introduction to spatiotemporal variations of ambient air pollutants and related public health impacts

1

Atin Adhikari

*Department of Biostatistics, Epidemiology and Environmental Health Sciences, Jiann-Ping Hsu
College of Public Health, Georgia Southern University, Statesboro, GA, United States*

There are myriad sources and processes of air pollutants in our atmosphere. These sources and processes are changing constantly. The understanding of these changes is important because the health effects of air pollutants are a product of the life cycle of specific pollutant, including the processes before emission, its emission, time in the air, followed by its time in other environmental media, and finally within humans and other living organisms through inhalation and other different routes of exposures. This section will cover recent findings on the seasonal and diurnal changes of common health-related air pollutants and changes in the media carrying these air pollutants, and finally the exposures to human beings and adverse public health outcomes. We will focus on six common air pollutants, which can harm human health and the environment, and may also cause property damage: (1) carbon monoxide (CO), (2) lead (Pb), (3) ground-level ozone (O_3), (4) particulate matter (PM), (5) nitrogen dioxide (NO_2, 6) sulfur dioxide (SO_2). The United States Environmental Protection Agency (EPA) has established US National Ambient Air Quality Standards (NAAQS) for these six air pollutants and these pollutants are referred as "criteria" air pollutants because EPA is regulating them by developing human health-based and environmental impact-based criteria or specific guidelines for setting permissible exposure levels. The US Clean Air act requires EPA to periodically review the NAAQS and revise or updated them if necessary for ensuring that the standards are providing the required amount of health and environmental protection.

1.1 Carbon monoxide

CO is a colorless and odorless gas, which is released when something is burned or oxidized, and this is a product of incomplete combustion when oxygen supply is insufficient. The density of CO is slightly lower than air. Atmospheric CO is a major sink of hydroxyl radicals in the troposphere. The radiative forcing potential of CO is approximately two times greater than CO_2 (Forster et al., 2007). The oxidation capacity of the atmosphere and the resulting concentrations of some atmospheric greenhouse gases are affected by the reactions of CO with hydroxyl radicals. Therefore, by these two different ways, atmospheric CO acts as both a direct and an indirect greenhouse gas (Thompson, 1992; Derwent, 1995). The most common outdoor sources of CO are cars, trucks, and other vehicles or machinery that burn fossil fuels. In indoors, a variety of household items, such as unvented kerosene and gas space heaters, leaking chimneys and furnaces, and gas stoves also release CO and can affect indoor air quality in homes (EPA, 2018a).

1.1.1 Spatiotemporal variations of CO

The natural concentration of CO in the clean atmosphere is about 0.2 ppm, and this level is generally not harmful for us. However, there are many changing sources behind the seasonal and diurnal variations of CO in the atmosphere, which can increase the atmospheric CO levels. In urban areas, the atmospheric CO levels are largely influenced by traffic-induced emissions. Based on the national trend analysis of EPA for CO levels (EPA, 2018b) in 51 cities, the average concentration of CO in the US was significantly higher than natural concentration level between 1980 and 2000, between 4 and 8 ppm. However, the average CO concentrations have decreased substantially over the years and the average concentration in 2017 was 1.3 ppm. Variations of CO levels were reported from other countries. For instance, measurements at fixed locations near heavily trafficked streets in seven UK cities found levels usually below 10 ppm, but peaks of up to 114 ppm were also recorded (Reed and Trott, 1971). A study conducted in rural areas of China near Beijing reported CO levels generally above 600 ppbv with a low level of 400 ppbv during the morning hours and a peak of up to 800 ppbv in the afternoon hours (Wang et al., 2008a). The authors of this study also found an increase in mean daytime mixing ratio of CO from 500 ppbv in June to 700 ppbv in July. Annual and diurnal variations of CO in 31 provincial capital cities in China based on air quality-monitoring data from China National Environmental Monitoring Center reported 1.0 to 2.5 mg/m^3 levels of CO in three different clusters of exposure levels (Zhao et al., 2016). A study conducted in Germany reported the seasonal cycle of CO, which showed a maximum concentration level in April and minimum in August (Thompson et al., 2009). These authors found a negligible trend in observed CO levels but found an interannual variability and attributed that to variations in local emissions. A study from the researchers of India reported CO levels in a Himalayan valley, which showed morning and evening peaks in all locations. They found that the CO levels dropped from their peak level of about 2000 ppbv in January to about 680 ppbv in June (Bhardwaj et al., 2018).

Several recent studies showed that sharp CO gradients exist near highways. Zhu et al. (2002a,b) measured wind speed and direction, traffic volume, and CO levels along transects downwind of Freeway 405 in Los Angeles, which is dominated by gasoline vehicles and also Freeway 710, which is dominated by high percentages of diesel vehicles and they found that relative concentrations of CO concentration decreased exponentially at 17–150 m downwind from the highways. Besides traffic, seasonal and interannual variations of CO are often influenced by wildfires. Some previous studies demonstrated such variations in tropical and southern hemisphere regions of the world (Bergamaschi et al., 2000; Langenfelds et al., 2002; Chen et al., 2010; Kopacz et al., 2010). The CO levels in coastal areas can be influenced by oceanic CO emissions. CO in seawater is mainly produced through photo-oxidation of color dissolved organic matter (Zuo and Jones, 1995; Zhang et al., 2008), and is lost through microbial consumption and emitted through sea-to-air releases (Bates et al., 1995; Xie et al., 2005). Large spatial and temporal variability is expected from oceanic emissions of CO because the processes driving the oceanic CO productions are related to the biological productivity in sea and ocean waters.

1.1.2 Public health impacts of CO

Breathing of air with very high concentrations of CO reduces the amount of oxygen that can be transported in our blood and important organs like the heart and brain. This type of high levels is possible in indoors or in other enclosed environments. CO can cause dizziness, confusion, unconsciousness, and death in these acute exposure levels. However, very high levels of CO are less likely to occur outdoors, as described earlier. But excess traffic emissions or wild fires, or oceanic emissions near coastal areas may elevate CO levels in outdoor environments. This increase in CO levels can be of particular concern for people suffering from different types of heart diseases. These patients have preexisted reduced ability for receiving oxygenated blood to their hearts in situations where heart has more oxygen requirement than usual. Therefore, these patients are vulnerable to the effects of CO when they are exercising or under increased stress. During these situations, short-term CO elevated levels may reduce the oxygen level in the heart of these patients accompanied by chest pain (Anderson et al., 1973; Volpino et al., 2004; Barn et al., 2018). This symptom is known as angina. Allred et al. (1989, 1991) reported decreased time-to-onset of angina and arrhythmia during exercises among coronary artery disease patients during 117 ppm CO exposure for one hour. The mechanism of toxicity from excess CO exposures is hypoxia induced by elevated carboxyhemoglobin (COHb) levels. COHb is the product of reaction between CO and hemoglobin. In healthy individuals, the levels of endogenous COHb are normally <1%–2% of total hemoglobin. When the COHb levels in bloods increase due to exposure to indoor or outdoor CO, then human body physiologically compensates that by increasing tissue oxygen levels through increased blood flow and blood vessel dilation. However, individuals with ischemic heart diseases have reduced oxygen delivery rate in their heart muscles and therefore CO exposure puts them at increased risk of hypoxia. According to the recent Agency for Toxic Substances and Disease Registry (ATSDR) report (ATSDR, 2012), enhanced myocardial ischemia and increased cardiac arrhythmias in coronary artery disease patients are possible at COHb level of 2.4%–6% in the blood at 14–40 ppm CO exposure levels. Several nonhypoxic mechanisms of action for CO toxicity are also proposed in the ATSDR (2012) report, which include binding of CO to heme proteins of blood other than hemoglobin and affecting several important physiological regulatory pathways, such as oxygen storage and utilization in brain and muscles, pathways for nitric oxide cell signaling and prostaglandin cell signaling, metabolic pathways for energy, and redox balance in cell.

According to the same ATSDR report (ATSDR, 2012), a number of neurological problems such as neurobehavioral or cognitive changes (including different visual and auditory sensory effects, such as decreased visual tracking, visual and auditory attentiveness, and visual perception), fine and sensorimotor acts of the nervous system, altered cognitive effects (learning performance, attention levels, driving performances), and changed brain electrical activities are possible at 5%–20% COHb in the blood and at 30–160 ppm CO exposure levels. Recent evidence suggests a possible link between CO exposure and neurocognitive impairment and behavioral disorders among children. CO can work as a neurotoxin because it can cross the placenta and reach the fetal circulation and the developing brain and thus functions as a potential public health threat (Greingor et al., 2001). The exact reasons and pathological pathways behind the CO-induced neurocognitive impairment and behavioral disorders, however, remain unclear.

Epidemiological studies on CO exposures and related health outcomes fall into two major categories. The relationships between long-term average ambient CO levels and health outcomes were examined in some studies, whereas in some other studies short-term exposures of <24 h were considered. End points of the epidemiological studies considering long-term CO exposures include mortality (Burnett et al., 2004; Dominici et al., 2003; Samoli et al., 2007), morbidly, and rates of medical assistance. Based on the estimated average CO levels of 1982–98, relative risks for mortality per 1 ppm increase in CO were approximately 0.97 (95% CI: 0.93, 1.0) for all causes of death, 0.95 (95% CI: 0.88, 0.99) for cardiopulmonary death, and 0.90 (95% CI: 0.83, 0.96) for lung cancer death (Pope III et al., 2002). Whereas examples of studies considering short-term CO exposures include rates of hospital admissions or emergency room visits and rates of medication use by asthmatic patients. Many epidemiological studies found that relatively low CO exposures (0.3–2 ppm) and consequent increased COHb levels in blood can be associated with exacerbated childhood asthma (Park et al., 2005; Rabinovitch et al., 2004; Rodriguez et al., 2007; Schildcrout et al., 2006; Silkoff et al., 2005; Slaughter et al., 2005; von Von Klot et al., 2002; Yu et al., 2000). However, these associations are confounded by coexposure to other air pollutants, such as NO_2, SO_2, O_3, and PM. Several previous studies on possible associations between inhalation exposures to CO and changes in pulmonary functions have shown mixed results (Chen and Fechter, 1999; Lagorio et al., 2006; Penttinen et al., 2001; Rabinovitch et al., 2004; Silkoff et al., 2005).

1.2 Lead

Lead is a heavy metal with low melting point and bluish-gray color, which is naturally occurring in the Earth's crust often combined with two or more other elements as lead compounds. The beneficial uses of lead are its use as an anticorrosion agent and development of alloys combining with other metals. These lead alloys or lead are present in pipes, storage batteries, cable covers, and radiation protective sheets. Lead compounds are often used in paints, dyes, ceramic glazes, and in caulking materials. Despite these beneficial uses, lead has significant harmful effects on human (see later). Therefore, the amount of lead use has been reduced in last two decades to minimize these harmful effects. Tetraethyl lead and tetramethyl lead were previously used to increase octane rating in the US as gasoline additives, but now lead has been banned for use in gasoline for motor vehicles since 1996.

Lead is present in all environmental media—the air, the soil, the water—in both outdoors and indoors. Past use of leaded gasoline and lead-based paint in homes and wide range of products, including batteries, ammunition, and cosmetics, pipes and plumbing materials, and previous lead-contaminated sites, such as former lead smelters, are acting as sources of lead in the environment. Current ambient sources of lead include aviation gasoline (for piston engine), smelters, foundries, and combustions of coals. Previously in the US—in 1979, cars released 94.6 million kilograms (208.1 million pounds) of lead in the air, which was reduced to 2.2 million kg (4.8 million pounds) in 1989, when the use of lead was limited but not banned, and since EPA banned the use of leaded gasoline in 1996, the amount of airborne lead levels decreased further (ATSDR, 2007). According to the EPA national trends of lead from 1980 to 2017 (EPA, 2018c), the airborne concentrations of lead dropped to $0.026\,\mu g/m^3$ in 2015 from the $1.855\,\mu g/m^3$ highest peak in 1988 and the concentration was $0.015\,\mu g/m^3$ in 2017. Based on the review of the air-quality criteria for lead, the EPA retained the existing 2008 standard, which is $0.15\,\mu g/m^3$ in three-month average concentration in total suspended particles.

1.2.1 Spatiotemporal variations of lead in air

Sources of lead emissions can vary from one place to another. In the US, major sources of lead in the air are ore and metals processing and piston-engine aircraft operating on leaded aviation fuel (EPA, 2018d). Other sources of lead in the US are waste incinerators, utilities, and lead-acid battery manufacturers and usually higher airborne lead concentrations found near lead smelters (EPA, 2018d). Lead emitted many years ago could contaminate soil through wet and dry deposition, and this soil lead can re-enter in the atmosphere as surface soil is disturbed by wind, vibrations, and other mechanical processes (Ehrman et al., 1992; Harris and Davidson, 2005). Usually lead levels in air are monitored with respect to the presence of lead in airborne PM. Sometimes lead containing fine particulate matter, such as $PM_{2.5}$ and PM_{10}, can travel from thousands of miles away from actual sources. In these situations, presence of lead isotopes is often used as chemical signatures to understand the origins of lead pollutants. For example, the coal and metal ores mined in eastern Asia have higher proportions of ^{208}Pb than the coal and ores used in USA. So presence of ^{208}Pb in PM of San Francisco can be monitored to understand the San Francisco area's airborne lead pollutants coming from eastern Asia. Many previous studies from all over the world studied the concentration and sources of Pb-containing PM

(Wang et al., 2006; Valavanidis et al., 2006; Okuda et al., 2008). Previous studies using single-particle mass spectrometry attributed 45% of the Pb-rich particles to coal combustions (Zhang et al., 2009). The level of lead in urban PM depends on the characteristics of the city (sources and intensity of lead emissions), its geographical location, and meteorological conditions (Dall'Osto et al., 2013). Hourly variations and spatial variations of elemental concentrations in $PM_{2.5}$ were investigated in Barcelona area by Dall'Osto et al. (2013), and they found higher concentrations of lead during night hours as industrial plumes were impacting Barcelona and these plumes were probably affected by wind direction and industrial cycles. Enhanced concentrations of lead in $PM_{2.5}$ were found to be influenced by Northerly Atlantic air masses. In another study, Li et al. (2010) examined the concentrations and origins of atmospheric lead and other trace species in northern China, and they employed the unmix receptor model, which resolved four factors in the aerosol composition data, including the sources of lead from biomass burning, industrial and coal combustion, dust, and a secondary source. The sources of lead from industrial and coal combustion and biomass burning were strongest in weak southerly winds before cold fronts, while the dust source was most active in strong northerly winds after cold fronts. The emissions from industrial processes and relatively small-scale coal burning activities were identified as the main source of ambient lead by Li et al. (2010).

Seasonal variations in blood lead levels in children were reported in several previous studies. Mean blood lead levels in the State of New York were found to be increased by 15%–30% in the late summer when compared with the mean values obtained during late winter and early spring (Haley and Talbot, 2004). Blood lead level monitoring of children of up to six years of ages in New York over a 48-month period showed yearly variations in blood lead concentrations with peak levels in the late summer times (Johnson and Bretsch, 2002). Although these seasonal variations in blood lead levels were primarily attributed to ingestion of lead in soil (Johnson and Bretsch, 2002), the seasonal variations of atmospheric lead could play an important role because atmospheric lead can be influenced by soil lead levels (Ehrman et al., 1992; Harris and Davidson, 2005).

1.2.2 Public health impacts of lead

The data from the US National Health and Nutrition Examination Surveys (NHANES) indicated that the mean blood levels of the US population dropped 78%, from 12.8 in 1976 to 2.8 µg/dL in 1991 (Brody et al., 1994). Data collected from NHANES III, phase II (1991–94), showed that 4.4% of children aged 1–5 years had blood lead levels ≥10 µg/dL, and the geometric mean blood lead level for children of up to 5 years ages was 2.7 µg/dL (Ballew et al., 1999). Although the blood lead levels are decreasing in recent years, the low exposure levels of lead may have significant public health impacts. Recent studies found that very low lead exposure levels (at blood lead level < 50 µg/L) can develop neurotoxic effects in childhood, and these effects can remain for many years in the later part of the life (Skerfving et al., 2015). A huge amount of information is available on the adverse health effects of lead on human health. Studies conducted in the past few decades were primarily focused on the health effects of low levels of lead exposures (blood lead level $< 20\,\mu g/dL$). The findings gathered from these studies came from studies of workers from a variety of industries and also from studies of children and adults in the general community. Most of the studies demonstrated that there are three sensitive targets for lead toxicity: (1) the developing nervous system, (2) the hematological and cardiovascular systems, and (3) the kidney.

Lead exposure in the human body causes damage to the peripheral and central nervous systems through several morphological effects, such as disruption of important molecules during neuronal differentiation and migration (Silbergeld, 1992), interfering with synapse development, through decreasing the reduction in neuronal sialic acid production (Bressler and Goldstein, 1991) and differentiation of glial cells prematurely (Cookman et al., 1987). Damages to nervous systems through pharmaceutical effects include substitution of calcium and zinc by exposed lead and inappropriate triggering on calmodulin (Goldstein, 1993). Several previous studies suggested that low-level lead exposure can significantly affect IQs, concentration ability, and attentiveness in exposed children (Koller et al., 2004). Young children are more vulnerable to lead poisoning because they can absorb more lead than adults from any source (WHO, 2018). Multiple deaths in young children were reported from Nigeria and Senegal due to exposure to lead-contaminated soil and dust among children (WHO, 2018). According to the WHO factsheet on lead and estimations of the Institute for Health Metrics and Evaluation (IHME), lead exposure in 2016 accounted for 540,000 deaths and 13.9 million years of healthy life lost (disability-adjusted life years or DALYs worldwide due to long-term effects of lead exposure on health) (WHO, 2018; IHME, 2018). The highest burden of this lead-related global health problem was observed in low-income and middle-income countries. Lead exposure worldwide was also accounted for global burden of 63.8% idiopathic developmental intellectual disability, 3% ischemic heart diseases, and 3.1% of strokes in the same IHME estimates of 2016 (WHO, 2018; IHME, 2018). Lead exposure can be associated with kidney disorders. Belgian Cadmibel Study (Staessen et al., 1992) examined the relationships between blood lead levels and renal creatinine and reported that covariate-adjusted creatinine clearance was significantly associated with blood lead levels in males. Lead exposure is also associated with blood pressure change. A meta-analysis of the papers published between 1980 and 2001 found a two-fold increase in blood lead concentration was significantly associated with the rise in systolic and diastolic pressures in both men and women (Nawrot et al., 2002).

1.3 Ozone

Ozone (O_3) is a gas molecule, which is composed of three oxygen atoms. This gas is highly reactive with strong oxidizing power. Tropospheric ozone can greatly influence public health and air quality. In addition, it is an important greenhouse gas. Ozone also plays an important role in photochemical processing of other atmospheric chemicals, and affect food security and viability of ecosystems. Ozone is not directly emitted in the air from any pollutant sources. It is primarily produced in the atmosphere through the catalytic oxidation of carbon compounds in the presence of oxides of nitrogen ($NO_x = NO + NO_2$). It is also generated through photochemical reactions between nitrogen oxides and volatile organic compounds (VOCs) (Trainer et al., 2000; Sillman, 1999). Because of these indirect generation pathways, ground-level ozone is considered as one of the most important secondary air pollutants in the atmosphere. A small amount of ozone can reach troposphere from stratospheric influx of ozone as well (Junge, 1962). Levels of ozone decrease through losses due to surface deposition and some ozone destroyed in the atmosphere by photochemical processes, primarily by photolysis and the subsequent chemical reactions between oxygen atoms and airborne water vapors. Production and destruction of ozone in certain atmospheric air mass depends largely on the short-wave radiations and the water vapor and concentrations of nitrogen oxides in air. Ozone can be classified as "good" or "bad" for human health and the environment depending on its presence in the atmospheric layers. Stratospheric ozone is considered as "good" because it protects living creatures on the Earth surface from hazardous ultraviolet radiations coming from the sun. Ground-level ozone is considered as "bad" because it can trigger several public health problems, particularly for children, the elderly, and other people who have respiratory diseases such as asthma (EPA, 2018e).

1.3.1 Spatiotemporal variations of ozone

Ozone concentrations in the atmosphere are influenced by solar radiation intensity and temperature gradients. Therefore, airborne ozone levels vary during different times of the day and in different seasons. Generally ozone concentrations are higher in summer when high pressure, low humidity, high temperature, and less ventilation cause stagnation and accumulation of air. High solar radiations at the same time trigger photochemical reactions forming ozone in the air. Tropospheric ozone near the Earth surface undergoes significant diurnal variation in summer months. The interactions and changes in the major determinants of surface ozone include precursor emissions, dry deposition, solar radiation, titration by NO_x (Sillman, 1999), vertical mixing rates in the planetary boundary layer, and mixing rates with the free troposphere (Zhang et al., 2006). The vertical mixing rates play an important role in redistributing ozone concentrations in the planetary boundary layer. During the diurnal variations of ozone in daytime, ozone is produced near the Earth surface with the maximum levels during afternoons through precursor reactions, which are influenced by solar radiations. Then this ozone is transported upward and mixed into the upper planetary boundary layer, which is often unstable in daytime. The planetary boundary layer is stable during the nighttime and therefore vertical mixing rate is low then. However, the surface ozone decreases to the minimum level during sunrise time due to different destruction processes, such as dry deposition and NO_x titration. The ozone levels in the afternoon hours were found to be several times greater than the minimum levels found in the early morning hours. Coyle et al. (2002) explained these diurnal variations as follows: the afternoon peak in ozone concentration develops when the atmosphere is most turbulent and UV levels are very high. Ozone levels drop during the night and early morning hours because the lower regions of the boundary layer become stable and thermally stratified as the ground surface cools, which reduces moving of ozone from the free troposphere. The ozone concentration decreases further as losses to dry deposition are not refilled by mixing from upper troposphere and photochemical reactions producing ozone do not occur. Therefore, the minimum concentrations of ozone are observed between midnight and dawn. The diurnal cycle of ozone is most pronounced during slow-moving warm anticyclonic weather when emissions of ozone precursors in great amounts can lead to widespread high ozone episodes. According to Ying et al. (2009), the increased levels of daytime O_3 concentrations can be attributed to increasing VOC emissions during daytime and decreasing NO_x emissions decreasing in the morning. This diurnal variation of ozone is important for diurnal and seasonal trends of ozone-related health hazards. Moreover, the information is important for evaluating the modeling uncertainties for climate change effects on different air pollutants and estimating long-range air pollution transport effects on air quality in the community, and estimating the effectiveness of pollution emission reductions programs. Ozone levels can differ in different countries of the world. For example, the ozone level in the US is continuously decreasing in urban sites in recent years (Pollack et al., 2013); however, long-term ozone level in China is increasing (Meng et al., 2009; Wang et al., 2008a,b,c) possibly due to increase of O_3 from stratosphere to troposphere (Xu and Lin, 2011; Lin et al., 2008), decadal circulation shifts (Ding et al., 2013), and large amount of VOCs and NO^x emissions in the atmosphere of China (Ma et al., 2016). Previous monitoring of ozone at European sites during the late 1900s and at the end of twentieth century indicate that average surface ozone levels have increased almost doubled in these European countries during this period (Volz and Kley, 1988; Anfossi and Sandroni, 1997). There are many reports available on the long-term trends and distributions of ozone in the troposphere (Carslaw, 2005; Fiore et al., 1998, 2002, 2005; Fusco and Logan, 2003; Gardner and Dorling, 2000; Jonson et al., 2006; Lelieveld and Dentener, 2000; Lu and Chang, 2005; Naja and Akimoto, 2004; Oltmans et al., 2006; Tarasova et al., 2003; Vingarzan and Taylor, 2003).

1.3.2 Public health impacts of ozone

Higher atmospheric concentration levels of ozone are serious threats for human health, ecosystems, and global climate (Fiore et al., 2009). Ozone can strongly react with the biomolecules in the human body to form ozonides and free radicals. The free radicals from this reaction can trigger inflammatory responses, which consequently increase systemic oxidative stress in the respiratory tract and other organs and may have adverse pulmonary and cardiovascular effects. Many epidemiological studies have reported that ground-level tropospheric ozone pollution can induce different types of adverse health effects, including mortality, respiratory disorders, and cardiovascular disorders (Bates, 2005; Bell et al., 2004; Ito et al., 2005; Jerrett et al.,

2009; Tao et al., 2011). Reports from WHO and previous meta-analysis by several researchers demonstrated short-term ozone exposure-related mortality and reported concentration-response factor for ozone and other air pollutants (WHO, 2001, 2004; Cairncross et al., 2007; Sicard et al., 2012; Bell et al., 2004, 2005; Ito et al., 2005). Jerrett et al. (2009) clearly found long-term health impact of ozone based on 1-h concentrations and their results were widely used later in several follow-up studies. Anenberg et al. (2010) estimated global premature death burden from anthropogenic ozone pollution using health impact functions based on long-term relative risk estimates for O_3 from the epidemiology literature and concluded that anthropogenic O_3 contributes significantly to global premature mortality. Lelieveld et al. (2013) estimated relative risks to calculate global, regional, and megacity premature mortality from ozone pollutions, and they found that age is most sensitive issue for ozone pollution-related health outcomes. They estimated that per 10 ppb increase in daily 8-h average ozone concentration, mortality risk for older persons was 1.27% (95% CI: 0.76, 1.78) in contrast to the mortality risk of younger persons, which was 0.60% (95% CI: 0.40, 0.80) and significantly lower in statistical analyses.

Ozone exposure can affect the patients with preexisting respiratory diseases by two ways: (1) the response to ozone pollution may interact with the pathological mechanisms of the underlying respiratory diseases or (2) these patients cannot tolerate the reduction in pulmonary functions or the increase in various respiratory symptoms (EPA, 2018f). Ozone can aggressively attack lung tissues through some chemical reactions as described in the paragraph earlier. When lung tissue is exposed to ozone, at the same time it is also exposed to other harmful pollutants created by the same photochemical processes that make ozone in the atmosphere. These combined exposures can aggravate ozone-induced respiratory symptoms, including asthma. People with asthma are especially susceptible to the hazardous effects of ozone exposures. Because the prevalence of asthma in children is particularly high in the US and because children are generally at risk of longer exposures due to their time spent in exercises and in other outdoor activities, they might be affected by ozone exposures. Sousa et al. (2013) reviewed recent epidemiological studies on ozone and childhood asthma and found evidence showing that short-term exposures to ozone can affect childhood asthma and ozone exposure can influence asthma development in children. According to EPA (2018f), people with asthma can experience following health hazards from excess ozone exposures: (1) increased respiratory symptoms, (2) lung function decrements, (3) increased asthma medication usage, (4) increased uses of healthcare services, and (5) increased asthma attack frequency. Ozone can constrict the airway muscles leading to wheezing and shortness of breath. This is interesting to note that studies that have assessed both CO and ozone found that ozone was either negatively or not associated with asthma indicators, while increases in average outdoor carbon monoxide levels of just 1 to 2 ppm were more positively associated than any other pollutant or particulates (Yu et al., 2000). Dai et al. (2018) conducted a time-series study in Hong Kong and examined the short-term effect of ambient ozone on emergency hospital admissions for subjects of different age groups. They concluded that the risk of ozone-related asthma hospitalization in elderly is greater than in children. Zu et al. (2017) recently conducted a study in Texas, USA, and found that Ambient ozone levels are positively associated with total asthma hospitalizations in Texas. The associations between ozone and asthma hospitalization are strongest in school-aged children. Anenberg et al. (2018) recently estimated that 9–23 million annual asthma emergency room visits globally in 2015 could be attributable to ozone exposure.

Besides the respiratory system, inhaling ozone may affect the heart as well. Previous epidemiological studies have proved that there is an association between O_3 and the increased incidence of morbidity and mortality from cardiovascular diseases (Brook et al., 2004). It is possible that pulmonary oxidant stress from ozone exposure can affect cardiovasculature and can initiate or aggravate cardiovascular disease developments. A previous study linked short-term exposures to high ozone levels to a particular type of cardiac arrhythmia, which can increase the risk of premature death and stroke (Sarnat et al., 2006). The authors found that increased levels of ambient ozone may increase the risk of supraventricular arrhythmia among elderly people. Another study of Jerrett et al. (2009) showed that effects of short-term variations in ozone on cardiovascular mortality were not conclusive. But a recent study of Raza et al. (2013) showed that short-term variations in ozone were significantly associated with out-of-hospital cardiac arrest.

1.4 Particulate matter (PM)

Any solid or liquid substances in the air that are not in the gas phase can be considered as particulate matter (PM). Generally airborne PM are classified according to the irregularly shaped and irregularly sized airborne particle's aerodynamic diameter, which is the diameter of the spherical particle with a density of $1000 \, kg/m^3$ (density of water) and the same settling velocity as the airborne irregularly sized and shaped particle. Total suspended particulate (TSP) matter includes all atmospheric aerosols, but generally we are concerned about smaller particles because they remain suspended for long time in the air, they are inhalable deeper into our lower respiratory tract, and they have the most adverse health impacts. Smaller particles are usually classified as the coarse fraction or PM_{10}, which includes particles of 2.5–10 μm aerodynamic diameters and the fine fraction or $PM_{2.5}$, which includes particles having ≤2.5 μm aerodynamic diameters. These particles can be made up of many different chemicals and biological components of airborne microorganisms and they may have different masses due to varieties of their direct emission sources, such as construction sites, unpaved roads, fields, smokestacks of factories, forest fires, and biomass and solid waste burning. Secondary particles can form in the atmosphere from complex reactions of chemicals, such as sulfur dioxide and nitrogen oxides, which can be emitted from automobiles, power plants, and industries. Therefore, whenever we study PM in the atmosphere, we usually focus on three primary characteristics of particulate pollutants, which are (1) total mass concentration, (2) size distribution of particles, and (3) chemical and biological (e.g., allergens, endotoxins) composition of particles. US NAAQS for PM pollution specify a maximum amount of outdoor airborne PM mass to protect human health, the environment, and public welfare. Recently, in 2012, the US EPA has revised the primary annual $PM_{2.5}$ standard based on particle mass from $15 \mu g/m^3$ to $12 \mu g/m^3$ and retained the 24-h $PM_{2.5}$ standard of $35 \mu g/m^3$ and 24-h PM_{10} standard of $150 \mu g/m^3$ (EPA, 2018g). The size distribution of different airborne particles is vital in understanding the respiratory deposition of particles in the human respiratory tract and their transport and removal in the atmosphere. The chemical and biological composition of particles determines the types of hazardous effects caused by PM on humans, vegetation, and different materials in the environment. Recently submicron particles of about $0.1 \mu m$ or $100 nm$ have been further classified as ultrafine particles. Although their mass concentrations in air is negligible compared to PM_{10} and $PM_{2.5}$, we are now focusing more on their health impacts because they have larger surface area, increased respiratory (particularly, alveolar) deposition, increased diffusion ability across the cell membranes and organelles in the human body, and increased transport to the human brain through the olfactory nerves. The composition of PM is important with respect to toxic effects of PM. The source apportionment analysis of various trace metal constituents and mass $PM_{2.5}$ data analyzed by Thurston et al. (2011) indicated that major U.S. $PM_{2.5}$ source categories and related elemental trace metals were motor vehicle traffic (EC); steel industry (Fe, Mn); metals industry (Pb, Zn); particles from soil (Ca, Si); coal combustions (As, Se); oil combustion (V, Ni); salt particles (Na, Cl); and biomass burning (K, OC). The $PM_{2.5}$ mass and speciation data have been previously characterized on a nationwide basis by Bell et al. (2007). They found strong seasonal and geographic variations in the compositions of $PM_{2.5}$ and seven out of the 52 components contributed $\geq 1\%$ to total PM2.5 masses, which were ammonium (NH_4^+), elemental carbon, organic carbon, nitrate (NO_3^-), silicon, sodium (Na^+), and sulfate (SO_4^{2-}).

Airborne particles can be directly emitted into the atmosphere (primary aerosols) or new aerosols can develop from interactions between primary aerosols and other gas-phase constituents in the atmosphere (secondary aerosol). The secondary aerosols are more important source of particles developing atmospheric haze. When gaseous molecules in the air become transformed to liquid or solid particles, there are three processes involved: absorption, nucleation, and condensation. When gas converts into solution in a liquid phase, then this is called absorption (e.g., absorption of atmospheric SO_2 in liquid H_2O droplets). When growth of clusters of airborne molecules forms a thermodynamically stable nucleus, then this process is called nucleation. This process is dependent on the vapor pressure of the condensable gas molecules and growth of molecular clusters occurs in supersaturated conditions of air. Condensation is the consequence of collisions between a gaseous molecule and an existing aerosol droplet in supersaturated atmospheric condition. Coagulation is the next process by which disconnected airborne particles come close to each other, contact, and join together by different surface forces and this process increase aerosol diameters. Several recent studies examined biological components of $PM_{2.5}$. For examples, bacterial endotoxin levels in $PM_{2.5}$ were examined by Heinrich et al. (2003) and Arteaga et al. (2015).

1.4.1 **Spatiotemporal variations of PM**

Studies on the spatiotemporal distribution of PM in the surface air are important because this distribution affects the level of PM exposure of a population and provides important information regarding the formation of PM in the atmosphere. Furthermore, simultaneous information about PM and other gaseous pollutants in the air can provide important formation about the generation of secondary atmospheric PM (Blanchard, 2003; Chan and Yao, 2008). Long-term trends in spatiotemporal variations of PM can inform whether appropriate regulatory steps are being implemented to mitigate health and environmental outcomes, and can motivate additional intervention and administrative actions.

Using a nationwide network of 429 monitoring sites for $PM_{2.5}$ and 312 monitoring sites for PM_{10}, the US EPA has developed ambient air quality trends for PM pollution (EPA, 2018h). These trends showing that average $PM_{2.5}$ concentrations have decreased over the years. For example, average $PM_{2.5}$ dropped from $13.02\,\mu g/m^3$ in 2000 to $8.50\,\mu g/m^3$ in 2017. Similarly, average PM_{10} concentrations have decreased over the years in the US. Average PM_{10} dropped from $65.55\,\mu g/m^3$ in 2000 to $56.31\,\mu g/m^3$ in 2017. In European countries, the evaluation of the status and trends of air quality is based on ambient air measurements reported by the countries to the European Environment Agency (EEA). Data reported for the period from 2002 to 2011, which are available in Airbase v. 7 (EEA, 2018), emissions of primary PM_{10} and $PM_{2.5}$ decreased by 14% and 16%, respectively, in the EU-27 member countries. The reductions in the same period for the EEA-32 member countries were 9% for PM_{10} and 16% for $PM_{2.5}$ (Guerreiro et al., 2014). Seasonal changes of PM in 100 US cities during 1987–2000 were analyzed by Peng et al. (2005) in a time-series study on mortality. The authors found that the daily mean of PM_{10} ranged from $13\,\mu g/m^3$ in Coventry, Rhode Island, to $49\,\mu g/m^3$ in Fresno, California.

Situations in developing countries such as China and India are somewhat different. For example, recent satellite-based spatiotemporal trends in $PM_{2.5}$ concentrations showed a mean annual increase of $1.97\,\mu g/m^3$ between 2004 and 2007 and then a decrease of $0.46\,\mu g/m^3$ between 2008 and 2013 (Ma et al., 2016). PM concentrations in India were found to be greater than WHO standards by Dey et al. (2012). These authors used Multiangle Imaging Spectro Radiometer (MISR)-retrieved columnar aerosol optical depths data to estimate surface $PM_{2.5}$ in India and found that 51% area of the country had $PM_{2.5}$ concentrations levels above $35\,\mu g/m^3$.

Zhang and Cao (2015) recently examined $PM_{2.5}$ data collected from 190 priority pollution monitoring cities in China and found that $PM_{2.5}$ concentrations were generally higher in northern cities than southern ones due to relative large PM emissions from various sources and unfavorable meteorological conditions for dispersion of $PM_{2.5}$.

Previous studies in the US showed clear seasonal trends in PM_{10} (Peng et al., 2005). The Southern California, Northwest, and Southwest regions had higher mean PM10 concentrations in the fall, while other US regions had their highest levels in the summer. Zhang and Cao (2015) recently found a clear seasonal variability of $PM_{2.5}$ in China, which was highest and lowest during the winter and summer, respectively.

Although the peaks were observed in winter and summer, high concentrations of $PM_{2.5}$ were also recorded during spring in Northwest and West Central China possibly due to excess dispersion of dust and also during autumn in East China possibly due to open biomass burning. When they examined diurnal variations of $PM_{2.5}$ in these cities, they found that the lowest and highest airborne $PM_{2.5}$ concentrations were observed during afternoon and evening, respectively. They attributed these diurnal variations to daily variation of the tropospheric boundary layer depth and emissions of particles from various anthropogenic sources. Xie et al. (2015) recently examined spatiotemporal variations of $PM_{2.5}$ and PM_{10} concentrations in 31 Chinese cities based on data from 286 monitoring sites obtained between 2013 and 2014. They found that highest monthly mean concentration of $PM_{2.5}$ was $>150\,\mu g/m^3$ in 10 Chinese cities and the worst concentrations were observed in two cities (Shijiazhuang and Xi'an; monthly mean concentrations $>200\,\mu g/m^3$). For PM_{10}, four cities showed highest monthly mean concentrations of $>250\,\mu g/m^3$ and again the worst in same two cities (Shijiazhuang and Xi'an; monthly mean concentrations $>300\,\mu g/m^3$). They also found highest concentrations of PM occurred in December, January, and February, and attributed these seasonal peaks to combinations of heating and industrial pollution in Chinese cities. In contrast, one study from India (Srimuruganandam and Nagendra, 2010) showed clear diurnal, weekly, and seasonal cycles of PM but highest concentrations were observed during post monsoon seasons (PM_{10}: $189\,\mu g/m^3$; $PM_{2.5} = 84\,\mu g/m^3$) compared to winter (PM_{10}: $135\,\mu g/m^3$; $PM_{2.5}$: $73\,\mu g/m^3$) and summer (PM_{10}: $102\,\mu g/m^3$; $PM_{2.5}$: $50\,\mu g/m^3$) seasons.

1.4.2 Public health impacts of PM

According to the US EPA (2018i), PM exposure can primarily affect both lungs and heart and PM exposures have been linked to a variety of health problems, such as aggravated asthma, decrease in lung functions, increased respiratory symptoms nonfatal heart attacks, irregular heartbeat (e.g., irritation of the airways, coughing, or difficulty breathing), and premature death in people with heart or lung diseases. $PM_{2.5}$ is now considered as one of the leading risk factors for premature mortality in the world. The recent Global Burden of Disease (GBD) assessment attributed 3.2 million premature deaths per year to ambient $PM_{2.5}$ exposures (Lim et al., 2012). The link between personal exposure to air pollution and severe cardiovascular and respiratory disorders has been investigated in numerous epidemiological studies (e.g., Langrish et al., 2012; Leitte et al., 2010; Zhou et al., 2010). Another study recently demonstrated that exposure to $PM_{2.5}$ and PM_{10} can aggravate chronic respiratory and cardiovascular diseases by altering host defenses and damaging the lung tissues (Chen and Zhao, 2011). Several previous studies found that chronic exposure to $PM_{2.5}$ was found to be associated with human morbidity and mortality (Beelen et al., 2007; Cao et al., 2011; Dockery et al., 1993; Eftim et al., 2008; Filleul et al., 2005; Gehring et al., 2006; Katanoda et al. 2011; Ostro et al., 2009; Pope III et al., 2002; Puett et al., 2009; Yorifuji et al., 2010). Some of these studies specifically considered lung-cancer mortality associations with $PM_{2.5}$ (Beelen et al., 2007; Dockery et al., 1993; McDonnell et al., 2000), and statistically significant ($P < .05$) relationship was observed in in the American Cancer Society study (Pope III et al., 2002; Turner et al., 2011). A recent multilevel meta-analysis of 22 research studies on the relationship between PM exposure and asthma (2000–16) showed that PM$_{2.5}$ exposure is significantly associated with asthma exacerbations but not PM$_{10}$ (Orellano et al., 2017). Another recent study conducted meta-analysis of case-control and cohort studies (1980–2015) that evaluated the association between pregnant mothers' exposure to PM and birth weight and preterm delivery-related problems and found that birth weight was negatively associated with 10 µg/m^3 increase in PM$_{10}$ and PM$_{2.5}$ (Lamichhane et al., 2015). The authors also found a significantly increased risk of preterm birth per 10 µg/m^3 increase in PM$_{10}$ and PM$_{2.5}$ exposure during the pregnancy periods. A US epidemiological study found a strong and consistent relationship between adult diabetes and PM exposure after adjusting for obesity and ethnicity (Pearson et al., 2010). Another recent study from Switzerland found that the odds of diabetes increased by 8% (95% CI: 2, 14%) per type 2 diabetes risk allele and by 35% (−8, 97%) per 10 µg/m^3 exposure to PM$_{10}$ (Eze et al., 2016). The authors attributed this finding of increased genetic risk for type 2 diabetes to alterations in insulin sensitivity. Another recent meta-analysis of 35 studies reported that increases in both PM2.5 and PM10 concentrations were associated with heart failure hospitalizations and related death. The authors also calculated a mean reduction in PM2.5 at of 3·9 µg/m^3 in the US may prevent 7978 heart failure hospitalizations (Shah et al., 2013). A systematic review and meta-analysis on the associations between PM$_{2.5}$ exposure and neurological disorders recently reported significant associations between PM$_{2.5}$ exposure and stroke, dementia, Alzheimer's disease, autism spectrum disorder, and Parkinson's disease (Fu et al., 2019). The authors found that both short- and long-term PM$_{2.5}$ exposure was associated with increased risks of stroke and mortality, and long-term PM$_{2.5}$ exposure was associated with increased risks of dementia, Alzheimer's disease, autism, and Parkinson's disease.

1.5 Nitrogen dioxide (NO$_2$)

Nitrogen dioxide (NO$_2$) is a gas with a strong harsh smell and this is one of the highly reactive oxides of nitrogen or nitrogen oxides (NO$_x$) (EPA, 2018j). Although all NO$_x$ have adverse impacts on human health and environment, NO$_2$ has the greatest concern among them. NO$_2$ primarily comes to the air from the burning of fuels, such as emissions from cars, trucks, and buses, power plants, and various off-road equipment using fuels. However, <1% of NO$_2$ in the air is also formed naturally during lightning or produced by plants, soil, and water. NO$_2$ is one of the most important ambient air pollutants because photochemical smog in the atmosphere forms from NO$_2$ and the smog has significant negative impacts on human health and environment. During NO$_2$ formation, first different anthropogenic activities increase the NO$_x$ levels in the urban atmosphere and oxides are diluted and chemically processed in the air for minutes to hours. In the presence of daylight, NO$_x$ is then converted into NO and

NO_2, and in the presence of OH radicals and VOCs, ozone forms from the chemical reactions of NO_2. NO_2 also interacts with atmospheric aerosols. The reactions of NO_2 with SO_2 on the surfaces of atmospheric aerosol particles can form sulfates in the atmosphere (Littlejohn et al., 1993). As stated earlier, ambient levels of NOx, NO, and NO_2 play an important role in the formation of ozone and also its destructions. NOx can contribute to the losses of ozone in stratosphere and thus influence global warming and climate changes. NO_x anthropogenic sources in urban areas and biogenic sources in rural areas are highly variable and atmospheric mixing ratios of NO_x gases vary in the atmosphere. In urban areas, ozone production is related to NO_2 but in contrast, in ozone production is limited by the available NOx, and therefore a linear relationship between atmospheric NO_2 and O_3 is not observed (Liu et al., 1987). NOx removal from the atmosphere occurs primarily through oxidation of NO_x to HNO_3 with ensuing loss through wet and dry depositions in the ambient environment. US EPA recently strengthened the health-based criteria for NO_2 for the NAAQS. According to the January 22, 2010 revisions, the current 1-h NO_2 standard was set at the level of 100 ppb and the annual average NO_2 standard was retained at the same level of 53 ppb.

1.5.1 Spatiotemporal variations of NO₂

Previous studies observed that the concentrations and speciations of various NO_x are strongly influenced by meteorological conditions in the atmosphere. Primary and secondary pollutants levels from NOx speciation are influenced by wind directions and seasons (Fahey et al., 1986; Munger et al., 1998; Hayden et al., 2003). During summer months, there is a strong correlation observed between ozone and NO_x (Trainer et al., 1993; Thornberry et al., 2001; Zaveri et al., 2003). In European countries, measurements of air pollutants from a high altitude site in the Swiss Alps found that NO_x concentrations are influenced by the contributions from polluted air coming from continental Europe and cleaner air coming from the free troposphere (Zellweger et al., 2003).

Atmospheric NO_x can be transported from continent to continent (Liang et al., 1998; Li et al., 2004) and affect global air quality. Thus, spatiotemporal variations of NO_2 in the atmosphere are influenced by chemical reactions related to NO_x, mixing of NO_x speciation products including NO_2, and losses of NO_x over different time periods and seasons.

1.5.2 Public health impacts of NO₂

Short-term exposures to high concentrations of NO_2 may cause following adverse health outcomes (EPA, 2018j): (1) irritation of airways in the respiratory system; (2) aggravation of respiratory diseases, particularly asthma, and increase of respiratory symptoms (such as coughing, wheezing, or difficulty breathing); (3) increased hospital admissions due to these disorders and visits to emergency rooms. Long-term exposures to high concentrations of NO_2 may contribute to asthma development and may also increase susceptibility to respiratory viral infections (EPA, 2018j). Therefore, asthmatic children and elderly subjects are usually at greater risk for the adverse health effects of atmospheric NO$_2$. As discussed earlier, NO$_2$ can also contribute to formation of particulate matter and ozone. Therefore, synergistic health effects of all these pollutants are anticipated.

Previous studies on the long-term effects of NO$_2$ exposure on mortality were limited and not conclusive. However, Faustini et al. (2014) recently reviewed 23 papers on the relationship between NO$_2$ and mortality, which were published between 2004 and 2013, and conducted meta-analyses of these published data. They concluded that long-term effect of NO$_2$ on mortality is evident and the effect is as great as that of PM$_{2.5}$. For increase of every 10 µg/m^3 in the annual NO$_2$ concentration, these authors found pooled effects of 1.04 (95% CI: 1.02–1.06) on overall mortality and 1.13 (95% CI: 1.09–1.18) on cardiovascular mortality. Another older systematic review of the health effects of NO$_2$ reported moderate evidences showing that short-term NO$_2$ exposure (24 h) of lower levels of <50 µg/m^3 increased both mortality and hospital admissions (Latza et al., 2009).

Some indoor studies and lab studies showed effects of NO$_2$ exposures on asthma and respiratory health issues (Belanger et al., 2006; Bauer et al., 1986), but epidemiological evidence for the association between ambient NO$_2$ exposure and respiratory symptoms did not result consistent findings. Methodological problems, confounding effects from other pollutants, and lack of prospective data were the reasons behind these inconsistences (Samet and Utell, 1990). A recent study of Greenberg et al. (2016) examined the effects of long-term NO$_2$ and/or SO$_2$ exposures on prevalence and severity of asthma in young adults and reported that high levels of SO$_2$ (13.3–592.7 µg/m^3) and high levels of NO$_2$ (27.2–43.2 µg/m^3) combination can increase the risk of asthma occurrence compared to the combination of high levels of NO$_2$ (27.2–43.2 µg/m^3) and moderate levels (6.7–13.3 µg/m^3) of SO$_2$ levels. Snowden et al. (2015) calculated the associations between ambient summertime NO$_2$ levels and lung function parameters in a cohort of asthmatic children in Fresno, California, and reported improved pulmonary functions among children during decrease in NO$_2$ levels. Truck drivers and others who spend longer time on roads can experience short-term elevated exposures to NO$_2$. People living near highways may have higher exposures to NO$_2$ as well. This population might be from the economically disadvantaged population groups. Thus, near-road environmental exposures to NO$_2$ may create some environmental justice issues, as well as threats to vulnerable populations like children and elderly people there with preexisting respiratory symptoms.

1.6 Sulfur dioxide (SO$_2$)

Sulfur dioxide is one of the common gases in the atmosphere. SO$_2$ is colorless, nonflammable, nonexplosive, toxic at high concentrations, and it has a nasty and sharp smell like a burnt match. It reacts easily with other substances in the atmosphere and forms acids in the atmosphere, especially sulfuric acid (H$_2$SO$_4$), which is the main component of acid rain and may damage the ecosystems, including crops and water bodies. SO_2 is one member of the group of gases called sulfur oxides or SO_x, and these gases generate from the burning of either sulfur or materials containing sulfur (EPA, 2018k). SO_2 is the most common in the atmosphere among different SO_x gases. Most of the SO_2 in air comes from various anthropogenic sources. Primary SO_2 emission sources are fossil fuel combustion at industries and power plants, as well as fuel combustion in automobiles, ships, and other equipment, which use fossil fuels. SO_2 is also released when some sulfur-containing mineral ores are processed in mine industries. There are some natural sources of SO_2 in the environment, which are oxidation of dimethyl sulfide emitted from phytoplankton and releases from volcanic eruptions (Seinfeld, 2003; Yang et al., 2007). SO_2 is a criteria pollutant under the US NAAQS because its exposure is related to several health effects and crop and environmental material damage is anticipated when acidic aerosols (both solid and liquid) generated from SO_2 are deposited on ground. Recently on May 25, 2018, based on a review of the most recent scientific literature on SO_2 exposure/health, the US EPA proposed to retain the existing primary NAAQS standard for SO_2, which is 75 ppb based on the 3-year average of the 99th percentile of the yearly distribution of 1-h daily maximum SO_2 concentrations in the outdoor air (EPA, 2018k). US EPA is also retaining the secondary NAAQS standard for SO_2, which is 500 ppb averaged over three hours, and this limit should not be exceeded more than once per year (EPA, 2018k). SO2 can also act as source of PM. Conversion of SO_2 to condensed phase sulfate (SO_4^{2-}) in the air can be a major source of $PM_{2.5}$ (Hand et al., 2012b,b; Tsigaridis et al., 2006).

1.6.1 Spatiotemporal variations of SO_2

Spatiotemporal variations of atmospheric SO_2 are an important research topic because these variations have significant impact on public health as well as atmospheric chemistry, radiation field, and finally on the climate. In the US, the US EPA is monitoring the trends of SO_2 at 42 monitoring stations since 1980 and they are using these data for setting and reviewing the national air quality standards for SO_2. These 42 air quality monitors measure concentrations of SO_2 throughout the US. Recent trends from 2000 to 2017 show that average SO_2 concentrations have decreased substantially over the years (EPA, 2018l). For example, the mean concentrations of SO_2 have dropped from 86.02 ppb in 2000 to 15.98 ppb in 2017. Seasonal variation of SO_2 in the lower atmosphere is anticipated because its atmospheric lifetime or residence time ranges from ~2 days in the winter to <1 day in the summer (Hains et al., 2008; Lee et al., 2011). Emissions from power plants demonstrate diurnal and seasonal variations according to the variations in electricity demand. A recent study of He et al. (2016) found changes in ambient SO_2 levels, which were matching to these power plant emission fluctuations in the eastern US. Previous studies also showed peak concentrations of ambient SO_2-related sulfate aerosol peaks in summer time in the US (Hidy et al., 1978). In addition to stationary networks, in recent years, satellite-based monitoring stations are observing changes in the tropospheric SO$_2$ levels. Some examples include the Global Ozone Monitoring Experiment (GOME) on ERS-2 (Khokhar et al., 2005; Thomas et al., 2005), the Ozone Monitoring Instrument (OMI) on EOS/Aura (Krotkov et al., 2006; Carn et al., 2007a, b; Yang et al., 2007), the Scanning Imaging Absorption Spectrometer for Atmospheric Cartography (SCIAMACHY) on ENVISAT (Lee et al., 2008; Richter et al., 2006), and the GOME-2 instrument on MetOp-A (Loyola et al., 2008). These measurements demonstrated columns of SO$_2$ in the troposphere and their variations. These satellite-based data have been analyzed to understand spatio-temporal variations of SO$_2$ in those areas where stationary monitors are not covering the entire areas due to logistical problems. For example, a recent study from China analyzed SCIAMACHY/ENVISAT satellite-based SO$_2$ data and reported that SO$_2$ levels in the eastern part of the country are decreasing since 2007 due to the regulatory measures taken for preparation of 2008 Olympic Games, while the SO$_2$ levels in the western part were increasing consistently during 2004–09, probably due to excess anthropogenic activities for the rapid economic development in these areas (Zhang et al., 2012).

1.6.2 Public health impacts of SO$_2$

Similar to other criteria air pollutants, SO$_2$ has significant adverse impacts on human health (Chen et al., 2007) because this is an irritant gas and it can constrict our lower respiratory tracts and may cause cardiovascular abnormalities, including changes in heart rates (Tunnicliffe et al., 2001).

Thus, short-term SO$_2$ exposures can harm our respiratory system and make breathing difficult (EPA, 2018k). Asthmatic children and elderly people are particularly sensitive to adverse effects of SO$_2$ (EPA, 2018k).

In addition to direct effects, SO$_2$ emissions may have secondary adverse health effects because high concentrations of SO$_2$ in the air also promote formation of other sulfur oxides (SO$_x$), which can react with other atmospheric compounds and form fine particles. These fine particles contribute to increased inhalable fine PM levels in the atmosphere and cause additional respiratory and cardiovascular health problems upon inhalation in lower airways (EPA, 2018k). Most of the previous studies on health effects of SO$_2$ can be categorized into two major outcomes: mortality and respiratory disorders.

Several previous studies showed associations between short-term exposures to SO$_2$ and increased risks of cardiorespiratory mortality and morbidity. For example, multicity studies conducted in Europe (Katsouyanni et al., 1997; Sunyer, Ballester, et al., 2003a) and Canada (Burnett et al., 2000) demonstrated such associations. However, independent health effect of SO$_2$ still remained inconclusive because SO$_2$ can promote development of atmospheric PM as described here. In fact, one European study initially found associations between ambient SO$_2$ and cardiovascular admissions for ischemic heart diseases, but this association disappeared after adjustment for PM$_{10}$ (Sunyer, Atkinson, et al., 2003b). One multicity Asian study (Public Health and Air Pollution in Asia or PAPA project), however, reported acute health effects of SO_2 (Kan et al., 2010; Wong et al., 2008). Chen et al. (2012) recently investigated the associations between short-term SO_2 exposures and daily mortality in 17 Chinese cities in a large-scale epidemiological study. This study suggested that short-term exposure to SO_2 is associated with increased mortality risk, but these associations may be attributable to other substances for which SO_2 can serve as precursors and therefore works as a surrogate. The authors of this study also found that the effects of SO_2 were attenuated when adjusted for NO_2.

A number of previous studies examined the associations between SO_2 exposures and asthma. For example, a recent study conducted in Wayne County, MI, in the US reported that daily 1-h peak SO_2 concentrations calculated from dispersion models were associated with increased odds of respiratory symptoms in asthmatic children, and this association was more prominent in low-income African-American and Latino/Hispanic asthmatic children (Lewis et al., 2016). Another large-scale study conducted in Israel assessed the effects of annual NO_2 and SO_2 exposures on asthma prevalence in 137,040 male subjects and found a weak but significant exposure response association in all models (Greenberg et al., 2017). Another study estimated the risk of asthma episodes in relation to short-term variations in SO_2 emitted from refinery stacks among children of 2–4 years of age living within 7.5 km of the stacks and found that short-term episodes of increased SO2 exposures were associated with children's asthma episodes (Smargiassi et al., 2008). A study from China reported an association between SO_2 levels and both asthma prevalence and current asthma symptoms among atopic children (Dong et al., 2011).

1.7 Challenges in epidemiological study designs and risk assessment for understanding air pollution-related health impacts

Recent findings on spatiotemporal variations of criteria air pollutants and their public health impacts are discussed here; however, measurements of air pollutants in most of the cited studies were conducted at a limited number of sites. Even US EPA measurements for NAAQS were conducted at a few hundred monitoring sites only. Sometimes these monitoring sites are away from the community who are suffering from pollutant-related health disorders, and they have limited geographic coverage. How these limited measurements represent actual human exposures is a questionable issue in air pollution epidemiology because personal exposures can differ from exposures measured at stationary monitoring locations. Moreover, for better understanding of the risks of air pollution-related diseases, long-term and large-scale cohort studies are required with long-term follow-up of subjects who live in different communities. Therefore, more studies are required addressing the variations in exposure generated at several spatial levels in different communities in different locations. For understanding global health burden of air pollutants, we need more exposure data, particularly from the developing countries of the world where along with the fastest-growing economies and increasing industrialization, the problems of air pollution and related health issues have recently worsened. Applications of satellite-based observations to surface air quality have been advanced in recent years (Hoff and Christopher, 2009; Martin, 2008). These observations can be utilized for improving the estimates of population exposure to various air pollutants, such as $PM_{2.5}$.

Beside this potential exposure measurement error, air pollution epidemiology may also suffer from confounding errors because, as discussed above, air pollutants have diurnal and seasonal variations and different long term trends as well. The determinants of these variations are therefore important determinants of adverse health outcomes as well and may function as important confounders. These confounders should be adequately considered in data analyses when determining the associations between air pollutants and adverse public health impacts. The spatial variations of other disease risk factors, such as diet, smoking, or sociodemographic factors, should be considered as confounders too in long-term cohort studies. Sometimes health outcomes from one air pollutant can be confounded by other pollutants too (e.g., SO_2 and $PM_{2.5}$; see earlier), and these errors should be adequately addressed in data analyses.

1.8 The significance of air monitoring networks and models for prediction of ambient air pollutants

Certainly we need more air monitoring networks covering more areas on the Earth surfaces. However, when establishing the new networks, we have to remember that we are designing these networks for protecting public health, and therefore besides collecting data on ambient concentrations air pollutants (including their dispersion patterns, secondary reactions, surface depositions) we have to collect data on human health parameters specific to air pollutants being monitored in different categories of subjects, such as children, adults, elderly people, and vulnerable population groups with preexisting conditions. Different types of computational tools are growing rapidly in recent years. Large databases are also available now focusing on air pollutant concentrations, air pollution exposures, and health status of populations, and these databases can be mined to provide newer insights on health effects of air pollutants at local to global levels. Newer informatics tools should be used for understanding patterns of air pollutants for understanding unknown associations between air pollutants and their effects on communities and ecosystems. New computational tools should be used not only for exposure assessment of air pollutants but also for explaining the kinetics and dynamics of these air pollutants and their metabolites in the human body. Computational and informatics tools should be applied for controlling the air pollutant emissions and better spatiotemporal statistical tools should be used for analyzing health effects of air pollutants in large-scale epidemiological studies. Use of all these tools may provide better understanding on the health effects of complicated air pollutant mixtures at local to global levels.

References

Agency for Toxic Substances and Disease Registry (ATSDR), 2007. U.S. Department of Health and Human Services, Public Health Service. Toxicological Profile for Lead. ATSDR, Atlanta, GA.

Agency for Toxic Substances and Disease Registry (ATSDR), 2012. U.S. Department of Health and Human Services, Public Health Service Toxicological Profile for Carbon Monoxide. ATSDR, Atlanta, GA.

Allred, E.N., Bleecker, E.R., Chaitman, B.R., Dahms, T.E., Gottlieb, S.O., Hackney, J.D., Pagano, M., Selvester, R.H., Walden, S.M., Warren, J., 1989. Short-term effects of carbon monoxide exposure on the exercise performance of subjects with coronary artery disease. N. Engl. J. Med. 321 (21), 1426–1432.

Allred, E.N., Bleecker, E.R., Chaitman, B.R., Dahms, T.E., Gottlieb, S.O., Hackney, J.D., Pagano, M., Selvester, R.H., Walden, S.M., Warren, J., 1991. Effects of carbon monoxide on myocardial ischemia. Environ. Health Perspect. 91, 89.

Anderson, E.W., Andelman, R.J., Strauch, J.M., Fortuin, N.J., Knelson, J.H., 1973. Effect of low-level carbon monoxide exposure on onset and duration of angina pectoris: a study in ten patients with ischemic heart disease. Ann. Intern. Med. 79 (1), 46–50.

Anenberg, S.C., Henze, D.K., Tinney, V., Kinney, P.L., Raich, W., Fann, N., Malley, C.S., Roman, H., Lamsal, L., Duncan, B., Martin, R.V., 2018. Estimates of the global burden of ambient PM 2.5, ozone, and NO 2 on asthma incidence and emergency room visits. Environ. Health Perspect. 126 (10), 107004.

Anenberg, S.C., Horowitz, L.W., Tong, D.Q., West, J.J., 2010. An estimate of the global burden of anthropogenic ozone and fine particulate matter on premature human mortality using atmospheric modeling. Environ. Health Perspect. 118 (9), 1189–1195.

Anfossi, D., Sandroni, S., 1997. Ozone levels in Paris one century ago. Atmos. Environ. 31 (20), 3481–3482.

Arteaga, V.E., Mitchell, D.C., Matt, G.E., Quintana, P.J., Schaeffer, J., Reynolds, S.J., Schenker, M.B., Mitloehner, F.M., 2015. Occupational exposure to endotoxin in PM2.5 and pre-and post-shift lung function in California dairy workers. J. Environ. Prot. 6 (05), 552–565.

Ballew, C., Khan, L.K., Kaufmann, R., Mokdad, A., Miller, D.T., Gunter, E.W., 1999. Blood lead concentration and children's anthropometric dimensions in the third National Health and nutrition examination survey (NHANES III), 1988-1994. J. Pediatr. 134 (5), 623–630.

Barn, P., Giles, L., Héroux, M.E., Kosatsky, T., 2018. A review of the experimental evidence on the toxicokinetics of carbon monoxide: the potential role of pathophysiology among susceptible groups. Environ. Health 17 (1), 13.

Bates, D.V., 2005. Ambient ozone and mortality. Epidemiology 16 (4), 427–429.

Bates, T.S., Kelly, K.C., Johnson, J.E., Gammon, R.H., 1995. Regional and seasonal variations in the flux of oceanic carbon monoxide to the atmosphere. J. Geophys. Res. Atmos. 100 (D11), 23093–23101.

Bauer, M.A., Utell, M.J., Morrow, P.E., Speers, D.M., Gibb, F.R., 1986. Inhalation of 0.30 ppm nitrogen dioxide potentiates exercise-induced bronchospasm in asthmatics. Am. Rev. Respir. Dis. 134 (5), 1203–1208.

Beelen, R., Hoek, G., van Den Brandt, P.A., Goldbohm, R.A., Fischer, P., Schouten, L.J., Jerrett, M., Hughes, E., Armstrong, B., Brunekreef, B., 2007. Long-term effects of traffic-related AIR pollution on mortality in a Dutch cohort (NLCS-AIR study). Environ. Health Perspect. 116 (2), 196–202.

Belanger, K., Gent, J.F., Triche, E.W., Bracken, M.B., Leaderer, B.P., 2006. Association of indoor nitrogen dioxide exposure with respiratory symptoms in children with asthma. Am. J. Respir. Crit. Care Med. 173 (3), 297–303.

Bell, M.L., Dominici, F., Ebisu, K., Zeger, S.L., Samet, J.M., 2007. Spatial and temporal variation in $PM_{2.5}$ chemical composition in the United States for health effects studies. Environ. Health Perspect. 115 (7), 989–995.

Bell, M.L., Dominici, F., Samet, J.M., 2005. A meta-analysis of time-series studies of ozone and mortality with comparison to the national morbidity, mortality, and air pollution study. Epidemiology 16 (4), 436–445.

Bell, M.L., McDermott, A., Zeger, S.L., Samet, J.M., Dominici, F., 2004. Ozone and short-term mortality in 95 US urban communities, 1987-2000. JAMA 292 (19), 2372–2378.

Bergamaschi, P., Hein, R., Heimann, M., Crutzen, P.J., 2000. Inverse modeling of the global CO cycle: 1. Inversion of CO mixing ratios. J. Geophys. Res. Atmos. 105 (D2), 1909–1927.

Bhardwaj, P., Naja, M., Rupakheti, M., Panday, A.K., Kumar, R., Mahata, K., Lal, S., Chandola, H.C., Lawrence, M.G., 2018. Variations in surface ozone and carbon monoxide in the Kathmandu Valley and surrounding broader regions during SusKat-ABC field campaign: role of local and regional sources. Atmos. Chem. Phys. 18, 11949–11971.

Blanchard, C.L., 2003. Spatial and Temporal Characterization of Particulate Matter. California Environmental Protection Agency, Air Resources Board, Research Division.

Bressler, J.P., Goldstein, G.W., 1991. Mechanisms of lead neurotoxicity. Biochem. Pharmacol. 41 (4), 479–484.

Brody, D.J., Pirkle, J.L., Kramer, R.A., Flegal, K.M., Matte, T.D., Gunter, E.W., Paschal, D.C., 1994. Blood lead levels in the US population: phase 1 of the Third National Health and Nutrition Examination Survey (NHANES III, 1988 to 1991). JAMA 272 (4), 277–283.

Brook, R.D., Franklin, B., Cascio, W., Hong, Y., Howard, G., Lipsett, M., Luepker, R., Mittleman, M., Samet, J., Smith Jr., S.C., Tager, I., 2004. Air pollution and cardiovascular disease: a statement for healthcare professionals from the expert panel on population and prevention science of the American Heart Association. Circulation 109 (21), 2655–2671.

Burnett, R.T., Brook, J., Dann, T., Delocla, C., Philips, O., Cakmak, S., Vincent, R., Goldberg, M.S., Krewski, D., 2000. Association between particulate-and gas-phase components of urban air pollution and daily mortality in eight Canadian cities. Inhal. Toxicol. 12 (Suppl 4), 15–39.

Burnett, R.T., Stieb, D., Brook, J.R., Cakmak, S., Dales, R., Raizenne, M., Vincent, R., Dann, T., 2004. Associations between short-term changes in nitrogen dioxide and mortality in Canadian cities. Arch. Environ. Health 59 (5), 228–236.

Cairncross, E.K., John, J., Zunckel, M., 2007. A novel air pollution index based on the relative risk of daily mortality associated with short-term exposure to common air pollutants. Atmos. Environ. 41 (38), 8442–8454.

Cao, J., Yang, C., Li, J., Chen, R., Chen, B., Gu, D., Kan, H., 2011. Association between long-term exposure to outdoor air pollution and mortality in China: a cohort study. J. Hazard. Mater. 186 (2–3), 1594–1600.

Carn, S.A., Krotkov, N.A., Yang, K., Hoff, R.M., Prata, A.J., Krueger, A.J., Loughlin, S.C., Levelt, P.F., 2007a. Extended observations of volcanic SO_2 and sulfate aerosol in the stratosphere. Atmos. Chem. Phys. Discuss. 7 (1), 2857–2871.

Carn, S.A., Krueger, A., Krotkov, N., Yang, K., Levelt, P., 2007b. Sulfur dioxide emissions from Peruvian copper smelters detected by the Ozone Monitoring Instrument. Geophys. Res. Lett. 34 (9), L09801.

Carslaw, D.C., 2005. On the changing seasonal cycles and trends of ozone at Mace head, Ireland. Atmos. Chem. Phys. 5 (12), 3441–3450.

Chan, C.K., Yao, X., 2008. Air pollution in mega cities in China. Atmos. Environ. 42 (1), 1–42.

Chen, G.D., Fechter, L.D., 1999. Potentiation of octave-band noise induced auditory impairment by carbon monoxide. Hear. Res. 132 (1–2), 149–159.

Chen, C., Zhao, B., 2011. Review of relationship between indoor and outdoor particles: I/O ratio, infiltration factor and penetration factor. Atmos. Environ. 45 (2), 275–288.

Chen, R., Huang, W., Wong, C.M., Wang, Z., Thach, T.Q., Chen, B., Kan, H., CAPES Collaborative Group, 2012. Short-term exposure to sulfur dioxide and daily mortality in 17 Chinese cities: the China air pollution and health effects study (CAPES). Environ. Res. 118, 101–106.

Chen, T.M., Kuschner, W.G., Gokhale, J., Shofer, S., 2007. Outdoor air pollution: nitrogen dioxide, sulfur dioxide, and carbon monoxide health effects. Am. J. Med. Sci. 333 (4), 249–256.

Chen, Y., Randerson, J.T., van der Werf, G.R., Morton, D.C., Mu, M., Kasibhatla, P.S., 2010. Nitrogen deposition in tropical forests from savanna and deforestation fires. Glob. Chang. Biol. 16 (7), 2024–2038.

Cookman, G.R., King, W., Regan, C.M., 1987. Chronic low-level lead exposure impairs embryonic to adult conversion of the neural cell adhesion molecule. J. Neurochem. 49 (2), 399–403.

Coyle, M., Smith, R.I., Stedman, J.R., Weston, K.J., Fowler, D., 2002. Quantifying the spatial distribution of surface ozone concentration in the UK. Atmos. Environ. 36 (6), 1013–1024.

Dai, Y., Qiu, H., Sun, S., Yang, Y., Lin, H., Tian, L., 2018. Age-dependent effect of ambient ozone on emergency asthma hospitalizations in Hong Kong. J. Allergy Clin. Immunol. 141 (4), 1532–1534.

Dall'Osto, M., Querol, X., Amato, F., Karanasiou, A., Lucarelli, F., Nava, S., Calzolai, G., Chiari, M., 2013. Hourly elemental concentrations in $PM_{2.5}$ aerosols sampled simultaneously at urban background and road site during SAPUSS—diurnal variations and PMF receptor modelling. Atmos. Chem. Phys. 13 (8), 4375–4392.

Derwent, R.G., 1995. Air chemistry and terrestrial gas emissions: a global perspective. Phil. Trans. R. Soc. A 351, 205–217.

Dey, S., Di Girolamo, L., van Donkelaar, A., Tripathi, S.N., Gupta, T., Mohan, M., 2012. Variability of outdoor fine particulate ($PM_{2.5}$) concentration in the Indian subcontinent: a remote sensing approach. Remote Sens. Environ. 127, 153–161.

Ding, A.J., Fu, C.B., Yang, X.Q., Sun, J.N., Zheng, L.F., Xie, Y.N., Herrmann, E., Nie, W., Petäjä, T., Kerminen, V.M., Kulmala, M., 2013. Ozone and fine particle in the western Yangtze river delta: an overview of 1 yr data at the SORPES station. Atmos. Chem. Phys. 13 (11), 5813–5830.

Dockery, D.W., Pope, C.A., Xu, X., Spengler, J.D., Ware, J.H., Fay, M.E., Ferris Jr., B.G., Speizer, F.E., 1993. An association between air pollution and mortality in six US cities. N. Engl. J. Med. 329 (24), 1753–1759.

Dominici, F., Sheppard, L., Clyde, M., 2003. Health effects of air pollution: a statistical review. Int. Stat. Rev. 71 (2), 243–276.

Dong, G.H., Chen, T., Liu, M.M., Wang, D., Ma, Y.N., Ren, W.H., Lee, Y.L., Zhao, Y.D., He, Q.C., 2011. Gender differences and effect of air pollution on asthma in children with and without allergic predisposition: Northeast Chinese children health study. PLoS One 6 (7), e22470.

EEA, E, 2018. AirBase-the European Air Quality DataBase. European Environment Agency. https://www.eea.europa.eu/data-and-maps/data/airbase-the-european-air-quality-database-8. (Accessed November 27, 2018).

Eftim, S.E., Samet, J.M., Janes, H., McDermott, A., Dominici, F., 2008. Fine particulate matter and mortality: a comparison of the six cities and American Cancer Society cohorts with a medicare cohort. Epidemiology 19 (2), 209–216.

Ehrman, S.H., Pratsinis, S.E., Young, J.R., 1992. Receptor modeling of the fine aerosol at a residential Los Angeles site. Atmos. Environ. 26 (4), 473–481.

EPA (US Environmental Protection Agency), 2018a. Carbon Monoxide (CO) Pollution in Outdoor Air. https://www.epa.gov/co-pollution. (Accessed November 27, 2018).

EPA (US Environmental Protection Agency), 2018b. National Trends in CO Levels. https://www.epa.gov/air-trends/carbon-monoxide-trends. (Accessed November 27, 2018).

EPA (US Environmental Protection Agency), 2018c. National Trends in Lead Levels. https://www.epa.gov/air-trends/lead-trends. (Accessed November 27, 2018).

EPA (US Environmental Protection Agency), 2018d. Basic Information About Lead Air Pollution. https://www.epa.gov/lead-air-pollution/basic-information-about-lead-air-pollution#how. (Accessed November 27, 2018).

EPA (US Environmental Protection Agency), 2018e. Ground-Level Ozone Pollution. https://www.epa.gov/ground-level-ozone-pollution. (Accessed November 27, 2018).

EPA (US Environmental Protection Agency), 2018f. Health Effects of Ozone in Patients With Asthma and Other Chronic Respiratory Disease. https://www.epa.gov/ozone-pollution-and-your-patients-health/health-effects-ozone-patients-asthma-and-other-chronic. (Accessed November 27, 2018).

EPA (US Environmental Protection Agency), 2018g. Revised Air Quality Standards for Particle Pollution and Updates to the Air Quality Index (AQI). https://www.epa.gov/sites/production/files/2016-04/documents/2012_aqi_factsheet.pdf. (Accessed November 27, 2018).

EPA (US Environmental Protection Agency), 2018h. National Air Quality: Status and Trends of Key Air Pollutants. https://www.epa.gov/air-trends. (Accessed November 27, 2018).

EPA (US Environmental Protection Agency), 2018i. Health and Environmental Effects of Particulate Matter (PM). https://www.epa.gov/pm-pollution/health-and-environmental-effects-particulate-matter-pm. (Accessed November 27, 2018).

EPA (US Environmental Protection Agency), 2018j. Nitrogen Dioxide (NO_2) Pollution. https://www.epa.gov/no2-pollution/basic-information-about-no2. (Accessed November 27, 2018).

EPA (US Environmental Protection Agency), 2018k. Sulfur Dioxide (SO_2) Pollution. https://www.epa.gov/so2-pollution. (Accessed November 27, 2018).

EPA (US Environmental Protection Agency), 2018l. Sulfur Dioxide Trends. https://www.epa.gov/air-trends/sulfur-dioxide-trends. (Accessed November 27, 2018).

Eze, I.C., Imboden, M., Kumar, A., von Eckardstein, A., Stolz, D., Gerbase, M.W., Künzli, N., Pons, M., Kronenberg, F., Schindler, C., Probst-Hensch, N., 2016. Air pollution and diabetes association: modification by type 2 diabetes genetic risk score. Environ. Int. 94, 263–271.

Fahey, D.W., Hübler, G., Parrish, D.D., Williams, E.J., Norton, R.B., Ridley, B.A., Singh, H.B., Liu, S.C., Fehsenfeld, F.C., 1986. Reactive nitrogen species in the troposphere: measurements of NO, NO_2, HNO_3, particulate nitrate, peroxyacetyl nitrate (PAN), O_3, and total reactive odd nitrogen (NO_y) at Niwot ridge, Colorado. J. Geophys. Res. 91 (D9), 9781–9793.

Faustini, A., Rapp, R., Forastiere, F., 2014. Nitrogen dioxide and mortality: review and meta-analysis of long-term studies. Eur. Respir. J. 44 (3), 744–753.

Filleul, L., Rondeau, V., Vandentorren, S., Le Moual, N., Cantagrel, A., Annesi-Maesano, I., Charpin, D., Declercq, C., Neukirch, F., Paris, C., Vervloet, D., 2005. Twenty five year mortality and air pollution: results from the French PAARC survey. Occup. Environ. Med. 62 (7), 453–460.

Fiore, A.M., Dentener, F.J., Wild, O., Cuvelier, C., Schultz, M.G., Hess, P., Textor, C., Schulz, M., Doherty, R.M., Horowitz, L.W., MacKenzie, I.A., 2009. Multimodel estimates of intercontinental source-receptor relationships for ozone pollution. J. Geophys. Res. 114 (D4). D04301-1-21.

Fiore, A.M., Horowitz, L.W., Purves, D.W., Levy, H., Evans, M.J., Wang, Y., Li, Q., Yantosca, R.M., 2005. Evaluating the contribution of changes in isoprene emissions to surface ozone trends over the eastern United States. J. Geophys. Res. 110 (D12). D12303-1-13.

Fiore, A.M., Jacob, D.J., Bey, I., Yantosca, R.M., Field, B.D., Fusco, A.C., Wilkinson, J.G., 2002. Background ozone over the United States in summer: origin, trend, and contribution to pollution episodes. J. Geophys. Res. 107 (D15). ACH 11-1-ACH 11-25.

Fiore, A.M., Jacob, D.J., Logan, J.A., Yin, J.H., 1998. Long-term trends in ground level ozone over the contiguous United States, 1980–1995. J. Geophys. Res. 103 (D1), 1471–1480.

Forster, P., Ramaswamy, V., Artaxo, P., Berntsen, T., Betts, R., Forster, P., Ramaswamy, V., Artaxo, P., Berntsen, T., Betts, R., Fahey, D.W., Haywood, J., Lean, J., Lowe, D.C., Myhre, G., Nganga, J., 2007. Changes in atmospheric constituents and in radiative forcing. In: Chapter 2. Climate Change 2007. The Physical Science Basis.

Fu, P., Guo, X., Cheung, F.M.H., Yung, K.K.L., 2019. The association between $PM_{2.5}$ exposure and neurological disorders: a systematic review and meta-analysis. Sci. Total Environ. 655, 1240–1248.

Fusco, A.C., Logan, J.A., 2003. Analysis of 1970–1995 trends in tropospheric ozone at Northern Hemisphere mid-latitudes with the GEOS-CHEM model. J. Geophys. Res. 108 (D15). ACH 4-1-25.

Gardner, M.W., Dorling, S.R., 2000. Meteorologically adjusted trends in UK daily maximum surface ozone concentrations. Atmos. Environ. 34 (2), 171–176.

Gehring, U., Heinrich, J., Krämer, U., Grote, V., Hochadel, M., Sugiri, D., Kraft, M., Rauchfuss, K., Eberwein, H.G., Wichmann, H.E., 2006. Long-term exposure to ambient air pollution and cardiopulmonary mortality in women. Epidemiology 17 (5), 545–551.

Goldstein, G.W., 1993. Evidence that lead acts as a calcium substitute in second messenger metabolism. Neurotoxicology 14 (2–3), 97–101.

Greenberg, N., Carel, R.S., Derazne, E., Bibi, H., Shpriz, M., Tzur, D., Portnov, B.A., 2016. Different effects of long-term exposures to SO_2 and NO_2 air pollutants on asthma severity in young adults. J. Toxicol. Environ. Health 79 (8), 342–351.

Greenberg, N., Carel, R.S., Derazne, E., Tiktinsky, A., Tzur, D., Portnov, B.A., 2017. Modeling long-term effects attributed to nitrogen dioxide (NO_2) and sulfur dioxide (SO_2) exposure on asthma morbidity in a nationwide cohort in Israel. J. Toxicol. Environ. Health 80 (6), 326–337.

Greingor, J.L., Tosi, J.M., Ruhlmann, S., Aussedat, M., 2001. Acute carbon monoxide intoxication during pregnancy. One case report and review of the literature. Emerg. Med. J. 18 (5), 399–401.

Guerreiro, C.B., Foltescu, V., De Leeuw, F., 2014. Air quality status and trends in Europe. Atmos. Environ. 98, 376–384.

Hains, J.C., Taubman, B.F., Thompson, A.M., Stehr, J.W., Marufu, L.T., Doddridge, B.G., Dickerson, R.R., 2008. Origins of chemical pollution derived from Mid-Atlantic aircraft profiles using a clustering technique. Atmos. Environ. 42 (8), 1727–1741.

Haley, V.B., Talbot, T.O., 2004. Seasonality and trend in blood lead levels of New York State children. BMC Pediatr. 4 (1), 8.

Hand, J.L., Gebhart, K.A., Schichtel, B.A., Malm, W.C., 2012b. Increasing trends in wintertime particulate sulfate and nitrate ion concentrations in the Great Plains of the United States (2000–2010). Atmos. Environ. 55, 107–110.

Hand, J.L., Schichtel, B.A., Pitchford, M., Malm, W.C., Frank, N.H., 2012a. Seasonal composition of remote and urban fine particulate matter in the United States. J. Geophys. Res. 117 (D5). D05209.

Harris, A.R., Davidson, C.I., 2005. The role of resuspended soil in lead flows in the California South Coast Air Basin. Environ. Sci. Technol. 39 (19), 7410–7415.

Hayden, K.L., Anlauf, K.G., Hastie, D.R., Bottenheim, J.W., 2003. Partitioning of reactive atmospheric nitrogen oxides at an elevated site in southern Quebec, Canada. J. Geophys. Res. 108 (D19), 4603.

He, H., Vinnikov, K.Y., Li, C., Krotkov, N.A., Jongeward, A.R., Li, Z., Stehr, J.W., Hains, J.C., Dickerson, R.R., 2016. Response of SO_2 and particulate air pollution to local and regional emission controls: a case study in Maryland. Earth's Future 4 (4), 94–109.

Heinrich, J., Pitz, M., Bischof, W., Krug, N., Borm, P.J., 2003. Endotoxin in fine ($PM_{2.5}$) and coarse ($PM_{2.5-10}$) particle mass of ambient aerosols. A temporo-spatial analysis. Atmos. Environ. 37 (26), 3659–3667.

Hidy, G.M., Mueller, P.K., Tong, E.Y., 1978. Spatial and temporal distributions of airborne sulfate in parts of the United States. In: Sulfur in the Atmosphere, pp. 735–752.

Hoff, R.M., Christopher, S.A., 2009. Remote sensing of particulate pollution from space: have we reached the promised land? J. Air Waste Manage. Assoc. 59 (6), 645–675.

Institute for Health Metrics and Evaluation (IHME), 2018. GBD Compare. https://vizhub.healthdata.org/gbd-compare/. (Accessed November 28, 2018).

Ito, K., De Leon, S.F., Lippmann, M., 2005. Associations between ozone and daily mortality: analysis and meta-analysis. Epidemiology 16, 446–457.

Jerrett, M., Burnett, R.T., Pope III, C.A., Ito, K., Thurston, G., Krewski, D., Shi, Y., Calle, E., Thun, M., 2009. Long-term ozone exposure and mortality. N. Engl. J. Med. 360 (11), 1085–1095.

Johnson, D.L., Bretsch, J.K., 2002. Soil lead and children's blood lead levels in Syracuse, NY, USA. Environ. Geochem. Health 24 (4), 375–385.

Jonson, J.E., Simpson, D., Fagerli, H., Solberg, S., 2006. Can we explain the trends in European ozone levels? Atmos. Chem. Phys. 6 (1), 51–66.

Junge, C.E., 1962. Global ozone budget and exchange between stratosphere and troposphere. Tellus 14 (4), 363–377.

Kan, H., Wong, C.M., Vichit-Vadakan, N., Qian, Z., 2010. Short-term association between sulfur dioxide and daily mortality: the public health and air pollution in Asia (PAPA) study. Environ. Res. 110 (3), 258–264.

Katanoda, K., Sobue, T., Satoh, H., Tajima, K., Suzuki, T., Nakatsuka, H., Takezaki, T., Nakayama, T., Nitta, H., Tanabe, K., Tominaga, S., 2011. An association between long-term exposure to ambient air pollution and mortality from lung cancer and respiratory diseases in Japan. J. Epidemiol. 1102090211–1102090211.

Katsouyanni, K., Touloumi, G., Spix, C., Schwartz, J., Balducci, F., Medina, S., Rossi, G., Wojtyniak, B., Sunyer, J., Bacharova, L., Schouten, J.P., 1997. Short term effects of ambient sulphur dioxide and particulate matter on mortality in 12 European cities: results from time series data from the APHEA project. BMJ 314 (7095), 1658.

Khokhar, M.F., Frankenberg, C., Van Roozendael, M., Beirle, S., Kühl, S., Richter, A., Platt, U., Wagner, T., 2005. Satellite observations of atmospheric SO_2 from volcanic eruptions during the time-period of 1996–2002. Adv. Space Res. 36 (5), 879–887.

Koller, K., Brown, T., Spurgeon, A., Levy, L., 2004. Recent developments in low-level lead exposure and intellectual impairment in children. Environ. Health Perspect. 112 (9), 987.

Kopacz, M., Jacob, D.J., Fisher, J.A., Logan, J.A., Zhang, L., Megretskaia, I.A., Yantosca, R.M., Singh, K., Henze, D.K., Burrows, J.P., Buchwitz, M., 2010. Global estimates of CO sources with high resolution by adjoint inversion of multiple satellite datasets (MOPITT, AIRS, SCIAMACHY, TES). Atmos. Chem. Phys. 10 (3), 855–876.

Krotkov, N.A., Carn, S.A., Krueger, A.J., Bhartia, P.K., Yang, K., 2006. Band residual difference algorithm for retrieval of SO_2 from the Aura Ozone Monitoring Instrument (OMI). IEEE Trans. Geosci. Remote Sens. 44 (5), 1259–1266.

Lagorio, S., Forastiere, F., Pistelli, R., Iavarone, I., Michelozzi, P., Fano, V., Marconi, A., Ziemacki, G., Ostro, B.D., 2006. Air pollution and lung function among susceptible adult subjects: a panel study. Environ. Health 5 (1), 11.

Lamichhane, D.K., Leem, J.H., Lee, J.Y., Kim, H.C., 2015. A meta-analysis of exposure to particulate matter and adverse birth outcomes. Environ Health Toxicol. 30, e2015011.

Langenfelds, R.L., Francey, R.J., Pak, B.C., Steele, L.P., Lloyd, J., Trudinger, C.M., Allison, C.E., 2002. Interannual growth rate variations of atmospheric CO_2 and its $\delta^{13}C$, H_2, CH_4, and CO between 1992 and 1999 linked to biomass burning. Glob. Biogeochem. Cycles 16 (3). 21-1–21-22.

Langrish, J.P., Li, X., Wang, S., Lee, M.M., Barnes, G.D., Miller, M.R., Cassee, F.R., Boon, N.A., Donaldson, K., Li, J., Li, L., 2012. Reducing personal exposure to particulate air pollution improves cardiovascular health in patients with coronary heart disease. Environ. Health Perspect. 120 (3), 367.

Latza, U., Gerdes, S., Baur, X., 2009. Effects of nitrogen dioxide on human health: systematic review of experimental and epidemiological studies conducted between 2002 and 2006. Int. J. Hyg. Environ. Health 212 (3), 271–287.

Lee, C.W., Lu, Z., Kwoun, O.I., Won, J.S., 2008. Deformation of the Augustine Volcano, Alaska, 1992–2005, measured by ERS and ENVISAT SAR interferometry. Earth Planets Space 60 (5), 447–452.

Lee, C., Martin, R.V., van Donkelaar, A., Lee, H., Dickerson, R.R., Hains, J.C., Krotkov, N., Richter, A., Vinnikov, K., Schwab, J.J., 2011. SO_2 emissions and lifetimes: estimates from inverse modeling using in situ and global, space-based (SCIAMACHY and OMI) observations. J. Geophys. Res. 116 (D6). D06304.

Leitte, A.M., Schlink, U., Herbarth, O., Wiedensohler, A., Pan, X.C., Hu, M., Richter, M., Wehner, B., Tuch, T., Wu, Z., Yang, M., 2010. Size-segregated particle number concentrations and respiratory emergency room visits in Beijing, China. Environ. Health Perspect. 119 (4), 508–513.

Lelieveld, J., Barlas, C., Giannadaki, D., Pozzer, A., 2013. Model calculated global, regional and megacity premature mortality due to air pollution. Atmos. Chem. Phys. 13 (14), 7023–7037.

Lelieveld, J., Dentener, F.J., 2000. What controls tropospheric ozone? J. Geophys. Res. 105 (D3), 3531–3551.

Lewis, T.C., Robins, T.G., Batterman, S.A., Mukherjee, B., Mentz, G.B., Parker, E.A., Israel, B.A., 2016. A16 epidemiology and health care costs of Pediatric and adult asthma: daily 1-hour peak levels of sulfur dioxide are associated with increased respiratory symptoms in Detroit children with asthma. Am. J. Respir. Crit. Care Med. 193, 1.

Li, Q., Jacob, D.J., Munger, J.W., Yantosca, R.M., Parrish, D.D., 2004. Export of NOy from the north American boundary layer: reconciling aircraft observations and global model budgets. J. Geophys. Res. 109 (D2). D02313.

Li, C., Wen, T., Li, Z., Dickerson, R.R., Yang, Y., Zhao, Y., Wang, Y., Tsay, S.C., 2010. Concentrations and origins of atmospheric lead and other trace species at a rural site in northern China. J. Geophys. Res. 115. D00K23.

Liang, J., Horowitz, L.W., Jacob, D.J., Wang, Y., Fiore, A.M., Logan, J.A., Gardner, G.M., Munger, J.W., 1998. Seasonal budgets of reactive nitrogen species and ozone over the United States, and export fluxes to the global atmosphere. J. Geophys. Res. 103 (D11), 13435–13450.

Lim, S.S., Vos, T., Flaxman, A.D., Danaei, G., Shibuya, K., Adair-Rohani, H., AlMazroa, M.A., Amann, M., Anderson, H.R., Andrews, K.G., Aryee, M., 2012. A comparative risk assessment of burden of disease and injury attributable to 67 risk factors and risk factor clusters in 21 regions, 1990–2010: a systematic analysis for the global burden of disease study 2010. Lancet 380 (9859), 2224–2260.

Lin, W., Xu, X., Zhang, X., Tang, J., 2008. Contributions of pollutants from North China plain to surface ozone at the Shangdianzi GAW Station. Atmos. Chem. Phys. 8 (19), 5889–5898.

Littlejohn, D., Wang, Y., Chang, S.G., 1993. Oxidation of aqueous sulfite ion by nitrogen dioxide. Environ. Sci. Technol. 27 (10), 2162–2167.

Liu, S.C., Trainer, M., Fehsenfeld, F.C., Parrish, D.D., Williams, E.J., Fahey, D.W., Hübler, G., Murphy, P.C., 1987. Ozone production in the rural troposphere and the implications for regional and global ozone distributions. J. Geophys. Res. 92 (D4), 4191–4207.

Loyola, D., Van Geffen, J., Valks, P., Erbertseder, T., Van Roozendael, M., Thomas, W., Zimmer, W., Wißkirchen, K., 2008. Satellite-based detection of volcanic sulphur dioxide from recent eruptions in Central and South America. Adv. Geosci. 14, 35–40.

Lu, H.C., Chang, T.S., 2005. Meteorologically adjusted trends of daily maximum ozone concentrations in Taipei, Taiwan. Atmos. Environ. 39 (35), 6491–6501.

Ma, Z., Xu, J., Quan, W., Zhang, Z., Lin, W., Xu, X., 2016. Significant increase of surface ozone at a rural site, north of eastern China. Atmos. Chem. Phys. 16 (6), 3969–3977.

Martin, R.V., 2008. Satellite remote sensing of surface air quality. Atmos. Environ. 42 (34), 7823–7843.

McDonnell, W.F., Nishino-Ishikawa, N., Petersen, F.F., Chen, L.H., Abbey, D.E., 2000. Relationships of mortality with the fine and coarse fractions of long-term ambient PM 10 concentrations in nonsmokers. J. Expo. Sci. Environ. Epidemiol. 10 (5), 427–436.

Meng, Z.Y., Xu, X.B., Yan, P., Ding, G.A., Tang, J., Lin, W.L., Xu, X.D., Wang, S.F., 2009. Characteristics of trace gaseous pollutants at a regional background station in northern China. Atmos. Chem. Phys. 9 (3), 927–936.

Munger, J.W., Fan, S.M., Bakwin, P.S., Goulden, M.L., Goldstein, A.H., Colman, A.S., Wofsy, S.C., 1998. Regional budgets for nitrogen oxides from continental sources: variations of rates for oxidation and deposition with season and distance from source regions. J. Geophys. Res. 103 (D7), 8355–8368.

Naja, M., Akimoto, H., 2004. Contribution of regional pollution and long-range transport to the Asia-Pacific region: analysis of long-term ozonesonde data over Japan. J. Geophys. Res. 109 (D21). D21306-1-15.

Nawrot, T.S., Thijs, L., Den Hond, E.M., Roels, H.A., Staessen, J.A., 2002. An epidemiological re-appraisal of the association between blood pressure and blood lead: a meta-analysis. J. Hum. Hypertens. 16 (2), 123.

Okuda, T., Katsuno, M., Naoi, D., Nakao, S., Tanaka, S., He, K., Ma, Y., Lei, Y., Jia, Y., 2008. Trends in hazardous trace metal concentrations in aerosols collected in Beijing, China from 2001 to 2006. Chemosphere 72 (6), 917–924.

Oltmans, S.J., Lefohn, A.S., Harris, J.M., Galbally, I., Scheel, H.E., Bodeker, G., Brunke, E., Claude, H., Tarasick, D., Johnson, B.J., Simmonds, P., 2006. Long-term changes in tropospheric ozone. Atmos. Environ. 40 (17), 3156–3173.

Orellano, P., Quaranta, N., Reynoso, J., Balbi, B., Vasquez, J., 2017. Effect of outdoor air pollution on asthma exacerbations in children and adults: systematic review and multilevel meta-analysis. PLoS One 12 (3), e0174050.

Ostro, B., Lipsett, M., Reynolds, P., Goldberg, D., Hertz, A., Garcia, C., Henderson, K.D., Bernstein, L., 2009. Long-term exposure to constituents of fine particulate air pollution and mortality: results from the California teachers study. Environ. Health Perspect. 118 (3), 363–369.

Park, J.W., Lim, Y.H., Kyung, S.Y., An, C.H., Lee, S.P., Jeong, S.H., JU, Y.S., 2005. Effects of ambient particulate matter on peak expiratory flow rates and respiratory symptoms of asthmatics during Asian dust periods in Korea. Respirology 10 (4), 470–476.

Pearson, J.F., Bachireddy, C., Shyamprasad, S., Goldfine, A.B., Brownstein, J.S., 2010. Association between fine particulate matter and diabetes prevalence in the United States. Diabetes Care 33, 2196–2201.

Peng, R.D., Dominici, F., Pastor-Barriuso, R., Zeger, S.L., Samet, J.M., 2005. Seasonal analyses of air pollution and mortality in 100 US cities. Am. J. Epidemiol. 161 (6), 585–594.

Penttinen, P., Timonen, K.L., Tiittanen, P., Mirme, A., Ruuskanen, J., Pekkanen, J., 2001. Ultrafine particles in urban air and respiratory health among adult asthmatics. Eur. Respir. J. 17 (3), 428–435.

Pollack, I.B., Ryerson, T.B., Trainer, M., Neuman, J.A., Roberts, J.M., Parrish, D.D., 2013. Trends in ozone, its precursors, and related secondary oxidation products in Los Angeles, California: a synthesis of measurements from 1960 to 2010. J. Geophys. Res. 118 (11), 5893–5911.

Pope III, C.A., Burnett, R.T., Thun, M.J., Calle, E.E., Krewski, D., Ito, K., Thurston, G.D., 2002. Lung cancer, cardiopulmonary mortality, and long-term exposure to fine particulate air pollution. JAMA 287 (9), 1132–1141.

Puett, R.C., Hart, J.E., Yanosky, J.D., Paciorek, C., Schwartz, J., Suh, H., Speizer, F.E., Laden, F., 2009. Chronic fine and coarse particulate exposure, mortality, and coronary heart disease in the nurses' health study. Environ. Health Perspect. 117 (11), 1697–1701.

Rabinovitch, N., Zhang, L., Murphy, J.R., Vedal, S., Dutton, S.J., Gelfand, E.W., 2004. Effects of wintertime ambient air pollutants on asthma exacerbations in urban minority children with moderate to severe disease. J. Allergy Clin. Immunol. 114 (5), 1131–1137.

Raza, A., Bellander, T., Bero-Bedada, G., Dahlquist, M., Hollenberg, J., Jonsson, M., Lind, T., Rosenqvist, M., Svensson, L., Ljungman, P.L., 2013. Short-term effects of air pollution on out-of-hospital cardiac arrest in Stockholm. Eur. Heart J. 35 (13), 861–868.

Reed, L.E., Trott, P.E., 1971. Continuous measurement of carbon monoxide in streets 1967–1969. Atmos. Environ. 5 (1), 27–39.

Richter, A., Wittrock, F., Burrows, J.P., 2006. SO$_2$ measurements with SCIAMACHY. In: Proc. Atmospheric Science Conference, pp. 8–12.

Rodriguez, C., Tonkin, R., Heyworth, J., Kusel, M., De Klerk, N., Sly, P.D., Franklin, P., Runnion, T., Blockley, A., Landau, L., Hinwood, A.L., 2007. The relationship between outdoor air quality and respiratory symptoms in young children. Int. J. Environ. Health Res. 17 (5), 351–360.

Samet, J.M., Utell, M.J., 1990. The risk of nitrogen dioxide: what have we learned from epidemiological and clinical studies? Toxicol. Ind. Health 6 (2), 247–262.

Samoli, E., Touloumi, G., Schwartz, J., Anderson, H.R., Schindler, C., Forsberg, B., Vigotti, M.A., Vonk, J., Košnik, M., Skorkovsky, J., Katsouyanni, K., 2007. Short-term effects of carbon monoxide on mortality: an analysis within the APHEA project. Environ. Health Perspect. 115 (11), 1578–1583.

Sarnat, S.E., Suh, H.H., Coull, B.A., Schwartz, J., Stone, P.H., Gold, D.R., 2006. Ambient particulate air pollution and cardiac arrhythmia in a panel of older adults in Steubenville, Ohio. Occup. Environ. Med. 63 (10), 700–706.

Schildcrout, J.S., Sheppard, L., Lumley, T., Slaughter, J.C., Koenig, J.Q., Shapiro, G.G., 2006. Ambient air pollution and asthma exacerbations in children: an eight-city analysis. Am. J. Epidemiol. 164 (6), 505–517.

Seinfeld, J.H., 2003. Tropospheric chemistry and composition: aerosols/particles. In: Holton, J.R., Curry, J.A., Pyle, J.A. (Eds.), Encyclopedia of Atmospheric Sciences. Vol. 6. Academic Press, London, UK, pp. 2349–2354.

Shah, A.S., Langrish, J.P., Nair, H., McAllister, D.A., Hunter, A.L., Donaldson, K., Newby, D.E., Mills, N.L., 2013. Global association of air pollution and heart failure: a systematic review and meta-analysis. Lancet 382 (9897), 1039–1048.

Sicard, P., Talbot, C., Lesne, O., Mangin, A., Alexandre, N., Collomp, R., 2012. The aggregate risk index: an intuitive tool providing the health risks of air pollution to health care community and public. Atmos. Environ. 46, 11–16.

Silbergeld, E.K., 1992. Mechanisms of lead neurotoxicity, or looking beyond the lamppost. FASEB J. 6 (13), 3201–3206.

Silkoff, P.E., Zhang, L., Dutton, S., Langmack, E.L., Vedal, S., Murphy, J., Make, B., 2005. Winter air pollution and disease parameters in advanced chronic obstructive pulmonary disease panels residing in Denver, Colorado. J. Allergy Clin. Immunol. 115 (2), 337–344.

Sillman, S., 1999. The relation between ozone, NOx and hydrocarbons in urban and polluted rural environments. Atmos. Environ. 33 (12), 1821–1845.

Skerfving, S., Löfmark, L., Lundh, T., Mikoczy, Z., Strömberg, U., 2015. Late effects of low blood lead concentrations in children on school performance and cognitive functions. Neurotoxicology 49, 114–120.

Slaughter, J.C., Kim, E., Sheppard, L., Sullivan, J.H., Larson, T.V., Claiborn, C., 2005. Association between particulate matter and emergency room visits, hospital admissions and mortality in Spokane, Washington. J. Expo. Sci. Environ. Epidemiol. 15 (2), 153.

Smargiassi, A., Kosatsky, T., Hicks, J., Plante, C., Armstrong, B., Villeneuve, P.J., Goudreau, S., 2008. Risk of asthmatic episodes in children exposed to sulfur dioxide stack emissions from a refinery point source in Montreal, Canada. Environ. Health Perspect. 117 (4), 653–659.

Snowden, J.M., Mortimer, K.M., Dufour, M.S.K., Tager, I.B., 2015. Population intervention models to estimate ambient NO2 health effects in children with asthma. J. Expo. Sci. Environ. Epidemiol. 25 (6), 567–573.

Sousa, S.I.V., Alvim-Ferraz, M.C.M., Martins, F.G., 2013. Health effects of ozone focusing on childhood asthma: what is now known—a review from an epidemiological point of view. Chemosphere 90 (7), 2051–2058.

Srimuruganandam, B., Nagendra, S.M.S., 2010. Analysis and interpretation of particulate matter–PM10, PM2.5 and PM1 emissions from the heterogeneous traffic near an urban roadway. Atmos. Pollut. Res. 1 (3), 184–194.

Staessen, J.A., Lauwerys, R.R., Buchet, J.P., Bulpitt, C.J., Rondia, D., Vanrenterghem, Y., Amery, A., Cadmibel Study Group, 1992. Impairment of renal function with increasing blood lead concentrations in the general population. N. Engl. J. Med. 327 (3), 151–156.

Sunyer, J., Atkinson, R., Ballester, F., Le Tertre, A., Ayres, J.G., Forastiere, F., Forsberg, B., Vonk, J.M., Bisanti, L., Anderson, R.H., Schwartz, J., 2003b. Respiratory effects of sulphur dioxide: a hierarchical multicity analysis in the APHEA 2 study. Occup. Environ. Med. 60 (8), e2.

Sunyer, J., Ballester, F., Tertre, A.L., Atkinson, R., Ayres, J.G., Forastiere, F., Forsberg, B., Vonk, J.M., Bisanti, L., Tenías, J.M., Medina, S., 2003a. The association of daily sulfur dioxide air pollution levels with hospital admissions for cardiovascular diseases in Europe (the Aphea-II study). Eur. Heart J. 24 (8), 752–760.

Tao, Y., Huang, W., Huang, X., Zhong, L., Lu, S.E., Li, Y., Dai, L., Zhang, Y., Zhu, T., 2011. Estimated acute effects of ambient ozone and nitrogen dioxide on mortality in the Pearl River Delta of southern China. Environ. Health Perspect. 120 (3), 393–398.

Tarasova, O.A., Elansky, N.F., Kuznetsov, G.I., Kuznetsova, I.N., Senik, I.A., 2003. Impact of air transport on seasonal variations and trends of surface ozone at Kislovodsk high mountain station. J. Atmos. Chem. 45 (3), 245–259.

Thomas, W., Erbertseder, T., Ruppert, T., Van Roozendael, M., Verdebout, J., Balis, D., Meleti, C., Zerefos, C., 2005. On the retrieval of volcanic sulfur dioxide emissions from GOME backscatter measurements. J. Atmos. Chem. 50, 295–320.

Thompson, A.M., 1992. The oxidizing capacity of the Earth's atmosphere: probable past and future changes. Science 256 (5060), 1157–1165.

Thompson, R.L., Manning, A.C., Gloor, E., Schultz, U., Seifert, T., Hänsel, F., Jordan, A., Heimann, M., 2009. *In-situ* measurements of oxygen, carbon monoxide and greenhouse gases from Ochsenkopf tall tower in Germany. Atmos. Meas. Tech. 2 (2), 573–591.

Thornberry, T., Carroll, M.A., Keeler, G.J., Sillman, S., Bertman, S.B., Pippin, M.R., Ostling, K., Grossenbacher, J.W., Shepson, P.B., Cooper, O.R., Moody, J.L., 2001. Observations of reactive oxidized nitrogen and speciation of NO y during the PROPHET summer 1998 intensive. J. Geophys. Res. 106 (D20), 24359–24386.

Thurston, G.D., Ito, K., Lall, R., 2011. A source apportionment of US fine particulate matter air pollution. Atmos. Environ. 45 (24), 3924–3936.

Trainer, M., Parrish, D.D., Buhr, M.P., Norton, R.B., Fehsenfeld, F.C., Anlauf, K.G., Bottenheim, J.W., Tang, Y.Z., Wiebe, H.A., Roberts, J.M., Tanner, R.L., 1993. Correlation of ozone with NOy in photochemically aged air. J. Geophys. Res. 98 (D2), 2917–2925.

Trainer, M., Parrish, D.D., Goldan, P.D., Roberts, J., Fehsenfeld, F.C., 2000. Review of observation-based analysis of the regional factors influencing ozone concentrations. Atmos. Environ. 34 (12–14), 2045–2061.

Tsigaridis, K., Krol, M., Dentener, F.J., Balkanski, Y., Lathiere, J., Metzger, S., Hauglustaine, D.A., Kanakidou, M., 2006. Change in global aerosol composition since preindustrial times. Atmos. Chem. Phys. 6 (12), 5143–5162.

Tunnicliffe, W.S., Hilton, M.F., Harrison, R.M., Ayres, J.G., 2001. The effect of Sulphur dioxide exposure on indices of heart rate variability in normal and asthmatic adults. Eur. Respir. J. 17 (4), 604–608.

Turner, M.C., Krewski, D., Pope III, C.A., Chen, Y., Gapstur, S.M., Thun, M.J., 2011. Long-term ambient fine particulate matter air pollution and lung cancer in a large cohort of never-smokers. Am. J. Respir. Crit. Care Med. 184 (12), 1374–1381.

Valavanidis, A., Fiotakis, K., Vlahogianni, T., Bakeas, E.B., Triantafillaki, S., Paraskevopoulou, V., Dassenakis, M., 2006. Characterization of atmospheric particulates, particle-bound transition metals and polycyclic aromatic hydrocarbons of urban air in the Centre of Athens (Greece). Chemosphere 65 (5), 760–768.

Vingarzan, R., Taylor, B., 2003. Trend analysis of ground level ozone in the greater Vancouver/Fraser Valley area of British Columbia. Atmos. Environ. 37 (16), 2159–2171.

Volpino, P., Tomei, F., La Valle, C., Tomao, E., Rosati, M.V., Ciarrocca, M., De Sio, S., Cangemi, B., Vigliarolo, R., Fedele, F., 2004. Respiratory and cardiovascular function at rest and during exercise testing in a healthy working population: effects of outdoor traffic air pollution. Occup. Med. 54 (7), 475–482.

Volz, A., Kley, D., 1988. Evaluation of the Montsouris series of ozone measurements made in the nineteenth century. Nature 332 (6161), 240.

Von Klot, S., Wölke, G., Tuch, T., Heinrich, J., Dockery, D.W., Schwartz, J., Kreyling, W.G., Wichmann, H.E., Peters, A., 2002. Increased asthma medication use in association with ambient fine and ultrafine particles. Eur. Respir. J. 20 (3), 691–702.

Wang, Q.G., Han, Z., Wang, T., Zhang, R., 2008c. Impacts of biogenic emissions of VOC and NOx on tropospheric ozone during summertime in eastern China. Sci. Total Environ. 395 (1), 41–49.

Wang, Y., McElroy, M.B., Munger, J.W., Hao, J., Ma, H., Nielsen, C.P., Chen, Y., 2008a. Variations of O_3 and CO in summertime at a rural site near Beijing. Atmos. Chem. Phys. 8 (21), 6355–6363.

Wang, X., Sato, T., Xing, B., 2006. Size distribution and anthropogenic sources apportionment of airborne trace metals in Kanazawa, Japan. Chemosphere 65 (11), 2440–2448.

Wang, J.L., Wang, C.H., Lai, C.H., Chang, C.C., Liu, Y., Zhang, Y., Liu, S., Shao, M., 2008b. Characterization of ozone precursors in the Pearl River Delta by time series observation of non-methane hydrocarbons. Atmos. Environ. 42 (25), 6233–6246.

WHO, 2001. Health Impact Assessment of Air Pollution in the WHO European Region. World Health Organization.

WHO, 2004. Meta-Analysis of Time-Series Studies and Panel Studies of Particulate Matter (PM) and Ozone (O3): Report of a WHO Task Group. World Health Organization.

Wong, C.M., Vichit-Vadakan, N., Kan, H., Qian, Z., 2008. Public health and air pollution in Asia (PAPA): a multicity study of short-term effects of air pollution on mortality. Environ. Health Perspect. 116 (9), 1195–1202.

World Health Organization (WHO), 2018. Lead Poisoning and Health. http://www.who.int/news-room/fact-sheets/detail/lead-poisoning-and-health. (Accessed November 27, 2018).

Xie, H., Zafiriou, O.C., Umile, T.P., Kieber, D.J., 2005. Biological consumption of carbon monoxide in Delaware Bay, NW Atlantic and Beaufort Sea. Mar. Ecol. Prog. Ser. 290, 1–14.

Xie, Y., Zhao, B., Zhang, L., Luo, R., 2015. Spatiotemporal variations of $PM_{2.5}$ and PM_{10} concentrations between 31 Chinese cities and their relationships with SO_2, NO_2, CO and O_3. Particuology 20, 141–149.

Xu, X., Lin, W., 2011. Trends of tropospheric ozone over China based on satellite data (1979–2005). Adv. Clim. Chang. Res. 2 (1), 43–48.

Yang, K., Krotkov, N.A., Krueger, A.J., Carn, S.A., Bhartia, P.K., Levelt, P.F., 2007. Retrieval of large volcanic SO2 columns from the Aura ozone monitoring instrument: comparison and limitations. J. Geophys. Res. 112 (D24). D24S43.

Ying, Z., Tie, X., Li, G., 2009. Sensitivity of ozone concentrations to diurnal variations of surface emissions in Mexico City: a WRF/Chem modeling study. Atmos. Environ. 43 (4), 851–859.

Yorifuji, T., Kashima, S., Tsuda, T., Takao, S., Suzuki, E., Doi, H., Sugiyama, M., Ishikawa-Takata, K., Ohta, T., 2010. Long-term exposure to traffic-related air pollution and mortality in Shizuoka, Japan. Occup. Environ. Med. 67 (2), 111–117.

Yu, O., Sheppard, L., Lumley, T., Koenig, J.Q., Shapiro, G.G., 2000. Effects of ambient air pollution on symptoms of asthma in Seattle-area children enrolled in the CAMP study. Environ. Health Perspect. 108 (12), 1209.

Zaveri, R.A., Berkowitz, C.M., Kleinman, L.I., Springston, S.R., Doskey, P.V., Lonneman, W.A., Spicer, C.W., 2003. Ozone production efficiency and NOx depletion in an urban plume: interpretation of field observations and implications for evaluating O_3-NO_x-VOC sensitivity. J. Geophys. Res. 108 (D14), 4436.

Zellweger, C., Forrer, J., Hofer, P., Nyeki, S., Schwarzenbach, B., Weingartner, E., Ammann, M., Baltensperger, U., 2003. Partitioning of reactive nitrogen (NOy) and dependence on meteorological conditions in the lower free troposphere. Atmos. Chem. Phys. 3 (3), 779–796.

Zhang, Y.L., Cao, F., 2015. Fine particulate matter ($PM_{2.5}$) in China at a city level. Sci. Rep. 5, 14884.

Zhang, Y., Liu, P., Queen, A., Misenis, C., Pun, B., Seigneur, C., Wu, S.Y., 2006. A comprehensive performance evaluation of MM5-CMAQ for the summer 1999 southern oxidants study episode—Part II: gas and aerosol predictions. Atmos. Environ. 40 (26), 4839–4855.

Zhang, X., van Geffen, J., Liao, H., Zhang, P., Lou, S., 2012. Spatiotemporal variations of tropospheric SO2 over China by SCIAMACHY observations during 2004–2009. Atmos. Environ. 60, 238–246.

Zhang, Y., Wang, X., Chen, H., Yang, X., Chen, J., Allen, J.O., 2009. Source apportionment of lead-containing aerosol particles in Shanghai using single particle mass spectrometry. Chemosphere 74 (4), 501–507.

Zhang, Y., Xie, H., Fichot, C.G., Chen, G., 2008. Dark production of carbon monoxide (CO) from dissolved organic matter in the St. Lawrence estuarine system: implication for the global coastal and blue water CO budgets. J. Geophys. Res. 113. C12020.

Zhao, S., Yu, Y., Yin, D., He, J., Liu, N., Qu, J., Xiao, J., 2016. Annual and diurnal variations of gaseous and particulate pollutants in 31 provincial capital cities based on in situ air quality monitoring data from China National Environmental Monitoring Center. Environ. Int. 86, 92–106.

Zhou, Y., Fu, J.S., Zhuang, G., Levy, J.I., 2010. Risk-based prioritization among air pollution control strategies in the Yangtze River Delta, China. Environ. Health Perspect. 118 (9), 1204.

Zhu, Y., Hinds, W.C., Kim, S., Shen, S., Sioutas, C., 2002a. Study of ultrafine particles near a major highway with heavy-duty diesel traffic. Atmos. Environ. 36 (27), 4323–4335.

Zhu, Y., Hinds, W.C., Kim, S., Sioutas, C., 2002b. Concentration and size distribution of ultrafine particles near a major highway. J. Air Waste Manage. Assoc. 52 (9), 1032–1042.

Zu, K., Liu, X., Shi, L., Tao, G., Loftus, C.T., Lange, S., Goodman, J.E., 2017. Concentration-response of short-term ozone exposure and hospital admissions for asthma in Texas. Environ. Int. 104, 139–145.

Zuo, Y., Jones, R.D., 1995. Formation of carbon monoxide by photolysis of dissolved marine organic material and its significance in the carbon cycling of the oceans. Naturwissenschaften 82 (10), 472–474.

Further reading

Bell, M.L., Zanobetti, A., Dominici, F., 2014. Who is more affected by ozone pollution? A systematic review and meta-analysis. Am. J. Epidemiol. 180 (1), 15–28.

Ma, Z., Hu, X., Sayer, A.M., Levy, R., Zhang, Q., Xue, Y., Tong, S., Bi, J., Huang, L., Liu, Y., 2015. Satellite-based spatiotemporal trends in PM2. 5 concentrations: China, 2004–2013. Environ. Health Perspect. 124 (2), 184–192.

Pirkle, J.L., Brody, D.J., Gunter, E.W., Kramer, R.A., Paschal, D.C., Flegal, K.M., Matte, T.D., 1994. The decline in blood lead levels in the United States: the National Health and nutrition examination surveys (NHANES). JAMA 272 (4), 284–291.

Statistical analysis for air pollution data

Jingjing Yin[a], Hao Zhang[b], Lixin Li[c]

[a]*Department of Biostatistics, Epidemiology, and Environmental Health Sciences, Jiann-Ping Hsu College of Public Health, Georgia Southern University, Statesboro, GA, United States*
[b]*Department of Information Technology, College of Engineering and Computing, Georgia Southern University, Statesboro, GA, United States*
[c]*Department of Computer Science, Georgia Southern University, Statesboro, GA, United States*

Recently, there is a dramatic increase in the necessity for statistical analysis of air pollution data. The collection of samples from the field and the laboratory analysis of the samples are generally very costly; therefore, we should be very cautious about choosing the correct statistical analysis to comprehend the sampled data and interpret the results.

There are many challenges in the analysis of air pollution data, such as potential measurement error and correlated sample data due to temporal and spatial clusters. The later sections of the chapter will discuss in particular the time series data as an example of correlated data.

This chapter is designed as an introductory tutorial of common statistical analysis for air pollution data using R (R Foundation for Statistical Computing, Vienna, Austria), a popular statistical software. Much of the statistical books give detailed theoretical basics for the statistical methods; therefore, in this chapter, we emphasize the simple applications of the statistical procedures. Especially, we explain in detail how to utilize R to carry out some statistical analyses, including the common time series analysis for air pollution data. This chapter is useful for readers who have minimal knowledge of statistical programming in R as programming codes are provided along with corresponding results and interpretations, so readers can easily adopt the R codes in this chapter for their own analysis. This chapter also explains the general concepts of popular statistical analysis methods for air pollution data, which can be useful for readers of nonstatistical backgrounds.

2.1 Descriptive and graphic summaries of data

Statistical analysis of data generally uses both descriptive and inferential statistics to draw conclusions. Descriptive analysis deals with the summarization, calculation, and graphical representation of data from a sample. Descriptive statistics summarize data so we can understand the distribution, relation, or trend of data. However, we cannot use descriptive analysis to infer a population, such as estimation, hypothesis testing, or prediction. In order to perform descriptive data analysis appropriately, we need to understand the types of data, i.e., numeric or categorical and the choices of descriptive statistics and graphics for different types of data.

2.1.1 Descriptive statistics

It is difficult to comprehend what pattern or relation the data presents if we merely look at the raw data, and it is more difficult for the environmental data as usually the study collects many observations. Descriptive analysis summarizes the data by simple summary statistics, frequency tables, or in graphics; thus, we can interpret the data more straightforwardly.

2.1.1.1 Univariate descriptive statistics

There are two types of statistics for summarizing the data distribution of a single variable: the center or location and the dispersion or spread. The center of the data distribution is usually presented by the sample mean/average, sample median, and sample mode. The spread of the data distribution can be measured by range, which is the difference between the maximum and the minimum, interquartile range (3rd quartile minus 1st quartile), and variance, which is the average of the squared distances between each data value to their sample average. If the data distribution is symmetric or close to, then sample mean and variance are commonly used. If the data is skewed (with extreme values and not distributed even close to symmetric), then sample median and interquartile range are usually used.

In order to analyze data and obtain the descriptive statistics in R, we first need to read the data (if it was saved in a raw data file) into R using the `scan()` function. For other types of data formats, you can choose from R functions: `read.csv()`, `read.table()`, or `read.delim()` depending on the data types. Also, you need to tell R the file directory by either setting up the working directory by `setwd()` or giving it directly in the above-mentioned functions.

```
setwd("C:\\Users\\... ")
mydat=read.csv("time_series.csv", header = TRUE)
```

Alternatively, just call:
```
mydat=read.csv("C:\\Users\\... \\time_series.csv", header = TRUE)
```

For using a specific column or variable in the data set, just call:
```
pmn=mydat$pmn
```

The pmn column gives the time series data of $PM_{2.5}$ for each month over years. For the source of the data, please refer to (Kersey et al., 2018; Li et al., 2014; Tong et al., 2018). In the same data set, there are also columns indicating the time (year, month) that we will use in later sections for time-series analysis.

Table 2.1 Summary statistics of PM$_{2.5}$

Min.	1st Qu.	Median	Mean	3rd Qu.	Max.
7.062	9.255	10.500	10.560	11.810	14.930

In R programming, you can obtain each descriptive statistic separately as `sd()` for standard deviation, `mean()` for mean, `var()` for variance, `range()` for range, and `quantile( , given_percentile=)` for quartile, etc. In addition, you can use `summary()` function. For example, the quartiles of monthly average value of PM$_{2.5}$ denoted as a variable "pmn," can be obtained by (Table 2.1):

```
summary(pmn)
```

The `summary ()` function also can be used for summarizing all variables (lcr stands for lung cancer rate, and pmn stands for PM$_{2.5}$ values) together in the data set (Table 2.2):

```
summary(mydat)
```

Alternatively, you can call `sapply()` function with a specified summary statistic such as mean, sd, var, min, max, median, range, and quartile. For example, to obtain the sample mean, we have (Table 2.3):

```
sapply(mydat, mean, na.rm=TRUE)
```

If you want group mean, then call `aggregate()` function (Table 2.4). For example, we can get the average PM$_{2.5}$ values for each month across all years as in Fig. 2.1:

```
aggregate(mydat$pmn, list(mydat$month), mean)
```

Plotting these values, we have:

```
a=aggregate(mydat$pmn, list(mydat$month), mean)
plot(a, type="l", xlab="months", ylab="PM2.5")
```

Table 2.2 Summary statistics of all variables

year	month	lcr	pmn
Min.: 1999	Min.: 1.00	Min.: 39.60	Min.: 7.062
1st Qu.: 2003	1st Qu.: 3.75	1st Qu.: 49.10	1st Qu.: 9.255
Median:2006	Median: 6.50	Median:51.65	Median:10.502
Mean: 2006	Mean: 6.50	Mean: 51.29	Mean:10.558
3rd Qu.: 2010	3rd Qu.: 9.25	3rd Qu.: 53.70	3rd Qu.: 11.812
Max.: 2014	Max.: 12.00	Max.: 60.50	Max.: 14.933

Table 2.3 Mean of all variables in the data set

year	month	lcr	pmn
2006.50000	6.50000	51.29375	10.55798

Table 2.4 Monthly means of PM$_{2.5}$

Group.1	x
1	1 11.103109
2	2 11.326965
3	3 10.934678
4	4 9.887844
5	5 9.187555
6	6 9.727839
7	7 10.963548
8	8 11.905695
9	9 11.536693
10	10 10.172143
11	11 9.666913
12	12 10.282784

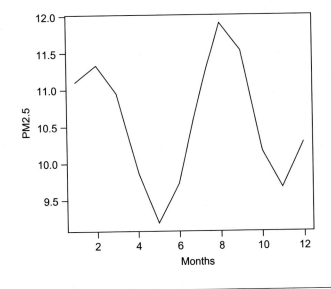

FIG. 2.1

Time series plot of PM$_{2.5}$ for each month across all years.

Later, we will learn that the ups and downs demonstrate some seasonality in the PM$_{2.5}$ data. Let us also check the changes over years by averaging across months as demonstrated in Fig. 2.2, and we observe a downward pattern over the years, which we will learn later is referred as trend-in-time-series;

```
a2=aggregate(mydat$pmn, list(mydat$year), mean)
plot(a2, type="l", xlab="year", ylab="PM2.5")
```

2.1.1.2 Sample correlation

Sample correlation is a descriptive statistic for describing the relationship between two continuous variables. If the data is symmetric and the relationship is linear, Pearson correlation coefficient estimate can be used. If the data is not symmetric, the nonparametric correlation estimate, Spearman correlation, can be used. We can call cor() function to obtain both correlations:

```
cor(pmny , lcr, method = "pearson")
[1] 0.9154907
cor(pmny , lcr, method = "spearman")
[1] 0.961001
```

We can see that $PM_{2.5}$ level and the lung cancer rate are highly correlated. Similarly, we can check if there is correlation between values from previous time and current time, which is referred as autocorrelation for time-series data.

2.1.2 Univariate plots

To observe the distribution of a numeric variable, generally we start with a histogram, density plot, or box plot. For categorical variable, we can use pie graph or bar chart. In inferential statistics, normality is an important distributional assumption for many statistical methods, and Q-Q plot is a common tool to help us to check normality of data. For the univariate plots, we applied the R package "Lattice," which gives better graphical presentation than the default base *R*graphics. So just call "library(lattice)" in R to use it.

2.1.2.1 Histogram

A histogram demonstrates the distribution of numerical data by counting the number of observations in each data range. For example, if we want to observe the distribution of the $PM_{2.5}$ levels in our data set, we can plot a histogram using the following R code:

```
histogram(~pmn, mydat, xlab="National PM2.5 level from 1999 to
2014")
```

The above R code produces Fig. 2.3. From Fig. 2.3, we can conclude that the $PM_{2.5}$ values are distributed symmetrically. To check the distribution of $PM_{2.5}$ for each year, simply plot histogram of pmn conditional on year as follows (Fig. 2.4):

```
histogram(~pmn | year, mydat, xlab="National PM2.5 level for each
year from 1999 to 2014")
```

We can always draw similar plots, such as histograms, for different levels of a grouping variable; as in the example by Fig. 2.4, we can draw histograms for each year. Similarly, we can also draw histograms for each month.

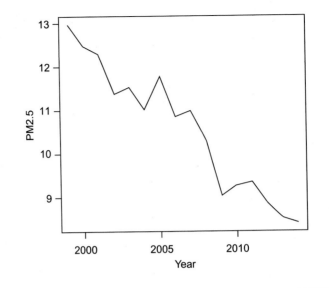

FIG. 2.2

Time series plot of $PM_{2.5}$ for each year across months.

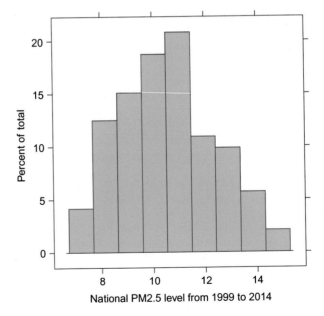

FIG. 2.3

Histogram of $PM_{2.5}$ values.

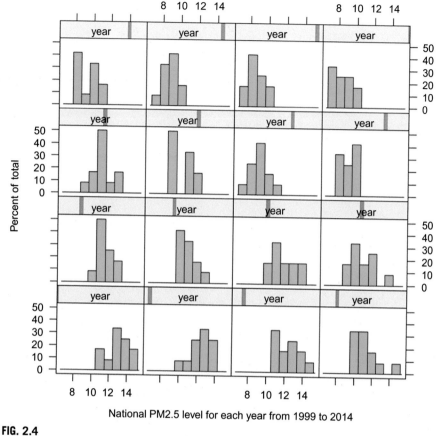

FIG. 2.4

Histogram of $PM_{2.5}$ values for each year.

2.1.2.2 Density plots

The Density plot looks like a smoothed version of the histogram, which applies kernel smoothing. The sample density plot estimates the population probability distribution of a continuous variable. We can draw an overall density plot and then draw the same density plots for different levels of a grouping variable (such as year or month). For example, we can use density plot to observe the distribution of $PM_{2.5}$ levels across all years (Fig. 2.5) and as well as for each year (Fig. 2.6) separately as follows:

```
densityplot(~pmn, mydat, xlab="National PM2.5 level from 1999 to 2014")
densityplot(~pmn | year, mydat, xlab="National PM2.5 level for each
year from 1999 to 2014")
```

We can conclude from the histograms and density plots that the $PM_{2.5}$ values are mostly symmetric for each year and also when pooling all years together.

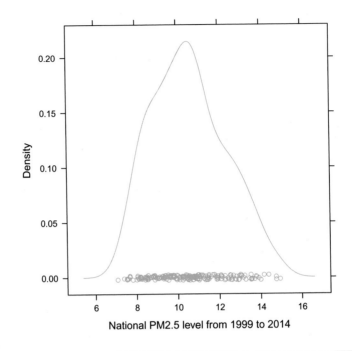

FIG. 2.5

Density plot of PM$_{2.5}$ values.

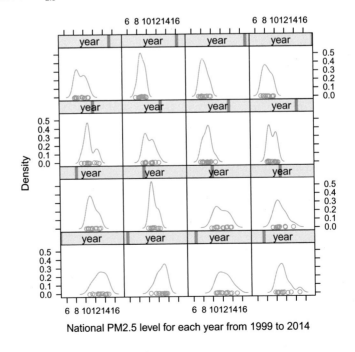

FIG. 2.6

Density plots of PM$_{2.5}$ values for each year.

2.1.2.3 Box plot

A box plot presents the minimum, first quartile, median, mean, third quartile, and maximum in one single graph. Compared to the histogram or density plot, the box plot is a more comprehensive presentation of the data. We can apply bwplot (group~value, data) in the "lattice" package to produce the box plot (Sarkar, 2008). For example, we can use box plot to observe the distribution of $PM_{2.5}$ levels across all years (Fig. 2.7) and as well as for each year (Fig. 2.8) or for each month (Fig. 2.9) separately as follows:

```
library(lattice)
bwplot(~pmn, mydat)
bwplot(year~pmn, mydat)
bwplot(month~pmn, mydat)
```

2.1.2.4 Q-Q plot

The Q-Q plot is used to assess normality of data as normality is a common statistical assumption for many procedures. If the sample data are normally distributed, the Q-Q plot should look like a rather straight line. For example, to see if the $PM_{2.5}$ values are normally distributed, we have plotted the Q-Q plot of $PM_{2.5}$ values for the entire data set as in Fig. 2.10. Fig. 2.10 depicts a fairly straight line on the diagonal; therefore, we conclude the $PM_{2.5}$ values are normally distributed.

```
qqmath(~pmn, mydat)
```

To check normality of $PM_{2.5}$ for each year, we can call the following to produce Q-Q plots for each year as in Fig. 2.11:

```
qqmath(~pmn | year, mydat)
```

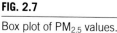

FIG. 2.7

Box plot of $PM_{2.5}$ values.

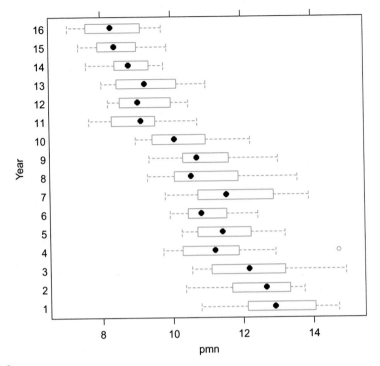

FIG. 2.8

Box plots of PM$_{2.5}$ values for each year.

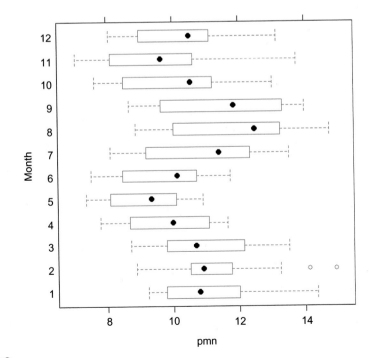

FIG. 2.9

Box plots of PM$_{2.5}$ values for each month.

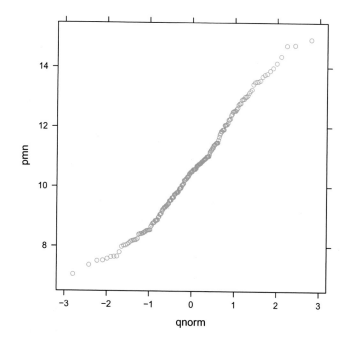

FIG. 2.10

Q-Q plot of PM$_{2.5}$ values.

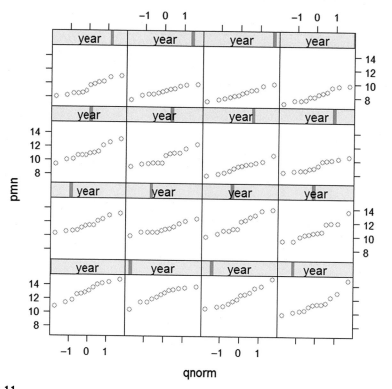

FIG. 2.11

Q-Q plots of PM$_{2.5}$ values for each year.

2.1.3 Scatterplot for relations between two variables

A scatterplot shows the relationship between two numerical variables. This plot is particularly useful for linear regression analysis. For example, to see if lung cancer rate changes along with the levels of $PM_{2.5}$, we can draw a scatterplot of $PM_{2.5}$ values against lung cancer rates as in Fig. 2.12. Note that $PM_{2.5}$ values should be placed on the X-axis and cancer rate values on the Y-axis as we assume the PM2.5 will affect lung cancer rate, but not the other direction. We always put predictor/independent variable on the X-axis and the response/dependent variable on the Y-axis. From Fig. 2.12, it is clear to see the two variables are positively correlated:

```
xyplot(lcr~pmny, mydat2,xlab="National PM2.5 level", ylab="National
Lung cancer rate")
```

2.2 Time series analysis

Most air pollution data are collected over time; therefore, the same sampling unit will have correlated measurements over time. In statistics, such data is referred as time series. A time series is a sequence of measurements of the same variable collected over time. The biggest challenge of time series data is that the data are not necessarily independent and identically distributed (i.i.d.); therefore, regular statistical inference methods assuming i.i.d. samples are not readily applied for time-series data.

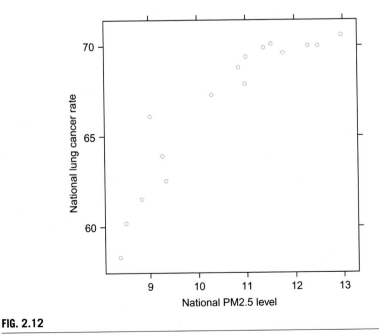

FIG. 2.12

Scatterplots of $PM_{2.5}$ values versus lung cancer rates.

2.2.1 Reading and plotting time series data

We can transform a data matrix or a vector into a time series object by using the `ts()` function. The potential dependency of the observations are related to the ordering of the observations; therefore, it is important to indicate the time when each of the measurements was connected. For data collected at regular time intervals less than a year, such as monthly or quarterly for each year, we can use "`frequency=`" parameter in the `ts()` function. For example, `frequency=12` for monthly data and `frequency=4` for quarterly data. This is common for air pollution data as usually the measurements were collected at more frequent time intervals.

The beginning time that the data was first collected can be specified in the "`start=`" parameter in the `ts()` function, and this can be set as a two-level time indicator. For example, for a start time of June 2003, you would set `start=c(2003,6)`. Similarly, you can specify the end time for the "`end=`" parameter. Therefore, to specify a time series object in R from the monthly collected $PM_{2.5}$ time series data from January 1999 to December 2014, we have

```
pmn<- ts(pmn, frequency=12, start=c(1999,1), end=c(2014,12))
```

	Jan	Feb	Mar	Apr	May	Jun	Jul	Aug	Sep	Oct	Nov	Dec
1999	14.36925	14.14525	13.06155	11.38569	10.80921	11.7454	13.52697	14.73047	13.98393	12.79078	12.51248	12.60173
2000	12.54458	13.2599	13.52996	11.42648	10.37418	11.47511	12.13434	12.79644	13.43586	11.91581	13.75579	13.14853
2001	13.79031	14.93347	12.91098	11.66088	10.91219	11.30363	12.6522	12.53382	13.50747	10.77485	10.54951	11.8598
2002	11.20342	11.48142	10.90932	10.23122	9.746394	10.08201	12.28464	14.74488	12.94047	11.2525	10.3397	11.23403
2003	12.08798	11.47446	11.41041	11.45282	10.26782	10.58273	12.39318	12.93682	13.20794	11.07849	10.68934	10.74287
2004	10.47984	11.31501	12.07211	10.78151	9.94276	10.50261	10.89145	12.44149	11.79669	11.00764	10.43278	10.43915
2005	11.02419	12.07763	12.25347	10.76055	9.80163	10.30553	12.75072	13.69066	13.88819	13.01653	10.71714	11.01645
2006	11.90824	10.69529	10.5008	9.946617	9.309125	10.21031	11.85846	13.56608	11.90908	10.33349	9.318469	10.59553
2007	10.57181	10.57195	10.58068	10.06146	9.359684	10.86304	12.06423	12.54694	13.01692	11.16801	9.988559	10.95289
2008	11.5693	11.00738	10.84224	9.541454	9.365464	9.387192	10.94673	12.24766	10.61466	9.537977	8.967015	9.555484
2009	9.43624	10.74025	9.803707	9.083268	7.987502	8.185597	8.793241	9.297602	9.641295	8.430727	7.653915	9.17981
2010	9.937665	10.48882	10.04669	8.862421	8.439003	8.550674	9.262722	10.26177	9.836926	8.540503	8.209449	8.800904
2011	10.21511	10.84596	9.807849	8.52381	8.217365	8.870305	10.10216	10.98488	9.62614	8.479632	8.02813	8.415697
2012	9.231998	8.887774	8.715934	8.198603	7.58579	8.396496	9.100993	9.78356	9.522488	8.533125	8.413299	9.753531
2013	9.643948	9.880725	8.773004	7.803394	7.373845	7.659537	8.545569	9.062344	8.992755	8.251906	8.033536	8.160556
2014	9.63588	9.426172	9.736149	8.48532	7.508916	7.525258	8.109185	8.865713	8.666258	7.642314	7.061515	8.067568

In order to comprehend the pattern of the time series data, usually the first step is to make a plot of the time series data by calling the `plot.ts()` function in R. The default labels for x and y axis are "Time" and the object name of the plotting time series. If you want to change the labels, just specify "`xlab=`" and "`ylab=`" accordingly. For example, to plot the time series of the above $PM_{2.5}$ data, we have

```
plot.ts(pmn, ylab="PM2.5")
```

From Fig. 2.13, we can see that there is a consistent downward trend over the entire time span. There seems some seasonality, but the pattern was not clear in this plot. One way to quickly check potential seasonality is to obtain the average for each month and compare the averages in a single plot. For example, Fig. 2.14 gives the

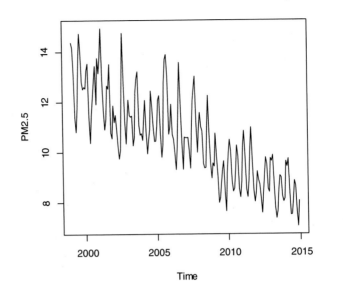

FIG. 2.13

Time series plot of overall trend of PM$_{2.5}$ values.

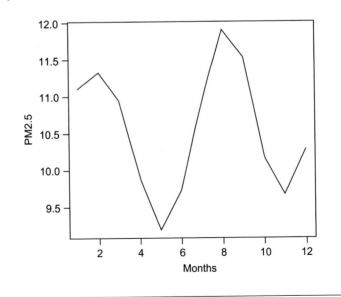

FIG. 2.14

Time series plot of seasonality of PM$_{2.5}$ values for each month.

monthly PM$_{2.5}$ averages collected together as a curve. We can obviously see that there is some seasonal pattern from Fig. 2.14.

In order to observe more clearly the seasonality, we need to decompose the time series with a seasonal pattern into four components: trend, season, and a remaining random error. Note that as in most cases, air pollution data is usually a seasonal time series; hence the discussion about time series analysis in this chapter will focus on the seasonal time series.

Usually the three components are additive if the seasonal pattern stays similar over time. The other is the multiplicative model, which is used when seasonal variation increases over time. For example, the seasonal variability of the above $PM_{2.5}$ time series plot looks more constant; therefore, an additive model would be better. To decompose the $PM_{2.5}$ time series, we call the `decompose()` function in R, and specify the parameter `type` = "additive" (or "multiplicative" if you want to fit a multiplicative model).

```
comp=decompose(pmn, type = "additive")
```

The decomposed $PM_{2.5}$ time series has four outputs: observed, trend, seasonal, and random; for the observed time series values, just call:

```
Comp$x
```

	Jan	Feb	Mar	Apr	May	Jun	Jul	Aug	Sep	Oct	Nov	Dec
1999	14.36925	14.14525	13.06155	11.38569	10.80921	11.7454	13.52697	14.73047	13.98393	12.79078	12.51248	12.60173
2000	12.54458	13.2599	13.52996	11.42648	10.37418	11.47511	12.13434	12.79644	13.43586	11.91581	13.75579	13.14853
2001	13.79031	14.93347	12.91098	11.66088	10.91219	11.30363	12.6522	12.53382	13.50747	10.77485	10.54951	11.8598
2002	11.20342	11.48142	10.90932	10.23122	9.746394	10.08201	12.28464	14.74488	12.94047	11.2525	10.3397	11.23403
2003	12.08798	11.47446	11.41041	11.45282	10.26782	10.58273	12.39318	12.93682	13.20794	11.07849	10.68934	10.74287
2004	10.47984	11.31501	12.07211	10.78151	9.94276	10.50261	10.89145	12.44149	11.79669	11.00764	10.43278	10.43915
2005	11.02419	12.07763	12.25347	10.76055	9.80163	10.30553	12.75072	13.69066	13.88819	13.01653	10.71714	11.01645
2006	11.90824	10.69529	10.5008	9.946617	9.309125	10.21031	11.85846	13.56608	11.90908	10.33349	9.318469	10.59553
2007	10.57181	10.57195	10.58068	10.06146	9.359684	10.86304	12.06423	12.54694	13.01692	11.16801	9.988559	10.95289
2008	11.5693	11.00738	10.84224	9.541454	9.365464	9.387192	10.94673	12.24766	10.61466	9.537977	8.967015	9.555484
2009	9.43624	10.74025	9.803707	9.083268	7.987502	8.185597	8.793241	9.297602	9.641295	8.430727	7.653915	9.17981
2010	9.937665	10.48882	10.04669	8.862421	8.439003	8.550674	9.262722	10.26177	9.836926	8.540503	8.209449	8.800904
2011	10.21511	10.84596	9.807849	8.52381	8.217365	8.870305	10.10216	10.98488	9.62614	8.479632	8.02813	8.415697
2012	9.231998	8.887774	8.715934	8.198603	7.58579	8.396496	9.100993	9.78356	9.522488	8.533125	8.413299	9.753531
2013	9.643948	9.880725	8.773004	7.803394	7.373845	7.659537	8.545569	9.062344	8.992755	8.251906	8.033536	8.160556
2014	9.63588	9.426172	9.736149	8.48532	7.508916	7.525258	8.109185	8.865713	8.666258	7.642314	7.061515	8.067568

The observed time series

Similarly, to list the estimated decomposed trend or seasonal component, we call `comp$trend` and `comp$seasonal`. For the remaining random noise, we call `comp$random`. We also can plot all the estimated component values in one single graph using `plot()` function and add X-axis labeling values by `axis()` function as in Fig. 2.15:

```
plot(comp)
time<-seq(1999,2015, by=1)
axis(1, at = time)
```

Fig. 2.15 gives the decomposed time series. After decomposing, we observe in Fig. 2.15 that the estimated seasonal factors vary from January to December but repeats the same pattern for each year. If we call for the decomposed seasonal component by calling `comp$seasonal` in R, we can see the seasonal component more clearly. It goes up slightly in Spring (February reaches to a value of 0.60); drops down in the Summer (May has the lowest value of −1.37), goes up again and peaks

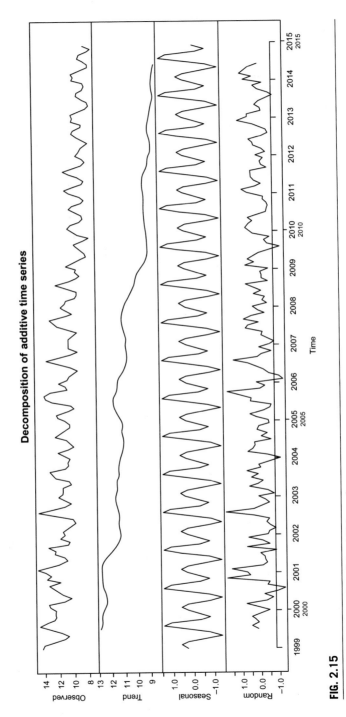

FIG. 2.15

Decomposed time series plots of $PM_{2.5}$ values.

in Fall (August has the largest value of 1.43), and finally drops down in the Winter (November reaches to a value of −0.78).

	Jan	Feb	Mar	Apr	May	Jun	Jul	Aug	Sep	Oct	Nov	Dec
1999	0.312183	0.597237	0.282105	-0.69372	-1.37282	-0.83119	0.449873	1.430651	1.072691	-0.29727	-0.78024	-0.16949
2000	0.312183	0.597237	0.282105	-0.69372	-1.37282	-0.83119	0.449873	1.430651	1.072691	-0.29727	-0.78024	-0.16949
2001	0.312183	0.597237	0.282105	-0.69372	-1.37282	-0.83119	0.449873	1.430651	1.072691	-0.29727	-0.78024	-0.16949
2002	0.312183	0.597237	0.282105	-0.69372	-1.37282	-0.83119	0.449873	1.430651	1.072691	-0.29727	-0.78024	-0.16949
2003	0.312183	0.597237	0.282105	-0.69372	-1.37282	-0.83119	0.449873	1.430651	1.072691	-0.29727	-0.78024	-0.16949
2004	0.312183	0.597237	0.282105	-0.69372	-1.37282	-0.83119	0.449873	1.430651	1.072691	-0.29727	-0.78024	-0.16949
2005	0.312183	0.597237	0.282105	-0.69372	-1.37282	-0.83119	0.449873	1.430651	1.072691	-0.29727	-0.78024	-0.16949
2006	0.312183	0.597237	0.282105	-0.69372	-1.37282	-0.83119	0.449873	1.430651	1.072691	-0.29727	-0.78024	-0.16949
2007	0.312183	0.597237	0.282105	-0.69372	-1.37282	-0.83119	0.449873	1.430651	1.072691	-0.29727	-0.78024	-0.16949
2008	0.312183	0.597237	0.282105	-0.69372	-1.37282	-0.83119	0.449873	1.430651	1.072691	-0.29727	-0.78024	-0.16949
2009	0.312183	0.597237	0.282105	-0.69372	-1.37282	-0.83119	0.449873	1.430651	1.072691	-0.29727	-0.78024	-0.16949
2010	0.312183	0.597237	0.282105	-0.69372	-1.37282	-0.83119	0.449873	1.430651	1.072691	-0.29727	-0.78024	-0.16949
2011	0.312183	0.597237	0.282105	-0.69372	-1.37282	-0.83119	0.449873	1.430651	1.072691	-0.29727	-0.78024	-0.16949
2012	0.312183	0.597237	0.282105	-0.69372	-1.37282	-0.83119	0.449873	1.430651	1.072691	-0.29727	-0.78024	-0.16949
2013	0.312183	0.597237	0.282105	-0.69372	-1.37282	-0.83119	0.449873	1.430651	1.072691	-0.29727	-0.78024	-0.16949
2014	0.312183	0.597237	0.282105	-0.69372	-1.37282	-0.83119	0.449873	1.430651	1.072691	-0.29727	-0.78024	-0.16949

In addition, Fig. 2.16 gives the Time series plot of the mean PM2.5 values for each month averaging out all years. From Fig. 2.16, we observe two peaks in a year for the $PM_{2.5}$ time series data, so we may consider the seasonal pattern repeats every six months (May to October and November to April). Alternatively, since the first peak (November to April) is slightly lower than the second peak (May to October), we can also consider the seasonal pattern has two peaks and repeats every year.

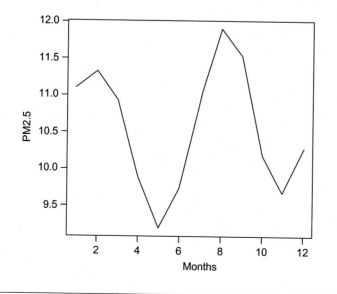

FIG. 2.16

Time series plot of $PM_{2.5}$ values averaging out all years.

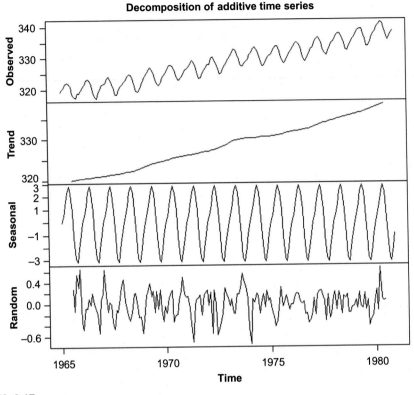

FIG. 2.17

Decomposed time series plot of CO_2 values.

Another example is the CO_2 emission data (Hyndman, 2014), and we decompose the data as shown in Fig. 2.17. We can see there is a constant upward trend, and the single peak in the seasonal components repeats each year. Therefore, this is a seasonal time series of same pattern every 12 months.

2.2.2 ARIMA time series models

ARIMA (Autoregressive Integrated Moving Average) models estimate the value at current time based on the past values (autoregressive) and past errors (moving average) of the sequence. AR(1) (autoregressive model of order 1) model is a linear model regressing the current value on the most recent past time. The order of the AR model indicates how many past values are used. Likewise, the order of MA indicates how many past error terms are used. Usually, we use p and q to note the orders of the AR and MA terms, respectively. In addition, ARIMA models are defined for stationary time series; i.e., the values do not change significantly over time. Therefore, for nonstationary time series, we need to difference the time series to make it stationary.

The number of times we need to difference the time series to be stationary determines the order of the difference, which is denoted as d. For example, a first order difference (d=1) is taken for linear trend. You can difference a time series using the "diff()" function in R.

The ARIMA model contains both AR and MA terms and differencing operations. AR model only uses autoregressive terms while MA model only uses moving average terms, and ARMA model has both terms but no differencing. We use p, d, q to denote the orders of AR, differencing, and MA terms respectively, and we use ARIMA(p,d,q) notation. For example, ARIMA(2,1,0) contains two AR terms and a first difference, but no MA term. If time series repeats with a certain pattern, we can use seasonal ARIMA model, and we use uppercase S to denote the number of time lags between two adjacent repeated patterns. For example, for monthly data, usually S=12, a seasonal AR(1) model would use values from the same month of last year to predict the current month. The seasonal ARIMA model can have both nonseasonal and seasonal terms and is denoted as ARIMA(p, d, q, P, D, Q)[S], with lowercase p, d, q representing the nonseasonal orders for AR, differencing and MA, while uppercase P, D, Q standing for the seasonal orders, and S=time span between repeated seasonal patterns.

2.2.2.1 Identifying ARIMA model

We generally eyeball the ACF (autocorrelation function) and the PACF (partial autocorrelation function) plots to "guess" some candidate ARIMA models for further evaluation. The ACF plot demonstrates the pattern of autocorrelations; each observation is correlated with its values in the previous times. If the correlation is positive, the ACFs are all positive. If the correlation is negative, the ACF values change between positive and negative. For example, Fig. 2.18 gives the scatterplots of current $PM_{2.5}$ values versus the corresponding previous values of lag 1, which is from a previous month as the data was collected monthly. Fig. 2.18 shows that $PM_{2.5}$ values are positively related with the most recent previous measurements with an estimated correlation of 0.82. The estimated regression line using the most recent previous values to predict current values is plotted as the red line in Fig. 2.18.

This scatterplot by Fig. 2.18 can be obtained by calling lag1.plot() function in the "astsa" package (Shumway and Stoffer, 2017). So first, you need to install the package and apply it using:

```
install.packages("astsa")
library(astsa)
lag1.plot(pmn, max.lag=1)
```

The "max.lag" parameter gives the maximum lagged values that you want to relate to the current value. For example, to obtain scatterplots of current value versus the previous four times (Fig. 2.19), we have:

```
lag1.plot(pmn, max.lag=4)
```

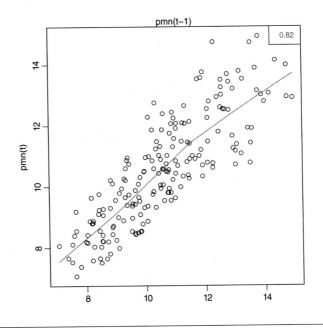

FIG. 2.18

Scatterplot of PM$_{2.5}$ values versus its past values of Lag 1.

Fig. 2.19 demonstrates that the correlations between PM2.5 and its past values of most recent four times are all positive but with a decreasing pattern as the lag time increases. To check such correlation for all times, we can use R functions `acf()` and `pacf()` to produce corresponding plots for autocorrelation and partial correlations, so that we can better identify the candidate time series models. Fig. 2.20 gives the ACF plot of PM$_{2.5}$ time series data; we can see that all correlations with the past values are positive, resulting in an ACF plot with all positive values. We use the following R code to produce Fig. 2.20:

```
acf(pmn)
```

In general statistics, the partial correlation is defined as the correlation between two variables conditioning on the values of other variables. For example, consider a regression model with y being the response variable and x1 and x2 predictor variables. The partial correlation between y and x1 is found by correlating the residuals from two different regressions: (1) regressing y on x2, (2) regressing x1 on x2. And then we correlate the remaining parts (residuals) of the two regression models. In the multiple regression, we usually refer to the regression coefficient (slope) as partial coefficient since the slope demonstrates the linear dependency between the variable of interest (x1) and the response (y) by holding the other predictors (such as x2) as constant. Similarly, for time-series analysis, the p-order PACF is defined as the partial correlation between values at current time and values from the previous time of p lags, conditioning on the values in between the two time points of interest. For example, the 4th order PACF of the monthly data calculates the partial correlation between current month (for example, the current is June) and the previous month of

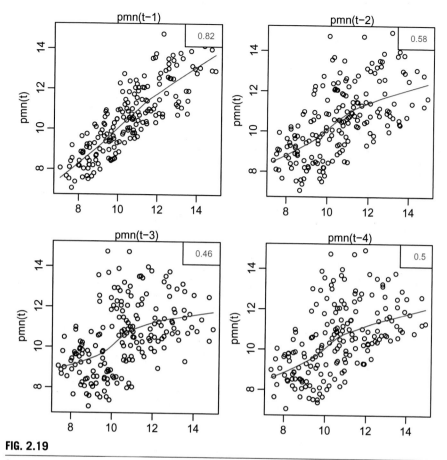

FIG. 2.19

Scatterplots of PM$_{2.5}$ values versus its past values of Lag 1, Lag 2, Lag 3, and Lag 4.

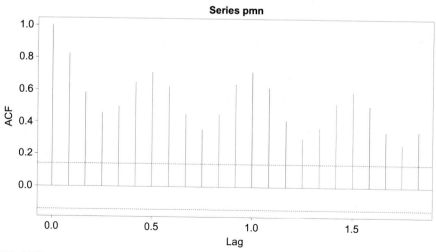

FIG. 2.20

ACF plot of PM$_{2.5}$ values.

four lags (which is February), conditioning on the months in between the current month and month of lag 4 (i.e., May, April, March).

Generally, we use the spikes in PACF plot to identify the order of the AR model and the spikes in ACF plot for the order of the MA model. If the plots are rather flat instead of exponentially decaying to zero, then the order for the respective model is zero. For example, the ACF plot of $PM_{2.5}$ series in Fig. 2.20 does not show any spikes but rather looks flat, which does not indicate a MA model (i.e., MA(0)). In addition, the ACF plot also demonstrates some seasonal pattern, which slowly decreases (without any spike) every six months, which suggests a seasonal model, and a seasonal differencing might be necessary (We will discuss this more later.). However, the PACF plot in Fig. 2.21 indicates an AR (1) model more likely due to the sharp spike on the time of lag 1. The R function `pacf()` gives PACF plot as shown in Fig. 2.21.

The orders of ARMA models sometimes are not so obviously seen from the ACF and PACF plots. If both ACF and PACF plots look rather flat, the series is usually nonstationary, and we need to take the difference of the data and evaluate the ACF and PACF plots again of the differenced data. For example, after taking a seasonal difference of the $PM_{2.5}$ values with twelve lags, the ACF and PACF plots depict some spikes, which indicates the differenced time series is more stationary, and the spikes can be used to determine the order of the ARIMA model. The following R code gives the differenced data by calling `diff()` function with a lag order of 12 and draws the ACF (Fig. 2.22) and PACF (Fig. 2.23) plots on the differenced data.

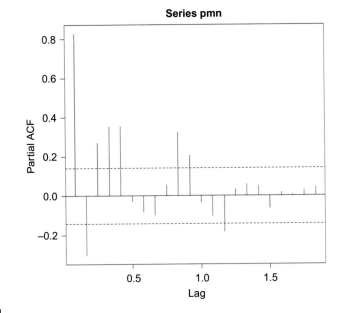

FIG. 2.21

PACF plot of $PM_{2.5}$ values.

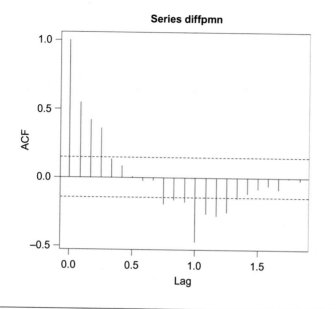

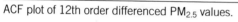

FIG. 2.22

ACF plot of 12th order differenced PM$_{2.5}$ values.

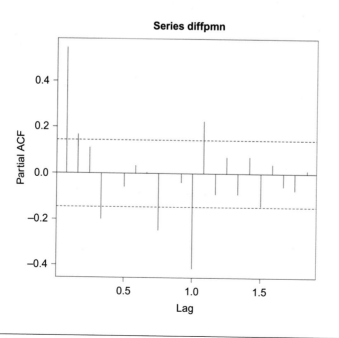

FIG. 2.23

PACF plot of 12th order differenced PM$_{2.5}$ values.

The spikes on ACF and PACF are more obvious in the seasonal differenced data, which suggests fitting a seasonal ARIMA model:

```
diffpmn=diff(pmn,12)
acf(diffpmn)
pacf(diffpmn)
```

Seasonal ARIMA models use values at time lags of multiples of S to estimate the time series, where S indicates by the number of time lags that the seasonal pattern will repeat. Generally, if the data is collected seasonally (such as monthly), it is helpful to use the time series plot to observe if the seasonal pattern appears at the corresponding time intervals. Also, the ACF plot demonstrates seasonality as if it decays to zero slowly at multiples of S (as demonstrated in Fig. 2.20). Seasonal time series is always nonstationary, but the values for the same months over the years are usually similar, which means the time series of the seasonal differenced data with time lags of multiples of S would be stationary. For example, the values in every January should be similar; therefore, the time series of the differenced data from January of the current year to the Januaries of previous years should be around zero; thus the time series of the seasonal differenced data is stationary.

Seasonal differencing removes seasonality and nonseasonal differencing removes nonseasonal trend. For seasonal time series with no trend, we generally take a difference of lag S. For example, the time series of 12th differenced $PM_{2.5}$ data is plotted in Fig. 2.24, which looks more stationary after the seasonal differencing compared with the original time series from Fig. 2.13.

Alternatively, if we think the seasonal pattern repeats every six months, we may choose $S = 6$ and reconstruct the time series data with frequency set as six.

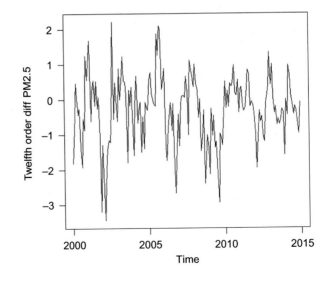

FIG. 2.24

Time series plot of 12th order differenced $PM_{2.5}$ values.

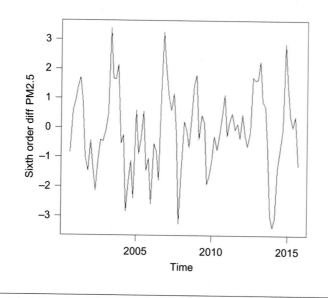

FIG. 2.25

Time series plot of 6th order differenced $PM_{2.5}$ values.

Then the time series of the 6th order differenced values are plotted in Fig. 2.25 by the following R code:

```
pmn2<- ts(pmn, frequency=6, start=c(1999,5), end=c(2014,11))
diff6pmn=diff(pmn2,6)
plot.ts(diff6pmn, ylab="sixth order diff PM2.5")
```

For a time series with trend and seasonal pattern at the same time, a seasonal difference is taken first, and the time series of the differenced data will be plotted to check if the trend pattern remains. Figs. 2.24 and 2.25 suggest the seasonal difference data is not yet stationary for some time intervals and may have additional trend pattern. If there is still trend pattern from the time series plot of the differenced data, we could take first differences and then evaluate again. In other words, we can obtain the difference of difference, i.e., a seasonal difference of a nonseasonal first-order difference. For example, for the month of June, we calculate the difference between the value of June minus May and the value of past June minus past May in the last year. After taking the difference of the difference for the $PM_{2.5}$ series, we have the time series plot in Fig. 2.26, which looks a lot more stationary compared with the original time series in Fig. 2.13 as well as compared with the seasonal differenced data in Figs. 2.24 and 2.25.

We usually examine the early lags to determine the type and order of nonseasonal terms and look for the seasonal pattern if it appears again at lags of multiples of S. For example, the ACF (Fig. 2.22) and PACF (Fig. 2.23) plots of the 12th order differenced data demonstrate some nonseasonal AR(1) (due to a sharp spike at the first lag in PACF from Fig. 2.23) and some seasonal MA(1) pattern (as the spike repeats at lag 13 in PACF from Fig. 2.23), so a potential candidate model could be

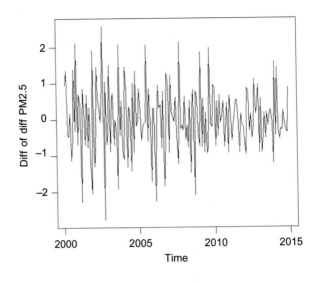

FIG. 2.26

Time series plot of the 12th order difference of the first order differenced PM$_{2.5}$ values.

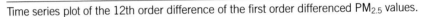

ARIMA(1,0,0)(0,1,1)[12]. We can use `sarima(series=,p=,d=,q=,P=,D=,Q=,S=)` to estimate the ARIMA model as follows:

```
sarima(pmn, 1,0,0,0,1,1,12)
```

$ttable
The estimates for ARIMA(1,0,0)(0,1,1)[6] model

	Estimate	SE	t value	P value
ar1	0.6218	0.0594	10.4649	0.0000
sma1	−0.9324	0.0914	−10.2063	0.0000
sma2	0.0488	0.0821	0.5950	0.5526
constant	−0.0256	0.0021	−11.9331	0.0000

$BIC

[1] −0.9435887

The modeling results indicate seasonal MA(2) is not necessary as its p-value is 0.5526, which is greater than the significance level 0.05, but the nonseasonal AR(1) and the seasonal MA(1) term are necessary due to small P values. And the diagnostic plots from Fig. 2.27 show the model fit is good: as the residuals are stationary "white noise" centered around zero with equal variability across time, and its distribution is approximately normal. In addition, the ACF plot of residuals are close to zero and do not have any spikes. The Ljung-Box test uses cumulative autocorrelations to see if autocorrelations for the residuals are close to zero.

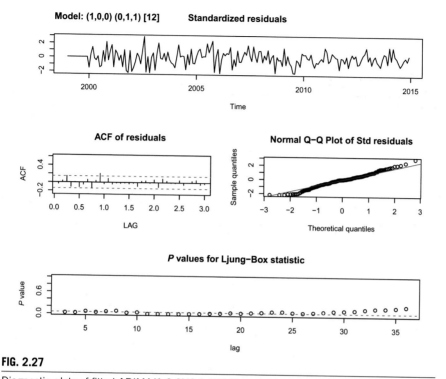

FIG. 2.27

Diagnostic plots of fitted ARIMA(1,0,0)(0,1,1)[12] model for PM$_{2.5}$ time series.

If we take a first difference for the trend and then a seasonal (S = 12) difference to cancel out the seasonality, we can have the ACF and PACF plots of the difference of difference as in Figs. 2.28 and 2.29. The plots demonstrate some nonseasonal AR(1) or MA(1) (due to a sharp spike at the first lag in both ACF and PACF) and some seasonal MA(1) or MA(2) pattern, so potential candidate models could be ARIMA(0,1,1)(0,1,2)[12], ARIMA(1,1,0)(0,1,2)[12], ARIMA(0,1,1)(0,1,1)[12], or ARIMA(1,1,0)(0,1,1)[12].

Again, we can use sarima() function to estimate the candidate ARIMA models one by one and choose the one with smallest AIC or BIC values. Compared with the above selected model using only seasonal differencing, all of the following models have larger BIC values, which indicates less compatibility of fit. Therefore, based on the BIC values, we finalize the best model to be ARIMA(1,0,0)(0,1,1)[12].

```
sarima(pmn, 0,1,1,0,1,1,12)
#$BIC
#[1] −0.8473227
sarima(pmn, 1,1,0,0,1,1,12)
#$BIC
#[1] −0.8360666
sarima(pmn, 0,1,1,0,1,2,12)
#$BIC
```

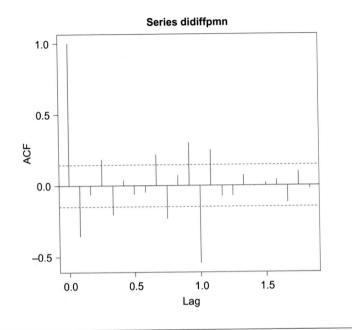

FIG. 2.28

ACF plot of the 12th order difference of the first order differenced PM$_{2.5}$ values.

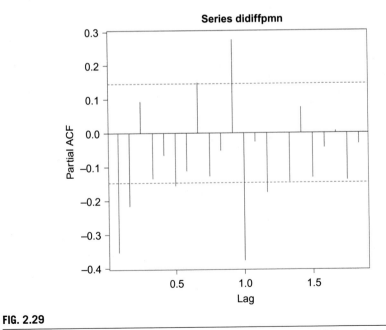

FIG. 2.29

PACF plot of the 12th order difference of the first order differenced PM$_{2.5}$ values.

```
#[1] −0.8154692
sarima(pmn, 1,1,0,0,1,2,12)
#$BIC
```
#[1] −0.8019518

Likewise, if we choose S=6, we can plot ACF and PACF of the 6th differenced data in Figs. 2.30 and 2.31.

Figs. 2.30 and 2.31 demonstrate some nonseasonal AR(1) (due to a sharp spike at the first lag in PACF) and some seasonal AR(1) or MA(1) or ARMA(1,1) pattern (as the spike repeats at lag 7 in ACF and PACF), so potential candidate models could be ARIMA(1,0,0)(0,1,1)[6], ARIMA(1,0,0)(1,1,0)[6], or ARIMA(1,0,0)(1,1,1)[6]. And based on BIC values, we select ARIMA(1,0,0)(0,1,1)[6] model; however, this model is still a poor fit compared to those models assuming S=12. This suggests even though there are two peaks repeated for each year, the seasonal pattern of 12 months will fit better if we consider the two peaks together as one pattern since they are not exactly the same:

```
sarima(pmn2, 1,0,0,1,1,1,6)
```
$BIC
[1] −0.5060954
```
sarima(pmn2, 1,0,0,0,1,1,6)
```
$BIC
[1] −0.5116944
```
sarima(pmn2, 1,0,0,1,1,0,6)
```
$BIC
[1] −0.1277012

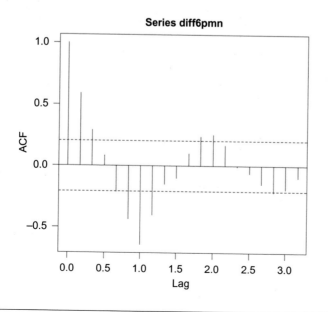

Series diff6pmn

FIG. 2.30

ACF plot of the 6th order differenced $PM_{2.5}$ values.

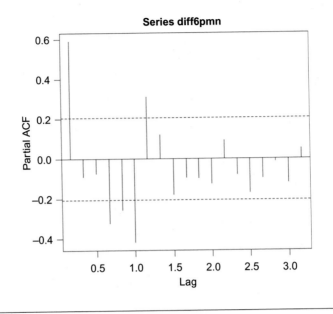

FIG. 2.31

ACF plot of the 6th order differenced $PM_{2.5}$ values.

We now choose the best ARIMA model for the CO_2 data. First, we know this is seasonal time series after decomposing in Fig. 2.17, so we would like to take seasonal differencing and plot the time series of the differenced data in Fig. 2.32. Fig. 2.32 demonstrates trend pattern after the seasonal differencing; therefore, we need to consider seasonal difference of a first-order difference; the times series plot of which looks more stationary (Fig. 2.33).

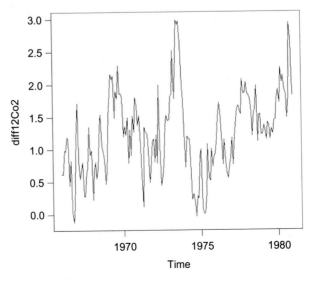

FIG. 2.32

Time series plot of the 12th order differenced CO_2 values.

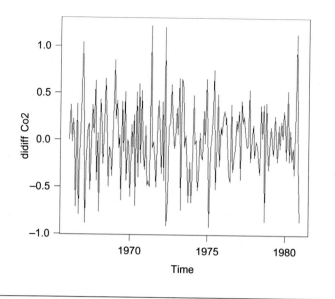

FIG. 2.33

Time series plot of the 12th order difference of the first order differenced CO_2 values.

Then, we check the ACF and PACF of the above-mentioned difference of difference. The ACF plot in Fig. 2.34 indicates a nonseasonal MA(1) (due to the peak at lag 1) and a seasonal MA(1) model (the peak repeats at lag 13). The PACF plot in Fig. 2.35 does not indicate any AR pattern. Therefore, we estimate the ARIMA (0,1,1)(0,1,1)[12] model.

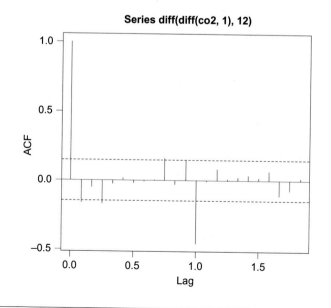

FIG. 2.34

ACF plot of the 12th order difference of the first order differenced CO_2 values.

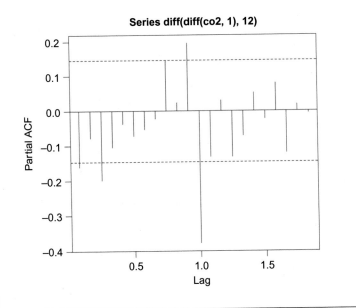

FIG. 2.35

PACF plot of the 12th order difference of the first order differenced CO_2 values.

$ttable
The estimates for ARIMA (0,1,1)(0,1,1)[12] model

	Estimate	SE	t value	P value
ma1	−0.3304	0.0823	−4.0169	1e−04
sma1	−0.8584	0.0887	−9.6813	0e+00

$BIC
[1] −2.24949

In summary, we use time series plot to examine initially if there is trend and seasonality. Then, we take appropriate differencing and re-examine the time series plot of the differenced data. In addition, we use ACF and PACF plots to determine type and orders of the potential ARIMA model(s). Then we fit the potential/candidate ARIMA models, compare model fit measures such as AIC and BIC values from each estimated models, and perform model diagnostics on the residuals. Based on model fit and diagnostics results, we choose the optimal ARIMA model that best fits the data.

2.2.2.2 Automatic ARIMA model selection

In this section, we present a straightforward automatic approach from the "forecast" R package, which is designed for automatic time series model selection and forecasting (Hyndman and Khandakar, 2007). This automatic approach yields the optimal

ARIMA model based on a search over all possible models and choose the one with best model diagnostic measures, such as lowest AIC or BIC values:

```
library("forecast")
auto.arima(pmn, ic = "bic")
```

The results from `auto.arima()` function suggest ARIMA(1,0,0)(0,1,2)[12] is the optimal ARIMA model selected by BIC criteria, but from the previous analysis for ARIMA(1,0,0)(0,1,1)[12] model, we knew SMA(2) term was not significant, thus not necessary in the model. Moreover, the BIC value of the auto-selected ARIMA(1,0,0)(0,1,2)[12] model is larger than our previously manual-selected model, i.e., ARIMA(1,0,0)(0,1,1)[12], thus is of less goodness of fit. The following gives the results of ARIMA(1,0,0)(0,1,2)[12], and we can see that the seasonal MA(2) is nonsignificant. Thus, it is not necessary to be in the model.

The estimates for ARIMA(1,0,0)(0,1,2)[12] model

ar1	sma1	sma2	drift
0.6218	−0.9324	0.0488	−0.0256
s.e. 0.0594	0.0914	0.0821	0.0021

δ^2 estimated as 0.3568

$ttable

	Estimate	SE	t value	P value
ar1	0.6218	0.0594	10.4649	.0000
sma1	−0.9324	0.0914	−10.2063	.0000
sma2	0.0488	0.0821	0.5950	**.5526**
constant	−0.0256	0.0021	−11.9331	.0000

$BIC

[1] −0.9435887

For the CO_2 data, the auto.arima results suggest a ARIMA(1,1,0)(1,1,1)[12] model with BIC value being −2.208315, which is larger than the manually selected model in the previous section (with a BIC value of −2.24949).

In conclusion, from the two examples, we see that automatic ARIMA model selection based on `auto.arima()` function does not necessarily give the "optimal" model, but it is close to the best model, and we can use it as a guideline or a starting model. Combined with eyeballing the ACF and PACF plots of the original or differenced time-series data, we can narrow down a few candidate models and then choose the optimal model based on model fit measures.

References

Hyndman, R., 2014. CO2 (ppm) mauna loa, 1965-1980. Retrieved from: https://datamarket.com/data/set/22v1/co2-ppm-mauna-loa-1965-1980#!ds=22v1&display=line.

Hyndman, R.J., Khandakar, Y., 2007. Automatic Time Series for Forecasting: The Forecast Package for R. Monash University, Department of Econometrics and Business Statistics.

Kersey, J., Yin, J., Adhikari, A., Zhou, X., Tong, W., Li, L., 2018. The impact of PM 2.5 on lung and bronchial cancers: regression and time series analysis in the US from 1999 to 2014. In: Paper Presented at the Proceedings of the 11th EAI International Conference on Mobile Multimedia Communications.

Li, L., Losser, T., Yorke, C., Piltner, R., 2014. Fast inverse distance weighting-based spatio-temporal interpolation: a web-based application of interpolating daily fine particulate matter $PM_{2.5}$ in the contiguous US using parallel programming and k-d tree. Int. J. Environ. Res. Public Health 11 (9), 9101–9141.

Sarkar, D., 2008. Lattice: Multivariate Data Visualization with R. Springer Science & Business Media.

Shumway, R.H., Stoffer, D.S., 2017. Time Series Analysis and its Applications: With R Examples. Springer.

Tong, W., Li, L., Zhou, X., Franklin, J., 2018. Efficient spatiotemporal interpolation with spark machine learning. Earth Sci. Inform. 12 (1), 87–96.

Case study: Does PM$_{2.5}$ contribute to the incidence of lung and bronchial cancers in the United States?

Jing Kersey[a,b], Jingjing Yin[b]

[a]*School of Mathematics and Natural Sciences, East Georgia State College, Statesboro, GA, United States*
[b]*Department of Biostatistics, Epidemiology, and Environmental Health Sciences, Jiann-Ping Hsu College of Public Health, Georgia Southern University, Statesboro, GA, United States*

Air quality has long been known to be strongly related to human health. Assessing the impact of air pollution on human health is a hot topic, and many studies have contributed to this body of research. This case study will illustrate how to conduct a statistical study related to air pollution and health issues using the statistical methods introduced in the previous chapter. To conduct a statistical study, we generally include the following parts:

- Identify the study background and the research question.
- Collect/obtain data.
- Conduct statistical data analyses.
- Present and discuss results.

3.1 Case study background

EPA, the US Environmental Protection Agency, was founded in 1970 to safeguard the environment and human health by reducing air pollution and other environmental risks. As an effort of fighting air pollution, the EPA has developed outdoor monitors across the US to collect and report air quality data, including criteria gases, particulates, meteorological toxics, and blanks (Air Data: Air Quality Data Collected at Outdoor Monitors Across the US, 2019). Particulate Matter (PM), also called particle pollution, includes PM$_{2.5}$ and PM$_{10}$ depending on the size of the inhalable particles. According to EPA, PM$_{2.5}$ is fine inhalable particles that generally have diameters of 2.5 micrometers or less. The largest fine particle of PM$_{2.5}$ is about one-thirtieth the size of a single human hair (Air Data: Air Quality Data Collected at Outdoor

Monitors Across the US, 2019). Due to the fine size of $PM_{2.5}$, it can be inhaled deeply into our lungs and even enter our bloodstream. $PM_{2.5}$ can also absorb gases or carry other fine toxic chemicals because of their large surface. Many researchers have conducted scientific studies and shown that exposure to particulate matter is associated with a variety of health issues, such as "premature death in people with heart or lung disease, nonfatal heart attacks, irregular heartbeats, aggravated asthma, and decreased lung function (Health and Environmental Effects of Particulate Matter (PM), 2019)."

3.2 Case description

This case study is to examine if $PM_{2.5}$ contributes to the incidence of lung and bronchial cancers in the United States. We obtained the necessary datasets for our study through available public databases. The datasets of lung cancer rates and geography of the urban area were downloaded from the National Program of Cancer Registries and United States Census Bureau, respectively. Based on the $PM_{2.5}$ data collected in the EPA monitoring sites, $PM_{2.5}$ data for any area (at national or state levels) were imputed using innovative machine learning and spatiotemporal interpolation. Then, we applied statistical analyses, such as descriptive statistics, scatter plots, time series analyses, generalized estimating equations, and lagged regression to explore the correlation between the lung and bronchial cancer annual rates and $PM_{2.5}$ values at both national and state levels. Part of the case study was published in the proceedings of the International Workshop on Environmental Health and Air Pollution in 2018 (Kersey et al., 2018).

3.3 Statistical study

3.3.1 Data

3.3.1.1 Lung and bronchial cancer dataset

Lung and bronchial cancer rates from 1999 to 2014 in the United States were downloaded from the National Program of Cancer Registries (NPCR) website. The dataset includes lung cancer rates for both national and state levels. Rates are age-adjusted to the 2000 US standard population and per 100,000 persons (US Cancer Statistics Working Group, 2017). See Table 3.1 for Lung and Bronchial cancer dataset description.

There are many public databases based on demographic characteristics (such as sex, age, and race) or based on tumor characteristics, such as site, year of diagnosis, stage, histology, and behavior. Those databases can be requested at the United States Cancer Statistics (USCS) site: https://www.cdc.gov/cancer/uscs/public-use/index. htm. An alternative way to obtain the lung and bronchial cancer dataset is to use the subsets of the above databases.

Table 3.1 Lung and bronchial cancer rates in the United States from 1999 to 2014

Variable name	Variable description
Area	Geographic area
All.Races.Rate	Cancer rate among all races
All.Races.LCI	Lower limit of 95% confidence interval for cancer rate among all races
All Races.UCI	Upper limit of 95% confidence interval for cancer rate among all races
Year	Year (1999-2014)

3.3.1.2 PM₂.₅ dataset

The $PM_{2.5}$ dataset is cited from Tong et al. (2019) and Li et al. (2014).

The $PM_{2.5}$ data was spatiotemporally interpolated and then aggregated at various spatial and temporal levels. Daily $PM_{2.5}$ data (1997-2015) in the contiguous United States were downloaded from the Environmental Protection Agency (EPA) Air Quality System and aggregated into monthly data with the schema (longitude, latitude, month, year, mean $PM_{2.5}$). Attributes longitude and latitude are to locate the centroids of a US census block group, and mean $PM_{2.5}$ is the average of daily values of $PM_{2.5}$ in the month at the centroid. Mean $PM_{2.5}$ values at every centroid were aggregated in each state to obtain the mean $PM_{2.5}$ value for each state at each month. Yearly and national data were aggregated accordingly for the illustration purpose of different statistical analysis.

The spatiotemporally interpolated $PM_{2.5}$ data in the contiguous United States were obtained using the following steps.

1) Daily $PM_{2.5}$ data (1997-2015) were downloaded in the contiguous United States from the Environmental Protection Agency (EPA) Air Quality System.
2) Daily $PM_{2.5}$ data were aggregated into monthly data with the schema (x, y, month, max $PM_{2.5}$, mean $PM_{2.5}$). Attributes x and y are the longitude and latitude at the centroid of a US census block group; $max\ PM_{2.5}$ is the highest daily value of $PM_{2.5}$ in the month at the centroid, and $mean\ PM_{2.5}$ is the average of daily values of $PM_{2.5}$ in the month at the centroid.
3) Using innovative machine learning and spatiotemporal interpolation methods, the aggregated monthly data were trained to find the optimal interpolation parameters, then interpolated at the centroids of census blocks in the contiguous United States and for every month between 1997 and 2015. The description of the resulted interpolation data is illustrated in Table 3.2.
4) In order to link $PM_{2.5}$ with the cancer data at the county level in the contiguous United States, the data described in Table 3.1 were aggregated into county level. The summary of the aggregated data at the county level is illustrated in Table 3.3.

Table 3.2 Spatiotemporally interpolated monthly $PM_{2.5}$ at census block group level in the contiguous United States from January 1997 to December 2015

Variable name	Variable description
FIPS Block GRP	Census 2000 FIPS block group code
Month	Month (1-12)
Year	Year (1997-2015)
Max $PM_{2.5}$	The highest daily value of $PM_{2.5}$ in the month at the census block
Mean $PM_{2.5}$	The average (arithmetic mean) of daily $PM_{2.5}$ value in the month at the census block

Table 3.3 Spatiotemporally interpolated monthly $PM_{2.5}$ at county level in the contiguous United States from January 1997 to December 2015

Variable name	Variable description
FIPS County	Census 2000 FIPS county code
Month	Month (1-12)
Year	Year (1997-2015)
Max $PM_{2.5}$	The highest daily value of $PM_{2.5}$ in the month at the county
Mean $PM_{2.5}$	The average (arithmetic mean) of daily $PM_{2.5}$ value in the month at the county

3.3.1.3 Urban area dataset
The dataset of urban area was obtained from the US Census Bureau at https://www.census.gov/geo/maps-data/data/cbf/cbf_ua.html. We calculated the urban percentage using the ratio between the urban area of cartographic boundary and the overall state area to represent the urban percentage (US Census Bureau, 2018). See Table 3.4 for the Urban area dataset description.

3.3.2 Statistical analysis
Time series analysis including ARIMA models and seasonal ARIMA models were used to predict the trend of lung and bronchial cancer rates (annually) and $PM_{2.5}$ values (both monthly and annually). The generalized estimating equation (GEE) was used to

Table 3.4 Urban area dataset description

Variable name	Variable description
State_Name	Name of states
State	Abbreviation of states
Urban_Area	Urban area of the state (square miles)
Total_Area	Total land area of the state (square miles)
Urban_Percent	Urban area percentage of the total area for states

estimate the parameters of the generalized linear regression models to determine the possible association of urban percentage with lung and bronchial cancer rates and $PM_{2.5}$ values, respectively. Furthermore, a lagged regression model regressing lung and bronchial cancer rates on its past time-series and $PM_{2.5}$ time-series was used.

The R code for the statistical analysis section is provided at the end of the chapter.

Step 1: Obtain the time series plots of $PM_{2.5}$ and lung cancer rate to see if there are trends over time. The time series plot of $PM_{2.5}$ levels for each month between Jan. 1999 and Dec. 2014 is shown in Fig. 3.1. The y-axis gives the values of $PM_{2.5}$ levels, and the x-axis gives the corresponding time point when the $PM_{2.5}$ was measured. It indicates that average national $PM_{2.5}$ has a trend over time, and it decreases in the long run. It seems the regularly repeating pattern of highs and lows related to months of the year is clear.

As shown in Fig. 3.2, the national lung and bronchial cancer rates dropped over the years. Particularly, it decreased greatly since 2005. Interestingly, it was also in 2005 that the EPA required that state and local governments have to meet the national ambient air quality standards for fine particulate matter ($PM_{2.5}$).

Step 2: Obtain a decomposition of the additive time series plot of $PM_{2.5}$ to see if there is a seasonal effect.

There is seasonality in monthly $PM_{2.5}$. Consistently in each year, $PM_{2.5}$ is low in the summer months and peaks between fall and winter (see Fig. 3.3).

Step 3: Develop ARIMA (Autoregressive Integrated Moving Average) models to see how the present value of $PM_{2.5}$ relates to past values and past prediction errors.

The optimal model by automatic model selection based on either Bayesian Information Criterion (BIC) or Akaike information criterion (AIC) is determined to be seasonal ARIMA $(1,0,0)$ $(0,1,2)[12]$. This model includes a non-seasonal AR(1) term, a seasonal MA(2) term, and seasonal difference of 1 without a non-seasonal difference, non-seasonal MA term, or seasonal AR terms, and the seasonal period is 12.

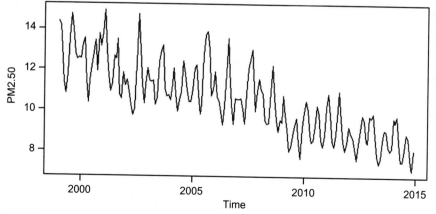

FIG. 3.1

Time series plot for national yearly $PM_{2.5}$.

The selected model ARIMA $(1,0,0)$ $(0,1,2)[12]$ can be used to predict future values of $PM_{2.5}$ if needed. For example, we predict the $PM_{2.5}$ values for each month of the first quarter of 2015 as shown in Table 3.5. Fig. 3.4 gives the prediction of monthly $PM_{2.5}$ values for 2015-2018.

Step 4: Conduct regression analysis with time series errors to confirm the model selected in Step 3 to see how year and percentage of urban area are related to $PM_{2.5}$.

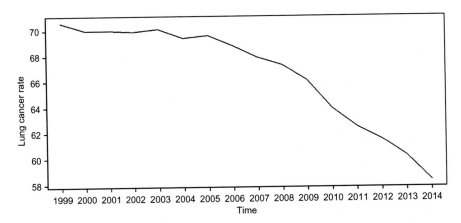

FIG. 3.2

Time series plot for national lung and bronchial cancer rates.

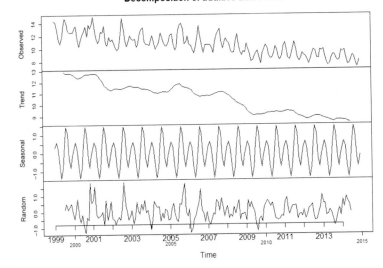

FIG. 3.3

Trend and seasonal effects of national $PM_{2.5}$.

Table 3.5 Predicted values of PM$_{2.5}$ of the first quarter of 2015

Time	Point forecast	80% Confidence interval		95% Confidence interval	
		LB	UB	LB	UB
Jan-15	8.57835	7.8332	9.34538	7.40527	9.75142
Feb-15	8.77782	7.87473	9.68091	7.39666	10.15898
Mar-15	8.34022	7.38973	9.29072	6.88657	9.79388
Apr-15	7.33197	6.36377	8.30017	5.85123	8.81270

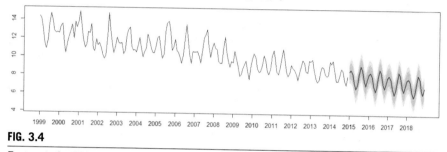

FIG. 3.4

Forecasts for PM$_{2.5}$ from 2015 to 2018.

Before conducting a regression analysis, it is necessary to examine the residuals for the selected ARIMA model. If the residuals have a time series structure, then adjustments for the estimated coefficients and standard errors will be needed.

For the optimal model, ARIMA (1,0,0) (0,1,2)[12] selected in Step 2, Fig. 3.5 shows the plots of residuals, ACF (autocorrelation function), and PACF(partial autocorrelation function). The residuals seem to be randomly distributed; in other words, residuals do not have an AR structure. Moreover, both the ACF and PACF show that not very much is going on except a spike at lag 3, which does not suggest the need to differentiate further.

Step 5: Conduct regression analysis to see how year and percentage of urban area related to PM2.5.

In the generalized linear regression model using the generalized estimating equation (GEE) (see Table 3.6), annual PM$_{2.5}$ value is strongly associated with years, and its value decreases each year consistently (estimate = −0.0272, P-value <.0001). This result is consistent with what we see with the ARIMA model. The regression analysis also discovers that the annual PM$_{2.5}$ for the individual state is significantly related to the urban percentage of that state—for every 1 percent increase in urban percentage, the PM$_{2.5}$ value increases by 29.87 percent (with P-value = .0336). As years pass by, the effect of the urban percentage becomes less important (estimate = −0.0148, P-value = .0346).

Similarly, analysis of GEE Parameter Estimates of national lung cancer rates (See Table 3.7) shows that the rate is strongly related to year, urban percentage, and year*urban percentage, with P-values of <.0001, .0018, and .0020 respectively.

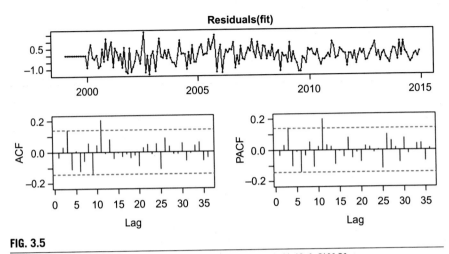

FIG. 3.5

Plots of residuals, ACF, and PACF for model ARIMA (1,0,0) (0,1,2)[12].

Table 3.6 Analysis of GEE parameters for $PM_{2.5}$ on year and urban area

			Time series analysis including ARIMA			
Parameter	**Estimate**	**Standard error**	**95% Confidence limits**		**Z**	**Pr > |Z|**
Intercept	56.8856	3.8758	49.2893	64.4820	14.68	<.0001
Year	−0.0272	0.0019	−0.0310	−0.0234	−14.14	<.0001
Urban percent	29.8661	14.0558	2.3173	57.4150	2.12	.0336
Year*urban percent	−0.0148	0.0070	−0.0284	−0.003	−2.3	.0346

Table 3.7 Analysis of GEE parameters for lung cancer rate on year and urban area

			Analysis of GEE parameter estimates			
Parameter	**Estimate**	**Standard error**	**95% Confidence limits**		**Z**	**Pr > |Z|**
Intercept	23.1645	2.2379	18.7783	27.5507	10.35	<.0001
Year	−0.0095	0.003	−0.037	−0.0073	−8.46	<.0001
Urban percent	28.0687	9.0068	10.4157	45.7216	3.12	0.0018
Year*urban percent	−0.0140	0.0045	−0.0228	−0.0051	−3.09	0.0020

Step 6: Conduct cross-correlation and lagged regression to explore the association between PM2.5 and lung cancer rate.

- **Cross-correlation function (CCF)**

To analyze the two potentially correlated time series—annual $PM_{2.5}$ and yearly lung cancer rates—we assume that $PM_{2.5}$ leads lung cancer. The CCF function (see Fig. 3.6) gives an initial estimate of h; the plot CCF plot shows h can be between −5 and 0. Moreover, 0 is most likely, which suggests $PM_{2.5}$ does not significantly lead to lung cancer in the United States during the years 1999-2014.

- **Lagged regression**

Since both $PM_{2.5}$ level and lung cancer rates are highly related to years, we use lagged regression to examine if the lung cancer rate is associated with $PM_{2.5}$ level.

First, we use scatter plots to identify possible lag patterns. Fig. 3.7 shows how the national yearly $PM_{2.5}$ (variable name: pmny) relates to its past values. Lag 1, 2, 3, and 6 seem to indicate a linear relation. For yearly lung cancer rate (variable name: lcr), lag 1, 2, 3, 4, and 5 may have a linear relation (see Fig. 3.8). We also consider lags of $PM_{2.5}$ over lung cancer rates. As shown in Fig. 3.9, the national lung cancer rates seem to have a strong association with national $PM_{2.5}$, particularly at lag 0 (no lag), 1, 2, and 4.

Second, we build lagged regression models for yearly lung cancer rates on national yearly $PM_{2.5}$ and other possible predictors. We start with three crude models without any lag:

Crude Model 1: Regression for lung cancer rate on $PM_{2.5}$
Crude Model 2: Regression for lung cancer rate on trend (year)
Crude Model 3: Regression for lung cancer rate on both $PM_{2.5}$ and trend (year)

The results for crude models (see Table 3.8) suggest that the trend over years is strongly related to lung cancer rate with P-value $= 2.83e-05$, but national $PM_{2.5}$ is not

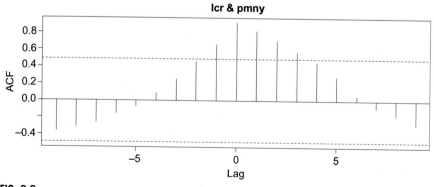

FIG. 3.6

CCF function for yearly $PM_{2.5}$ and yearly lung cancer rates.

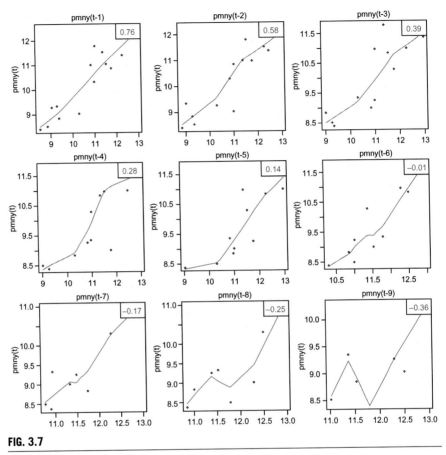

FIG. 3.7

Lags of national yearly $PM_{2.5}$.

(P-value = .0958). The strong association showed in both Fig. 3.3 and Crude Model 1 is due to collinearity between $PM_{2.5}$ and trend (year).

Furthermore, we build lagged regression models with lag effect, including lag effects from both lung cancer rate and $PM_{2.5}$.

For example,

Lagged Model 1: Regression for lung cancer rate on cancer rate lag 1 (variable name: lcrlag1)

Lagged Model 2: Regression for lung cancer rate on $PM_{2.5}$ lag 1 (variable name: pmnylag1)

Lagged Model 3: Regression for lung cancer rate on cancer rate lag 1 (variable name: lcrlag1) and $PM_{2.5}$ lag1

After fitting all the possible lagged regression models, the final selected optimal model only includes lag1 of cancer rate (cancer rate from previous year), lag1 of $PM_{2.5}$, and lag5 of $PM_{2.5}$ (see Table 3.9), suggesting that the time trend of the adjacent year for lung and bronchial cancer rates plays a major role leading to its current rate.

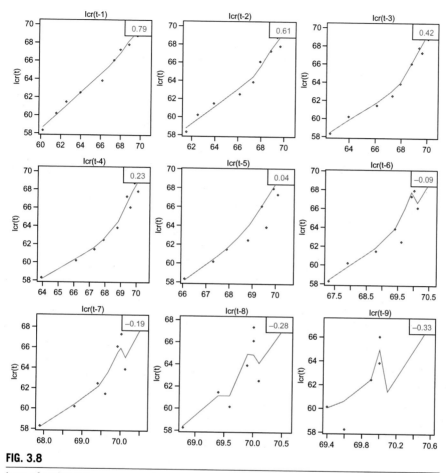

FIG. 3.8

Lags of national lung cancer rates.

Step 7: State Level Analysis

Lagged regression analysis with the national data suggests that $PM_{2.5}$ does not have a strong causal effect on lung and bronchial cancer. However, we notice that $PM_{2.5}$ is strongly associated with the urban-percentage of states. So, we continue the analysis with the following states:

The most rural state	Wyoming (WY)
The most urban state	New Jersey (NJ)
Fully urban	District of Columbia (DC)
The state with highest $PM_{2.5}$	California (CA)

All the statistical analysis for state-level follows the same steps (step 1 to step 6) as for the national level.

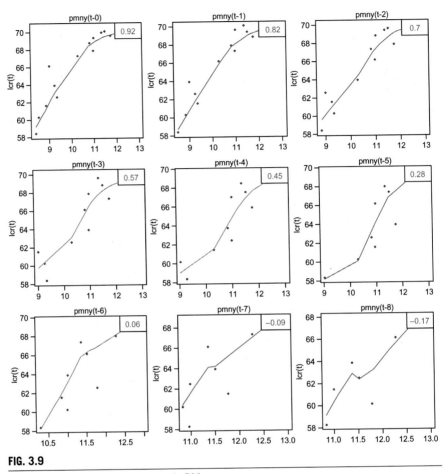

FIG. 3.9

Lags of lung cancer rates over yearly PM$_{2.5}$.

Table 3.8 Summary of results: crude models for yearly lung cancer rate (***Pr < 0.001, **Pr < 0.01)

Model	Coefficients	Estimate	Std. error	t value	Pr(>|t|)
Crude Model 1	Intercept	32.057	7.449	4.304	0.00507**
	pmny	3.365	0.795	4.233	0.00548**
Crude Model 2	Intercept	2885.343	120.031	24.040	3.40e-07***
	trend	−1.404	0.060	−23.51	3.89e-07***
Crude Model 3	Intercept	3303.167	225.804	14.628	2.70e-05***
	trend	−1.609	0.31	−14.487	2.83e-05***
	pmny	−0.627	0.306	−2.049	0.0958

Table 3.9 The optimal lagged regression model for lung cancer rate (***Pr < 0.001, *Pr < 0.05)

Coefficients	Estimate	Std. error	t value	Pr(>|t|)
Intercept	3.33170	1.30957	2.544	0.063701
lcrlag1	0.70962	0.04289	16.544	7.82e-05***
pmnylag1	1.13608	0.10918	10.405	0.000482***
pmnylag5	0.29613	0.09368	3.161	0.034147*

Table 3.10 Summary of results: regression analysis for WY, NJ, DC, and CA (***Pr < 0.001, *Pr < 0.05)

State	Coefficients	Estimate	Std. error	t value	Pr(>|t|)
Wyoming (WY)	Intercept	2218.273	810.8791	2.736	0.0170 *
	pmy	−1.5421	2.0033	−0.770	0.4552
	trend	−1.0753	0.3987	−2.697	0.0183 *
New Jersey (NJ)	Intercept	915.9432	449.7654	2.036	0.0626
	pmy	1.0516	0.5040	2.087	0.0572
	trend	−0.4310	0.2214	−1.947	0.0735
District of Columbia (DC)	Intercept	3445.5756	1413.9248	2.437	0.0313 *
	pmy	−1.356	1.3817	−0.807	0.4351
	trend	−1.6788	0.6962	−2.43	0.0328 *
California (CA)	Intercept	3063.9788	351.5695	8.715	8.65e-07 ***
	pmy	−0.4258	0.2927	−1.455	0.17
	trend	−1.4981	0.1734	−8.641	9.52e-07 ***

Statistical analysis for those four areas carries similar results (see Table 3.10) as the national level analysis. Lung cancer rates in all four areas have a significant trend over time, but they are not significantly affected by $PM_{2.5}$. Notice that for the most urban state—New Jersey (NJ) and highest $PM_{2.5}$ state—California (CA), the p-value is 0.0572 and 0.17, respectively, for $PM_{2.5}$ effect when considering both trend effect and $PM_{2.5}$ effect.

3.4 Discussion

At both the national level and state level, we found that there was no statistical evidence that $PM_{2.5}$ contributed to the lung and bronchial cancer rates in the United States based on the data of a 15-year span (1999–2014) in the current study. However, both the lung cancer rates data and the $PM_{2.5}$ data are highly correlated to time. We then consider the following discussion questions.

- What are the strengths of the case study?

How to obtain good data for a study is always a concern when we perform a statistical analysis. Good data sources should be original, comprehensive, current, and reliable. Public databases from the agencies in the federal government are good sources. For example, this case study obtained lung cancer data from the CDC (Centers for Disease Control and Prevention), $PM_{2.5}$ data from EPA, and urban area data from the US Census Bureau, EPA.

Using innovative machine learning and spatiotemporal interpolation methods to impute $PM_{2.5}$ data at the centroids of census blocks highlights the strengths of the study. The available $PM_{2.5}$ data are daily measures collected through EPA monitoring sites across the United States. However, the available lung cancer rates are at the national and state level. It is necessary to obtain the $PM_{2.5}$ measures for a US census block such as a state. The detailed procedure of imputing needed $PM_{2.5}$ data is described in the data.

This case study demonstrated how to conduct a statistical study of the possible relation between air pollution ($PM_{2.5}$) and health problems (lung and bronchial cancer) in detail. It serves as an example to guide our readers to conduct statistical studies on similar topics.

- What are the limitations of the study, and how can we improve it?

One of the limitations of the study is the method whereby the monthly and yearly PM2.5 were obtained. In this study, it assumed that centroid data in our study was a stratified random sample from the contiguous United States, and then the monthly and yearly $PM_{2.5}$ were calculated for each state in the contiguous United States using the arithmetic mean of concentration values of $PM_{2.5}$ at the centroids of census block groups for each state. Since the assumption may not be the case, the national estimates of $PM_{2.5}$ values using the arithmetic mean of $PM_{2.5}$ may not be accurate. A different approach, such as weighted mean based on the density of centroids or population, may provide better estimates.

The available lung and bronchial cancer data consisted of a 15-year span (1999–2014) at the time when the study was performed, and no significant evidence of association between $PM_{2.5}$ and lung cancer rates was identified. However, it may take decades to develop lung cancer. Historical data of lung and bronchial cancers and $PM_{2.5}$ over a longer time frame will help future research to see if $PM_{2.5}$ is associated with lung cancer.

- When can the other related study be done in the future?

In the state level analysis, we emphasized that for the most urban state—New Jersey (NJ) and the highest $PM_{2.5}$ state—California (CA), the P-value is .0572 and .17, respectively, for $PM_{2.5}$ effect in the regression on lung cancer rates considering trend effect and $PM_{2.5}$ effect. This result indicates that the most urban state—New Jersey (NJ) and the highest $PM_{2.5}$ state—California (CA) had a relative tendency to have a significant $PM_{2.5}$ effect among all contiguous United States. It suggests that when the $PM_{2.5}$ level is higher, there might be some association between lung and bronchial cancer rates and $PM_{2.5}$. Therefore, similar studies can be conducted in the regions with remarkable $PM_{2.5}$ concentrations, such as in China or India.

The urban percentage was found to be strongly related to $PM_{2.5}$. In addition to looking at national and state levels such as in the case study, we can also try different geographical levels for future work. For example, we can group states with similar characteristics, such as by population sizes, urban percentage, or/and geographical adjacency.

3.5 **R code**

```
#Change working directory
rm(list = ls())
setwd( "C:/datasets")
mydat=read.csv("national time series month.csv", header = TRUE)
mydat2=read.csv("national timeseries year.csv", header = TRUE)
mydat3=read.csv("urban_area.csv", header=TRUE)
summary(mydat)
library(astsa)

pmny=mydat2$pmny
lcr=mydat2$lcr
pmn=mydat$pmn

pmn<- ts(pmn, frequency=12, start=c(1999,1), end=c(2014,12))
lcr<- ts(lcr, frequency=1, start=1999, end=2014)
pmny<- ts(pmny, frequency=1, start=1999, end=2014)

# filling NA values using seasonal Kalman filter
#install.packages("zoo")
#time series plot for national yearly PM2.5
ilits1 <- ts(pmn, frequency=12, start=c(1999,1), end=c(2014,12))
library(zoo)
ilitsa=na.StructTS(ilits1)
plot.ts(ilitsa,xlab="time", ylab="PM2.50")

#time series plot for national yearly Lung and Bronchus cancer rate
plot.ts(lcr, ylab="Yearly Lung and Bronchus cancer rate")
time2<-seq(1999,2014, by=1)
axis(1, at = time2)

#plot of trend and seasonal effects of national PM2.5
ilicomp=decompose(ilitsa)
plot(ilicomp)
time<-seq(1999,2014, by=1)
axis(1, at = time)

#install.packages("forecast")
library("forecast")
```

```
fit=auto.arima(ilits1, ic = "aicc")
fit
tsdisplay(residuals(fit))
# the same as bic

fit2=auto.arima(ilits1, ic = "bic")
fit2
x11()
tsdisplay(residuals(fit2))

#predicted values of PM2.5 of the first quarter of 2015
ili4cast=forecast(fit2, h=4)
ili4cast

#predicted values of PM2.5 during 2015 and 2018
pmpred=forecast(fit2, h=48)
write.csv(pmpred, file="PM25 2015-2018 pred.csv")

#plot of prediction of monthly PM2.5 values during 2015 and 2018:
plot(pmpred)
time2<-seq(1999,2018, by=1)
axis(1, at = time2)

#regression analysis
reg1=lm(pmny~year,data=mydat2)
summary(reg1)

reg2=lm(pmny~year+urban_percent,data=mydat2)
summary(reg2)

########################

pm=mydat2$pmny
lcr=mydat2$lcr
mort=ts(pm)
mortdiff=diff(mort,1) # creates a variable = x(t) x(t-1)
x11()
plot(mortdiff,type="o") # plot of first differences
acf(mortdiff,xlim=c(1,14)) # plot of first differences, for 14 lags
mortdifflag1=lag(mortdiff,-1)
y=cbind(mortdiff,mortdifflag1) # bind first differences and lagged
first differences
mortdiffar1=lm(y[,1]~y[,2]) # AR(1) regression for first differences
summary(mortdiffar1) # regression results
```

```
    acf(mortdiffar1$residuals, xlim = c(1,14)) # ACF of residuals for
14 lags.

    ########################
    ar(lh)
    ar(lh, method = "burg")
    ar(lh, method = "ols")
    ar(lh, FALSE, 4) # fit ar(4)

    (sunspot.ar <- ar(sunspot.year))
    predict(sunspot.ar, n.ahead = 15)
    ## try the other methods too

    ar(ts.union(BJsales, BJsales.lead))
    ## Burg is quite different here, as is OLS (see ar.ols)
    ar(ts.union(BJsales, BJsales.lead), method = "burg")
    ########################

    #Cross Correlation and lagged regression
    #1)crude model lungcancer~pm2.5 (significant)

    ccfvalues =ccf(lcr, pmny)
    ccfvalues
    lag1.plot(lcr,9)
    lag1.plot(pmny,9)
    #lag2.plot (lcr, pmny, 10)
    lag2.plot (pmny,lcr,8)
    trend=time(lcr)
    alldata=ts.intersect(lcr,pmny,trend,pmnylag1=lag(pmny,-1),
pmnylag2=lag(pmny,-2), pmnylag1=lag(pmny,-1),
            pmnylag3=lag(pmny,-3),pmnylag4=lag(pmny,-4),
pmnylag5=lag(pmny,-5),pmnylag6=lag(pmny,-6),
            pmnylag7=lag(pmny,-7),
            pmnylag8=lag(pmny,-8),
            lcrlag1=lag(lcr,-1), lcrlag2=lag(lcr,-2),
lcrlag3=lag(lcr,-3), lcrlag4=lag(lcr,-4))
    crude<-lm(lcr~pmny, data=alldata)
    summary (crude)
    acf2(residuals(crude))

    crude1<-lm(lcr~trend+pmny, data=alldata)
    summary (crude1)
    acf2(residuals(crude1))
```

```
crude2<-lm(lcr~trend, data=alldata)
summary (crude2)
acf2(residuals(crude2))

#2)with lag effect: nothing significant except lcrlag1

tryit = lm(lcr~lcrlag1+lcrlag2+lcrlag3+lcrlag4, data = alldata)
#not significant
summary (tryit)
acf2(residuals(tryit))

tryit1 = lm(lcr~lcrlag1, data = alldata)
summary (tryit1)
acf2(residuals(tryit1))

tryitpm = lm(lcr~pmnylag6, data = alldata)
summary (tryitpm)
acf2(residuals(tryitpm))

tryit2 = lm(lcr~pmny+lcrlag1, data = alldata) #lcrlag1, lcrlag2,
lcrlag3, lcrlag4
summary (tryit2)# each univariate significant
acf2(residuals(tryit2))

tryit3 = lm(lcr~lcrlag1+lcrlag2, data = alldata)
summary (tryit3) #lcrlag2 is not significant
acf2(residuals(tryit3))

tryit4 = lm(lcr~lcrlag1+lcrlag3, data = alldata)
summary (tryit4) #lcrlag3 is not significant
acf2(residuals(tryit4))

tryit5 = lm(lcr~lcrlag1+lcrlag4, data = alldata)
summary (tryit5) #lcrlag4 is not significant
acf2(residuals(tryit5))

tryit6= lm(lcr~pmnylag1+pmnylag2+pmnylag3+pmnylag4+pmnylag5+
pmnylag6, data = alldata)
summary (tryit6)
acf2(residuals(tryit6))

tryit7= lm(lcr~lcrlag1+pmnylag1+pmnylag2+pmnylag5, data = alldata)
summary (tryit7)
acf2(residuals(tryit7))
```

```
tryit8= lm(lcr~lcrlag1+pmnylag1+pmnylag5, data = alldata)
summary (tryit8)
acf2(residuals(tryit8))


tryit9= lm(lcr~pmny+lcrlag1+pmnylag1+pmnylag5, data = alldata)
summary (tryit9)
acf2(residuals(tryit9))


tryit10= lm(lcr~trend+lcrlag1+pmnylag1+pmnylag5, data = alldata)
summary (tryit10)
acf2(residuals(tryit10))


tryit11= lm(lcr~lcrlag1+pmnylag1+pmnylag5, data = alldata)
summary (tryit11) #final optimal model
acf2(residuals(tryit11))


#State Level analysis (most rural state-Wyoming (WY), most urban
state-New Jersey (NJ), DC,CA)
#lungcancer~pm2.5+urban%
pm25<-read.csv("pm25_state_year_month.csv", header=TRUE)
lung<-read.csv("lungcancer.csv",header=TRUE)
names(lung)[names(lung) == 'All.Races.Rate'] <- 'rate'


#1)Wyoming
WYlcr<-subset(lung, Area=="Wyoming",select = 'rate')
WYpmy<-subset(pm25, state=="WY" & year>"1998" & year<"2015" )
WYpmy<- aggregate(x = WYpmy['pm_mean'],
        by = list(year = WYpmy$year),
        FUN = mean)
WYpmy<-ts(WYpmy,frequency=1, start=1999, end=2014)
WYpm<-WYpmy[,2]
WYlcr<-ts(WYlcr,frequency=1, start=1999, end=2014)
WYtrend=time(WYlcr)


modelWY1<-lm(WYlcr~WYpm)
summary(modelWY1)


modelWY2<-lm(WYlcr~WYpm+WYtrend)
summary(modelWY2)


modelWY3<-lm(WYlcr~WYtrend)
summary(modelWY3)
```

```
#2)New Jersey
NJlcr<-subset(lung, Area=="New Jersey",select = 'rate')
NJpmy<-subset(pm25, state=="NJ" & year>"1998" & year<"2015" )
NJpmy<- aggregate(x = NJpmy['pm_mean'],
         by = list(year = NJpmy$year),
         FUN = mean)
NJpmy<-ts(NJpmy,frequency=1, start=1999, end=2014)
NJpm<-NJpmy[,2]
NJlcr<-ts(NJlcr,frequency=1, start=1999, end=2014)
NJtrend=time(NJlcr)

modelNJ1<-lm(NJlcr~NJpm)
summary(modelNJ1)

modelNJ2<-lm(NJlcr~NJpm+NJtrend)
summary(modelNJ2)

modelNJ3<-lm(NJlcr~NJtrend)
summary(modelNJ3)

#The above State Level analysis can also be done with the following code.
state1<-c('WY', 'NJ', 'DC', 'CA')
statename<-c('Wyoming','New Jersey','District of Columbia','
  California')
state1cr<-subset(lung, Area==statename[1],select = 'rate') #staten
ame[1]=Wyoming;statename[2]=New Jersey...
  statepmy<-subset(pm25, state==state1[1] & year>"1998" &
year<"2015" )
  statepmy<- aggregate(x = statepmy$pm_mean,
            by = list(year = statepmy$year),
            FUN = mean)
  statepmy<-ts(statepmy,frequency=1, start=1999, end=2014)
  statepm<-statepmy[,2]
  state1cr<-ts(state1cr,frequency=1, start=1999, end=2014)
  statetrend=time(state1cr)

  statedata=ts.intersect(state1cr,statepm,statetrend,
            statepmlag1=lag(statepm,-1), statepmlag2=lag
              (statepm,-2),
            statepmlag3=lag(statepm,-3),statepmlag4=lag
              (statepm,-4), statepmlag5=lag(statepm,-5),
            statepmlag6=lag(statepm,-6),
            statepmlag7=lag(statepm,-7),
            statepmlag8=lag(statepm,-8),
```

```
          statelcrlag1=lag(statelcr,-1), statelcrlag2=lag
             (statelcr,-2), statelcrlag3=lag(statelcr,-3),
          statelcrlag4=lag(statelcr,-4))

  model1<-lm(statelcr~statepm)
  summary(model1)

  model2<-lm(statelcr~statetrend)
  summary(model2)

  model3<-lm(statelcr~statepm+statetrend)
  summary(model3)

  model4<-lm(statelcr~statepm+statepmlag1+statetrend,
data=statedata)
  summary(model4)

  model5<-lm(statelcr~statelcrlag1+statepm+statepmlag1+
          statepmlag2+statepmlag3+statepmlag4, data=statedata)
  summary(model5)
```

References

Air Data: Air Quality Data Collected at Outdoor Monitors Across the US, 2019. EPA. Retrieved from: https://www.epa.gov/outdoor-air-quality-data. (Accessed 28 May, 2019).

Health and Environmental Effects of Particulate Matter (PM), 2019. EPA. Retrieved from: https://19january2017snapshot.epa.gov/pm-pollution/health-and-environmental-effects-particulate-matter-pm_.html. (Accessed 30 January, 2019).

Kersey, J., Yin, J., Adhikari, A., Zhou, X., Tong, W., Li, L., 2018. The impact of $PM_{2.5}$ on lung and bronchial cancers: regression and time series analysis in the U.S. from 1999 to 2014. In: Workshop on Environmental Health and Air Pollution. EAI, Qindao, China, https://doi.org/10.4108/eai.21-6-2018.2276588.

Li, L., Losser, T., Yorke, C., Piltner, R., 2014. Fast inverse distance weighting-based spatio-temporal interpolation: a web-based application of interpolating daily fine particulate matter $PM_{2.5}$ in the contiguous U.S. using parallel programming and k-d tree. Int. J. Environ. Res. Public Health 3 (9), 9101–9141.

Tong, W., Li, L., Zhou, X., Franklin, J., 2019. Efficient spatiotemporal interpolation with Spark machine learning. Earth Sci. Inform. 12 (1), 87–96.

US Cancer Statistics Working Group, 2017. United States Cancer Statistics: 1999-2014 Incidence and Mortality Web-based Report. U.S. Department of Health and Human Services, Centers for Disease Control and Prevention and National Cancer Institute, Atlanta. Retrieved from: https://www.cdc.gov/cancer/uscs/public-use/index.htm. (Accessed 28 May, 2019).

US Census Bureau, EPA. Retrieved from: https://www.census.gov/programs-surveys/geography/data.html. (Accessed 28 May, 2019).

Bayesian hierarchical modeling for the linkages between air pollution and population health

4

Yongping Hao[a], Hope Landrine[b], Michele Ver Ploeg[c], Xingyou Zhang[c]

[a]*Department of Housing and Urban Development, Washington, DC, United States*
[b]*Center for Health Disparities Research, East Carolina University, Greenville, NC, United States*
[c]*USDA Economic Research Service, Washington, DC, United States*

Air pollution is a leading environmental risk factor associated with population health, especially with respiratory diseases and their related deaths. According to a World Health Organization (WHO) report, 91% of the world's population was living in places where the WHO air quality guideline levels were not met and ambient (outdoor) air pollution in both cities and rural areas was estimated to cause 4.2 million premature deaths worldwide in 2016 (World Health Organization, 2016). Numerous studies have been conducted to explore and examine the associations between air pollution and population health outcomes. However, the complexity of the linkages between air pollution and health remains challenging epidemiologically as well as statistically. This chapter introduces a Bayesian hierarchical modeling framework to develop appropriate statistical models to better understand the complex relationships between air pollution and health. We demonstrate this robust and flexible modeling approach with a case study in the United States.

4.1 Introduction: GLMMs and Bayesian estimation via MCMC

Bayesian hierarchical modeling benefits from two most significant advances in modern applied statistics: generalized linear mixed models (GLMMs) (McCulloch et al., 2008) and Bayesian estimation via Markov chain Monte Carlo (MCMC) simulation (Gilks et al., 1996).

GLMMs have two significant advantages: (1) like generalized linear models (GLMs) (McCullagh and Nelder, 1989), GLMMs can handle both continuous and categorical outcomes (also called dependent or responsible variables in literature. To be consistent, we use outcomes in this chapter) and (2) like linear mixed models, GLMMs can handle complex correlated structure or cluster effects among individual observations in a dataset. GLMs and linear mixed models are extensions of classic linear regression models that model continuous outcomes, assuming normal distribution of and statistical independency among individual observations. However, such assumptions might not be true in reality. In reality, in addition to continuous outcomes, binary and count outcomes are quite common, collected or observed data tend to be correlated because multiple observations might come from the same geographic units, or close proximity over space, or from the same units over multiple time periods, or combinations of them.

A GLMM has three basic components (Stroup, 2013):

- a linear predictor (η),
- a link function ($g(.)$), and
- a probability distribution for the outcome (Y) from the exponential family of distributions.

A linear predictor is the linear combination of regression coefficients associated with fixed effects as well as random effects:

$$\eta = X\beta + Z\gamma$$

where β is a vector of regression coefficients associated with a matrix of fixed effects (X) and γ is a vector of regression coefficients associated with a matrix of random effects (Z), often called design matrix. Both X and Z are known; both β and γ are unknown and have to be estimated.

The example below highlights fixed effects vs random effects and their corresponding regression coefficients:

Imagine that we have collected health data from 1000 individuals who reside in 50 different neighborhoods in a city, and we are interested in neighborhood air pollution exposure impact on the development of individual chronic obstructive pulmonary disease (COPD), a binary outcome (has COPD, yes vs no). Like others, we could use the census tract as the geographic proxy for neighborhood in the United States. Then, the fixed effects (X) could include individual characteristics (e.g., age, sex, race/ethnicity, income, and other individual-level factors) and neighborhood-level factors (e.g., air pollution in terms of ozone concentration level and neighborhood poverty rate). Thus, the fixed effects are the same as the explicit predictors or covariates included in classic linear regression models. Hence, we could use the regression coefficients (β) to quantify the exact and explicit fixed effects (X) on the risk of developing COPD; likewise, we could use the regression coefficients (γ) of the random effects (Z) and their variance to quantify the neighborhood-level variations in the risk of developing COPD. The random effects are neighborhood indicators that show an individual from a specific neighborhood. Here we want to know if neighborhood context has an impact on the risk of developing COPD.

Epidemiologically, we want to know if there are substantial neighborhood-level variations in COPD prevalence, or if neighborhood contexts have substantial influence on COPD risk. Statistically, we want to control the potential correlations between individuals from the same neighborhood. If we ignore such potential correlations, then the fitted statistical model could produce smaller standard errors for the regression coefficients (β) of the fixed effects (X). These underestimated standard errors for the regression coefficients (β) would inflate the significance of the fixed effects (X), produce more narrow confidence intervals for β, and could result in inappropriate or even misguided conclusions about the associations between COPD and the fixed effects (X). Thus, the inclusion of the random effects in the statistical models is also to account for the potential statistical dependency (correlations) among individuals and allow us to draw appropriate statistical inference (conclusions) for the fixed effects in the model.

As we mentioned earlier, because the datasets in the real world are often correlated, the random effects concept provides us with a statistical tool to address the complex correlations or statistical dependency in the datasets under study. Our usual assumption that individually observed records in a dataset are statistically independent is often violated. The potential correlations among individual outcomes in a dataset could be due to repeated measures for the same individual over time, or for multiple individuals from the same geographic units or socioeconomic clusters, or their combinations. For any data analysis, it is beneficial to explore and examine if there are substantial correlations in the outcomes of interest, geographically, temporally, by group (from the same person or community), or their combinations. To make appropriate statistical inference about the exact effects of fixed effect predictors (also called explanatory variables or covariates) on the outcomes under study, we must address these potential correlations in our datasets. We include fixed effects in the model to quantify how the outcome variables respond to changes in those explanatory variables, including those explicit contextual variables; we include random effects in the model to account for complex statistical dependency in the outcome data and to quantify the variations among random effect levels and their implicit contextual influence on the outcomes.

The link function ($g(.)$) maps the monotonic relationship between the expected data mean of the outcome of interest ($E(Y)$) and the linear predictor (η):

$$g(E(Y)) = \eta$$

In our example, the COPD outcome has two values ($Y=1$ or $Y=0$). The link function for the binary outcome is the logit function:

$$g\left(E(Y)\right) = \mathrm{logit}\left(p|y = 1\right) = \log\left(\frac{p}{1-p}\right) = \eta = X\beta + Z\gamma$$

where p is the expected probability of having COPD ($Y=1$). Now, we see how the outcome of interest (Y), COPD, links with the fixed effects (X) and random effects (Z) that compose of the linear predictor (η) via a logit link function. Thus, a GLMM regression model allows the linear predictor component related to the outcome variable via a link function. For normally distributed continuous outcome variables, the link function is an identity function and the expected mean (μ) of Y is directly equal to the linear combinations of the fixed effects and random effects and their regression coefficients:

$$g\left(E(Y)\right) = E(Y) = \mu = X\beta + Z\gamma$$

Thus, the classic linear or linear mixed regression model is a special case of GLMM. The link functions vary by the probability distribution of outcome variables, that is, identity function for normal continuous outcomes, logit function for binary and binomial outcomes, and log function for Poisson count outcomes. GLMMs could model outcomes (Y) from the exponential family of distributions that include normal (Gaussian), binary, binomial or multinomial, Poisson, negative binomial, beta, exponential, gamma, geometric, inverse Gaussian, and inverse gamma distributions. Each probability distribution for an outcome has its unique link function; common link functions include identity, logit, log, probit, complementary log-log, and inverse. These standard exponential family probability distributions could handle the most common outcomes of interest: continuous, categorical, and counts (see detailed examples of the commonly used link functions in appendix 2A in Stroup's book of *Generalized Linear Mixed Models*; Stroup, 2013).

GLMMs provide a very flexible and powerful statistical framework to specify the regression models that model the associations between risk factors (predictors) and outcomes. However, estimating the model parameters of GLMMs is challenging because the optimization of maximum likelihood of GLMMs, especially for nonnormal outcomes, has become analytical intractable. Thus, numerical approximation has often been used in model fitting and estimation of GLMMs. Bayesian estimation via MCMC recently has become a more popular alternative approach for the estimation of GLMMs.

Bayesian estimation via MCMC makes Bayesian inference via complex statistical models such as GLMM practical (Gilks et al., 1996). The basic idea of Bayesian inference or Bayesian learning is straight forward: update our belief or knowledge about a statistical/epidemiological hypothesis via available evidence, information, or data. Actually, we are Bayesian statisticians in our daily lives insofar as we constantly update our current knowledge about things as we acquire more evidence or experience about them. Our learning, logically, is a Bayesian learning process. The original idea of Bayesian inference came from Thomas Bayes (died in 1761)' posthumous publication in *Philosophical Transactions by the Royal Society of London* in 1763 (Bayes, 1763); the basic idea of Bayesian inference was developed more than 150 years earlier than popular classic frequentist statistics developed in the early 20th century mainly by R.A. Fisher, Jerzy Neyman, and E.S. Pearson (https://en.wikipedia.org/wiki/Frequentist_probability). But the Bayesian inference for most real problems has become analytically intractable. Bayesian inference remained a beautiful, but impractical statistical inference concept until the breakthrough in MCMC computer simulation in the 1990s made Bayesian inference practical for complex data analysis.

Bayesian inference has three basic components (Gelman et al., 2013):

- prior probability distribution ($p(\theta)$),
- likelihood function of the observed data ($p(y|\theta)$), and
- posterior probability distribution ($p(\theta|y)$).

Following the Bayes' theorem, the posterior probability distributions ($p(\theta|y)$) of model parameters (θ) are mathematically equal to the likelihood function of the observed data ($p(y|\theta)$) multiplied by the prior probability distributions ($p(\theta)$) over the normalizing constant ($\int p(y|\theta)p(\theta)$) so that the posterior probability distributions ($p(\theta|y)$) could integrate to one:

$$p(\theta|y) = \frac{p(y|\theta)p(\theta)}{\int p(y|\theta)p(\theta)} \propto p(y|\theta)p(\theta)$$

The Bayesian inference process is to update priors ($p(\theta)$) with the data likelihood ($p(y|\theta)$) and to obtain the posterior probability distributions ($p(\theta|y)$) of the unknown statistical model parameters for statistical inference.

Prior probability distributions ($p(\theta)$), often called priors, are used to quantify our understanding or knowledge or belief about a system or process or event in terms of probability before we take any evidence (observed data) into account. For example, when we construct a GLMM statistical model, the regression coefficients for fixed effects are unknown model parameters; we assume normal distributions with zero means and very large variances to express our limited knowledge or lack of knowledge about these model parameters; and these normal distributions are called noninformative priors. If we have certain knowledge about one parameter that could only have positive values, then we could introduce a left-truncated normal distribution as its prior; and the prior like this has incorporated some of knowledge or known rules and is often called informative prior. By nature, the prior setting-up reflects our uncertainties about the model parameters.

The likelihood function of the observed data ($p(y|\theta)$), often called likelihood, is the joint probability distribution of the observed data expressed as a function of these unknown statistical model parameters (θ). Posterior probability distributions ($p(\theta|y)$) of unknown statistical model parameters could be used to make statistical inferences about statistical hypotheses of interest or the characteristics of a process or a system under study. Despite the mathematical beauty of Bayes' theorem, for most real data analysis, the computation for the integral in the denominator ($\int p(y|\theta)p(\theta)$) is often analytically intractable; and thus, we could not produce accurate analytic closed forms for our posterior probability distributions ($p(\theta|y)$), like those common standard probability distributions. The breakthrough development of MCMC computer simulation in the 1990s has made Bayesian inference practical for complex real data analysis: MCMC methods allow us to sample from the conditional probability distributions based on the production of likelihood and priors ($p(y|\theta)p(\theta)$) and estimate the posterior distribution ($p(\theta|y)$) for each model parameter.

MCMC is a set of stochastic simulation tools for generating random variables from univariate or multivariate probability distribution functions (Gilks et al., 1996). For a random variable, we can express its probability distribution in two basic forms: (1) analytical closed-form formula and (2) a large number (1000 or more) of random values that follow its probability distribution. For a normally distributed random variable (y) with a mean of μ and a variance of δ^2, it could be accurately defined as the following formula:

$$y = f(x) = \frac{1}{\delta\sqrt{2\pi}} e^{-\frac{(x-\mu)^2}{2\delta^2}}, \quad -\infty \leq x \leq \infty$$

On the other hand, we can represent the distribution of y in terms of 1000 realized values that are randomly generated from this normal distribution. Although the normal distribution formula is accurate, a large number of realized values only approximately characterizes the probability distribution of y but has great computational flexibility and efficiency. In Bayesian inference, posterior probability distributions ($p(\theta|y)$) of statistical model parameters are often analytically intractable and have no accurate analytical formula like a normal distribution. The alternative way, a large number of randomly realized values of these model parameters, has become the only choice for us to represent their probability distributions. The MCMC methods could simulate a stationary ergodic Markov chain whose samples asymptotically follow the posterior densities ($p(\theta|y)$) that are only based on the production of likelihood of observed data and priors ($p(y|\theta)p(\theta)$). With a large sample (1000 or more) drawn from a model parameter's posterior probability density via MCMC methods, we could conveniently make inferences about the model parameter probability distribution summary statistics (mean, median, mode, variance, and standard errors) and percentiles to obtain credible intervals. Of course, we could also use the MCMC sample of posteriors for prediction or forecasting. The most common MCMC method for Bayesian estimation is Metropolis-Hastings algorithm, including its special case Gibbs sampling [see more details in Gilks et al., 1996]. The great improvement of computational efficiencies of MCMC methods has contributed to today's popularity of Bayesian statistics. The importance of MCMC methods in Bayesian statistics and its applications could never be overestimated. Bayesian estimation via MCMC will further promote Bayesian analysis for its growing applications.

Bayesian hierarchical modeling makes full use of the flexibility of GLMMs for statistical model construction and Bayesian estimation via MCMC for model estimation and inference. For example, to better understand the complex linkages between air pollution and population health outcomes, we are often required to combine multiple data sources at various geographic levels. Statistical models are the main tools to rigorously integrate these data together and evaluate the systematic linkages between air pollution exposure and health outcomes. Bayesian hierarchical modeling provides a robust and flexible statistical modeling framework for studying the linkages between air population and health outcomes, which often involves complex spatial and spatiotemporal correlations.

4.2 Bayesian hierarchical modeling

The Bayesian hierarchical modeling framework involves three basic stages or steps (Arab et al., 2008): (1) data model; (2) process model; and (3) parameter model.

The purpose of the data model is to specify an appropriate probability distribution model for the observed outcome and to explore the possible stochastic process generating the observed data. We treat the observed data as a realization of a stochastic process, and one major goal of statistical modeling is to identify an appropriate representation of this data-generating process for our observed data. For example, a Bernoulli distribution is often assumed for a binary outcome, a Poisson distribution for count data, and a normal distribution for continuous health outcome data. Exploratory data analysis could be applied at this stage to evaluate the basic distribution characteristics of the observed outcomes: mean, median, minimum, maximum, range, quartiles, and histogram shape (normality and skewness). Such measures provide further information to determine which probability model is appropriate for the observed data. Thus, the goal of the data model stage is to determine the probability distribution for the observed outcome; this probability distribution could be one of the common standard probability distributions. We then apply the link function ($g(.)$) to the mean parameter (μ) of this probability distribution for the observed data and form a linear predictor as done in GLMMs:

$$g\left(E(Y)\right) = g(\mu) = \eta$$

The specified probability distribution for our health outcome is, by nature, a conditional probability distribution: the observed data are independently distributed, conditioning on all the model parameters we include in the statistical model at the later stages of process model and parameter model.

The process model stage commonly involves constructing the linear relationships between the linear predictor of an outcome and its potential predictors or risk factors or covariates as in a GLMM statistical model. Subject-oriented knowledge is critical at this stage. Statistical modeling is about how we translate our subject matter knowledge-based conceptual model into a numerical statistical model (Cox and Donnelly, 2011). A good statistical model should not only be statistically well specified but also epidemiologically make sense from a subject perspective. Here, we could follow the flexible GLMMs statistical modeling framework to specify the structural relationship between health outcomes and their associated risk factors. In the context of air pollution and health outcome linkage studies, the relationships between health outcome, risk factor, and exposure to air pollution, as we do for a GLMM statistical model, could be intuitively expressed as follows:

$$\eta = X\beta + Z\gamma = x\beta' + e * b + Z\gamma$$

where we divide the fixed effects (X) into two components, a set of non-air-pollution predictors, risk factors, or explanatory variables (x) that associated with health outcome (y) and a set of air-pollution exposure metrics (e), quantifying individual or population exposure to air pollution, and β' and b are their corresponding vector of regression coefficients. In air pollution and health studies, we are mainly interested in whether b is statistically significant; if so, it suggests that air pollution exposure is linked to the occurrence and/or development of the health outcomes under study. The random effect Z is used to specify the complex correlations among the observed data under study. The observed data records are commonly correlated, which means they are not statistically independent. The inclusion of random effects accounts for the potential statistical dependency among the observed data. In the context of air pollution and health studies, spatial proximity or adjacency between areas or individuals could result in substantial spatial correlations among health outcomes. We will use a case study to highlight how to handle these common and complex spatial correlations in the following section. Temporal correlations or spatial and temporal correlations could arise because of multiple records from the same areas or from the same individuals over time.

In the process model stage, we need to construct the design matrix (Z) to account for all these known correlations among health data records. Thus, the process model specification requires us to consider at least two aspects for model formulation: (1) what factors and air pollution exposure metrics (including their transformation and interactions) should be included in a statistical model? These factors are specified as fixed effects in a GLMM model; (2) what is the hierarchy of observed data structure for health outcomes of interest? This hierarchy is specified as random effects in a GLMM model. Sometimes a factor could be specified as fixed effect or random effect, depending on your research objectives or other model choice concerns. For example, when multiple records for each city are obtained for a multicity study; the cities could be specified as fixed effects in the model if only city-specific effects in the sample are concerned; alternatively, the city factor could be specified as random effects if the cities are viewed as a sample of the entire city population from a state or a region or a country.

The purpose of the parameter model is to further specify statistical assumptions about the model parameters from the data model and process model stages. Model parameters are unknown and need to be estimated later. But in a hierarchical model specification process, we need to specify a statistical probability distribution for each of these model parameters. In practice, we often assume noninformative normal distributions with means of zeroes and variances of large values for fixed effects regression coefficients (β' and b); we also assume normal distribution with means of zeroes and variance parameters (δ^2) for random effects regression coefficients (γ); these normal distributions are called priors (prior probability distributions for the unknown model parameters). We further assign an inverse gamma distribution for these variance parameters (δ^2). Various prior distributions (e.g., half-t distribution) have been explored for variance parameters in hierarchical models (Gelman, 2006). The prior probability distribution for the model parameter priors is often called hyperprior. From a Bayesian perspective, all model parameters are random variables, the differences between random effects and fixed effects are not themselves and more about their priors' setting-up: each fixed effect model parameter has its independently specified prior, whereas a group of model parameters for a random effect has a common prior. In the previous example, if we treat cities as fixed effects, then each city will have a model parameter with a normal distribution prior with a mean of zero and a large variance; if we treat the city as a random effect, then a normal distribution prior with a mean of zero and a common variance parameter of δ^2_{city} will be assumed for all city-specific model parameters, that is, all city-specific model parameters are drawn from a common normal probability prior.

The multistage Bayesian hierarchical model specification approach is flexible and we could conveniently build up a statistical model that represents complex dependency structures in the observed data, just like simple Lego blocks forming a complex object. Once a Bayesian hierarchical model is set up, we need to estimate the model parameters via MCMC. Bayesian estimation via MCMC is available in most common statistical software, such as SAS, R, Stata, Matlab, and Mplus. There are also some specialized Bayesian statistical software programs, such as WinBUGS and JAGS. WinBUGS in particular is a very flexible and powerful Bayesian statistical program, and it could handle complex spatial or spatiotemporal models that are often needed to evaluate the linkages between air pollution exposure and population health outcomes.

For most data analyses, we formulate multiple models for the same health outcome due to various concerns, which inevitably raises a challenging task of model selection. Deviance information criterion (DIC) is routinely used for Bayesian model selection and comparison (Spiegelhalter et al., 2002). Deviance is a measure of overall fit of a statistical model, defined as -2 times the log likelihood:

$$D(\theta) = -2\log\{p(y|\theta)\}$$

In a Bayesian framework, we could calculate two deviance measures:
- the posterior mean of the deviances based on model parameter posterior distributions (samples):

$$\overline{D(\theta)} = E\{D(\theta)|y\} = E_{\theta|y}\left[-2\log\{p(y|\theta)\}\right]$$

- the deviance based on the posterior means of model parameter posterior distributions:

$$D(\bar{\theta}) = D\{E(\theta|y)\} = -2\log\left[p\{y|\bar{\theta}(y)\}\right]$$

The difference between the posterior mean of the deviance $(\overline{D(\theta)})$ and the deviance of posterior means $(D(\bar{\theta}))$ is the estimate of the effective number of model parameters in a Bayesian statistical model:

$$p_D = \overline{D(\theta)} - D(\bar{\theta})$$

Then, DIC is the sum of the posterior mean of the deviance $(\overline{D(\theta)})$ and the estimate of effective number of model parameters in a Bayesian statistical model (p_D):

$$\text{DIC} = \overline{D(\theta)} + p_D = D(\bar{\theta}) + 2p_D$$

The form of DIC is quite similar to classic Akaike's information criterion (AIC) (Akaike, 1974):

$$\text{AIC} = -2\log\{p(y|\hat{\theta})\} + 2p$$

where $\hat{\theta}$ is the maximum likelihood estimate of model parameters (θ) and p is the number of model parameters (p). DIC is a "generalized AIC" under noninformative priors.

Like AIC, DIC balances model fit (the posterior means of the deviance) and model complexity (the number of effective model parameters). Smaller DIC values suggest better model fit of the observed data. In general, if DIC values between two models differ by 10, the model with higher DIC should be ruled out; if they differ by 5–10, the model with lower DIC should be preferred; if they differ by <5, then these two make little difference in terms of model fitting of data. DIC could be conveniently calculated from MCMC posterior samples and is available from Bayesian statistical software or programs, such as WinBUGS and SAS Proc MCMC.

4.3 Case study: Bayesian hierarchical spatial modeling

Our case study explores the linkages between air population (ozone and particular matters) and chronic lower respiratory disease (CLRD) mortality in the United States. The main objective was to evaluate county level associations between long-term exposure to ambient ozone and fine particulate matter (PM$_{2.5}$, particles with an aerodynamic diameter of 2.5 µm or less), and CLRD mortality in the contiguous United States (48 states and District of Columbia). The full study has been published in *American Journal of Respiratory Critical Care Medicine* (Hao et al., 2015). As a case study for this chapter, we summarize only the Bayesian hierarchical spatial modeling process, including Bayesian hierarchical spatial model setup (data model, process model, and parameter model), Bayesian model estimation via MCMC, Bayesian model comparison, and Bayesian model inference.

4.3.1 Data model

The population health outcome of interest is the number of CLRD deaths among adults 45 years of age or older during 2007–2008 for each of 3109 counties from 48 states and the District of Columbia in the contiguous United States. We excluded Alaska and Hawaii because air pollution data are not available. The outcome of interest is county-level death count, and hence a Poisson distribution was assigned for these county-level death counts in all 3109 counties in the contiguous United States:

$$y_i \sim \text{Poisson}(\lambda_i), \ i = 1,\ldots,3109$$

where λ_i is the expected death count for a county (i) under Poisson distribution. The variance parameter of Poisson distribution is its mean (λ_i). Log function is the link function for Poisson distribution, thus we have the linear predictor (η_i) for county (i):

$$\eta_i = g\left(E\left(y_i\right)\right) = \log\left(\lambda_i\right)$$

4.3.2 **Process model**

At this stage, we need to specify the fixed effects and random effects for the statistical models of interest. In this study, we have

$$\log(\lambda_i) = \log(\text{Pop}_i) + x_i\beta' + e_i * b + Z_i\gamma$$

where Pop_i is the population size for adults aged 45 years old and above for county (i) in 2007–2008, and serves as an offset variable to scale the modeled Poisson mean (λ_i), and $\log(\text{Pop}_i)$ is treated like a fixed effect regression variable with a fixed parameter of one; x_i is a set of the six fixed effect non-air-pollution variables, including regression intercept ($x=1$), county-level percentage of adults 65 years old and above in 2007–2008, percentage of adult population aged 18 years old and above under federal poverty level in 2007–2008, percentage of lifetime smokers among adults 18 years old and above in 2007–2008, percentage of being obese among adults 18 years old and above in 2007–2008, and average annual extremely hot days ($\geq 90°F$) during 2001–2008; e_i is a set of the two fixed effect air pollution exposure variables, including county-level 8-h maximum zone and 24-h average $PM_{2.5}$ concentration during 2001–2008. Thus, we have six regression coefficients (β_k', $k=1, 2, 3, 4, 5, 6$) for non-air-pollution fixed effect variables and two regression coefficients (b_k', $k=1, 2$) for air-pollution exposure fixed effect variables.

The models contain three potential random effects (Z_i): (1) *state random effects* that specify the possible correlations between counties from the same state and represent the possible state-level contextual influence on CLRD mortality: $ST_{j[i]}$, $j=1, ..., 49$; (2) *county unstructured random effects* that specify the county heterogeneous effects (spatial heterogeneity) on CLRD mortality and represent county-level unspecified implicit contextual effects: U_i, $i=1, ..., 3109$; and (3) *county spatially structured random effects* (often called spatial random effects in the literature) that specify the potential spatial correlations among neighboring counties (spatial homogeneity) and represent the possible implicit county adjacent environment's influence on CLRD mortality: S_i, $i=1, ..., 3109$. Here are five models with different random effects specification:

- Model I: $\log(\lambda_i) = \log(\text{Pop}_i) + x_i\beta' + e_i * b + ST_{j[i]}$
- Model II: $\log(\lambda_i) = \log(\text{Pop}_i) + x_i\beta' + e_i * b + ST_{j[i]} + U_i$
- Model III: $\log(\lambda_i) = \log(\text{Pop}_i) + x_i\beta' + e_i * b + ST_{j[i]} + S_i$
- Model IV: $\log(\lambda_i) = \log(\text{Pop}_i) + x_i\beta' + e_i * b + ST_{j[i]} + U_i + S_i$
- Model V: $\log(\lambda_i) = \log(\text{Pop}_i) + x_i\beta' + e_i * b + ST_{j[i]} + \rho_i * U_i + (1 - \rho_i) * S_i$

Model I (state random effect model) is the simplest one with only state random effects and assumes county-level random effects ignorable for this dataset. Model II (county random effect model) includes both state random effects and county-level unstructured random effects and assumes spatial correlations among neighboring counties are ignorable. Model III (county spatial random effect model) includes both state random effects and county-level spatially structured random effects and assumes county-level unstructured random effects ignorable. Model IV (county spatial convolution model) includes state random effects and both county-level unstructured and structured random effects. Model V (county spatial mixture model) includes state random effects and both county-level unstructured and structured random effects but assumes the balance between spatial heterogeneity and spatial homogeneity could change across the contiguous United States; for some counties, county-level spatial heterogeneity could be dominant, while for other counties, county-level spatial homogeneity could be dominant; thus, a balance parameter $(0 < \rho_i < 1, \quad i = 1, \ldots, 3109)$ is introduced into model V. The model complexity increases from models I to V.

When we specify different models, we usually start with the simplest one then gradually add complexity for additional models. For each model formulation, we should have some appropriate statistical assumptions and epidemiological justifications. Our goal is to build a model that accounts for and balances both statistical concerns and epidemiological insights; we also avoid constructing a model that is too complex for our data to support.

4.3.3 Parameter model

The prior specification (parameterization) is critical for Bayesian model setup and has direct impact on model converging as well as on DIC. At this stage, we assign priors for unknown statistical model parameters, including the possible variance parameters for the probability distribution of the health outcome, the regression coefficients of fixed effects, the regression coefficients and their variance parameters of random effects, and any other model parameters revealed in the prior stages of data model and process model.

The variance parameter of the Poisson distribution for the county-level CLRD deaths is equal to its mean parameter (λ_i). $\log(\lambda_i)$ has already been assigned to a linear combination of fixed effects and random effects and their corresponding regression coefficients. If we deal with a normally distributed outcome, we need to assign a prior for its variance parameter at this stage. We assign noninformative priors for fixed effect regression coefficients with a series of normal distributions with means of zero and large variances of 100,000.

$$\beta_k' \sim \text{Normal}\left(0, 100,000\right), \quad k = 1, 2, 3, 4, 5, 6$$

$$b_i \sim \text{Normal}\left(0, 10,0000\right), \quad k = 1, 2$$

We next assign a normal distribution with a mean of zero and a variance parameter for regression coefficients of the state random effects and county-level unstructured random effects:

$$ST_j \sim \text{Normal}\left(0, \delta_{st}^2\right)$$

$$U_i \sim \text{Normal}\left(0, \delta_u^2\right)$$

Then, we assign the county-level spatial random effects a common intrinsic conditional autoregressive (ICAR) prior (Banerjee et al., 2014). Under this specification, it is assumed that

$$S_i \mid S_j \sim \text{Normal}\left(\overline{S}_i, \frac{\delta_s^2}{n_i}\right), \ j \in \Omega_i$$

where Ω_i is the set of neighbors for county (i), n_i is the number of neighbors for county (i), and $\overline{S}_i$ is the mean of spatial random effects (S_j) of county (i) neighbors ($j \in \Omega_i$). The conditional variance (δ_s^2) for spatial random effects determines the amount of spatial correlations across neighboring counties.

Then, we assign noninformative gamma distributions for the inverse of variance parameters of random effects in this study:

$$\frac{1}{\delta_{st}^2} \sim \text{Gamma}\left(0.001, 0.001\right)$$

$$\frac{1}{\delta_u^2} \sim \text{Gamma}\left(0.001, 0.001\right)$$

$$\frac{1}{\delta_s^2} \sim \text{Gamma}\left(0.001, 0.001\right)$$

We also assign a noninformative beta distribution, equivalent to uniform prior, for the spatial mixture parameter (ρ_i):

$$\rho_i \sim \text{Beta}\left(1, 1\right)$$

We keep consistent prior specifications for all model parameters for five models in this case study and assure that their DIC values are comparable, since DIC is sensitive to prior parameterization.

4.3.4 **Bayesian model estimation and comparison**

Bayesian estimation of hierarchical spatial models is quite challenging. In particular, model V (spatial mixture model) is so complex that it is hard to implement using conventional statistical software, such as SAS. Hence, we implemented all five models in the most flexible and powerful WinBUGS version 1.4.3 that uses Gibbs sampling to generate MCMC samples for model parameters and calculate DIC values. Table 4.1 displays the DIC values for all five models.

When we compare DIC values between different models, DIC values for models I and II are much higher than those DIC values for models III–V, which means that models III–V better fit the data than models I and II. Among models III–V, the DIC differences are <5 and they have quite similar fit to the data. However, model III (county spatial random effect model) has smallest number of random effects parameters and is the most parsimonious model, thus model III was selected as the final model.

Table 4.1 DIC for models with different random effects

Model	DIC
I (state random effect model)	21,606
II (county random effect model)	21,607
III (county spatial random effect model)	21,475
IV (county spatial convolution model)	21,475
V (county spatial mixture model)	21,479

4.3.5 Bayesian inference

We used model III (county spatial random effect model) and estimated all the model parameters and their credible intervals. In this study, we were most interested in the linkages between population exposure to ambient air pollution (ozone and $PM_{2.5}$). Our final estimates of the adjusted rate ratios and their credible intervals are 1.05 (1.01–1.09) per 5 ppb for ozone and 1.07 (0.99–1.14) per $5 \, \mu g/m^3$ for $PM_{2.5}$. Thus, this national case study suggests that long-term exposure to ambient air pollution, especially exposure to ozone, could substantially increase the risk of population CLRD mortality in the contiguous United States. Bayesian inference could also address more complex and interesting epidemiological questions. For example, we could use a model like model III to estimate how many CLRD premature deaths could be prevented if all counties with higher ambient air pollution would have reduced their ozone and $PM_{2.5}$ concentration to their corresponding median levels.

4.4 Summary and conclusions

It is quite challenging to apply Bayesian hierarchical modeling to analyze the potential linkages between air pollution and population health, especially when complex spatial or spatiotemporal dependency occurs in the observed data. We hope that this chapter might be a useful guide for both epidemiologists and applied statisticians who wish to start this challenging data-analytic journey.

GLMMs could easily be extended to handle more complex and multiple correlated outcomes under a broader generalized linear latent and mixed models (GLLAMMs) (Skrondal and Rabe-Hesketh, 2004). Some observed data distributions are so complex that a single probability distribution might not be appropriate and a mixed probability distribution should be specified. For example, an excessively large amount of zeros could happen to count data. In this case, a zero-inflated Poisson distribution (a mixture of binomial and Poisson probability models) would be more appropriate. On the other hand, we should avoid building a complex hierarchical model that could not be supported by the data. The growing popularity of Bayesian methods and their applications have been encouraging more computationally efficient MCMC methods and their combination with numerical approximation and optimization methods for Bayesian analysis of complex Big Data.

Disclaimer

The views expressed in this publication are those of the authors and do not necessarily represent the views of the Department of Housing and Urban Development. The findings and conclusions in this publication are those of the authors and should not be construed to represent any official USDA or U.S. Government determination or policy.

References

Akaike, H., 1974. A new look at the statistical model identification. IEEE Trans. Automat. Contr. 19 (6), 716–723.

Arab, A., Hooten, M.B., Wikle, C.K., 2008. Hierarchical spatial models. In: Shekhar, S., Xiong, H. (Eds.), Encyclopedia of Geographical Information Science. Springer, New York, pp. 425–431.

Banerjee, S., Carlin, B.P., Gelfand, A.E., 2014. Hierarchical Modeling and Analysis for Spatial Data. Chapman and Hall/CRC.

Bayes, T., 1763. An essay towards solving a Problem in the Doctrine of Chances. Philos. Trans. R. Soc. Lond. 53, 370–403.

Cox, D.R., Donnelly, C.A., 2011. Principles of Applied Statistics. Cambridge University Press.

Gelman, A., 2006. Prior distributions for variance parameters in hierarchical models. Bayesian Anal. 1 (3), 515–533.

Gelman, A., Carlin, J.B., Stern, H.S., Rubin, D.B., 2013. Bayesian Data Analysis. CRC Press, Boca Raton, FL.

Gilks, W.R., Richardson, S., Spiegelhalter, D.J., 1996. Markov Chain Monte Carlo in Practice. Chapman and Hall, London.

Hao, Y., Balluz, L., Strosnider, H., Wen, X.J., Li, C., Qualters, J.R., 2015. Ozone, fine particulate matter, and chronic lower respiratory disease mortality in the United States. Am. J. Respir. Crit. Care Med. 192 (3), 337–341.

McCullagh, P., Nelder, J.A., 1989. Generalized Linear Models. Chapman & Hall/CRC Press.

McCulloch, C.E., Searle, S.R., Neuhaus, J.M., 2008. Generalized, Linear, and Mixed Models. Wiley.

Skrondal, A., Rabe-Hesketh, S., 2004. Generalized Latent Variable Modeling: Multilevel, Longitudinal, and Structural Equation Models. Chapman & Hall/CRC Press.

Spiegelhalter, D.J., Best, N.G., Carlin, B.P., Linde, A.v.d., 2002. Bayesian measures of model complexity and fit (with discussion). J. R. Stat. Soc. Ser. B: Stat. Methodol. 64 (4), 583–639.

Stroup, W., 2013. Generalized Linear Mixed Models: Modern Concepts, Methods and Applications. CRC Press, Boca Raton, FL.

World Health Organization, 2016. Ambient (Outdoor) Air Quality and Health. https://www.who.int/en/news-room/fact-sheets/detail/ambient-(outdoor)-air-quality-and-health.

Machine learning for spatiotemporal big data in air pollution

5

Weitian Tong

Department of Computer Science, Eastern Michigan University, Ypsilanti, MI, United States

5.1 Introduction

Air pollution remains as a significant environmental public health concern and has attracted much attention from geographers, environmental scientists, and public health professionals such as epidemiologists. Typical air pollutants such as $PM_{2.5}$, PM_{10}, NO_2, SO_2, O_3, etc., are mainly produced by the industrial and traffic emission. Among the pollutants, *particulate matter* (PM) has attracted immense attention. PM is a mixture of microscopic solids and liquid droplets that vary in origin, size, and composition. Usually, PM is so small that it can get deep into the lungs and cause serious health problems (Zanobetti and Schwartz, 2009). PM in air pollution with diameters of $<2.5\,\mu m$ ($PM_{2.5}$) is thought to pose a particular greater risk to health, because it is more likely to be toxic and can be breathed more deeply into lungs (Li et al., 2014). This was verified by the report from the European Study of Cohorts for Air Pollution Effects (Raaschou-Nielsen et al., 2013) that $PM_{2.5}$ is associated with a risk for the development of lung cancer. Recently, PM in outdoor air pollution was even designated a Group I carcinogen to humans by the International Agency for Research on Cancer (2013). The association between the risk of overall lung cancer and the concentrations of $PM_{2.5}$ has also been studied in Hystad et al. (2013), Puett et al. (2014), Beelen et al. (2008), Pope et al. (2002), Næss et al. (2006), Dockery et al. (1993), and Attfield et al. (2012). Moreover, acute stroke mortality (Hong et al., 2002), visibility reduction (Ghim et al., 2005; Sloane et al., 1991), and daily mortality in many US cities (Laden et al., 2000) are all attributed to $PM_{2.5}$ exposure.

With the advancement of geospatial technologies, especially geographic information systems (GIS), environmental exposure analysis has made significant progress. An accurate understanding of $PM_{2.5}$ in a continuous space-time domain is critical for meaningful assessment of the quantitative relationship between the adverse health effects and the concentrations of $PM_{2.5}$, which can help to identify, monitor, and evaluate interventions, such as establishment and enforcement of air quality standards, reduction of industry or auto emissions, and local community-based efforts. Since air pollution data, including the $PM_{2.5}$ data, are commonly recorded at scattered or localized sampling locations, it is often necessary to predict or estimate the air pollution concentration at new data points within the range of a discrete set of known data points, which is known as *interpolation* in numerical analysis. Spatial interpolation methods have already been extensively investigated over the years based on the assumption that the nearer two points are, the higher correlation they are. Common spatial interpolation methods, such as kriging (Krige, 1951), inverse distance weighting (IDW) (Shepard, 1968), trend surface (Zurflueh, 1967), and splines (De Boor, 1978), are available in major GIS software packages. Interested reader can also refer to the excellent surveys on spatial interpolation techniques by Li and Heap (2008, 2011).

Nowadays, modern sensors are able to monitor $PM_{2.5}$ concentrations and other related data at an increasing temporal resolution, and thus producing extremely large volume of spatiotemporal data sets. However, the data collection is only carried out at a limited number of monitoring locations and in a noncontinuous manner. Therefore, advancing spatiotemporal interpolation methodology to gain a better understanding of the observed $PM_{2.5}$ data is urgent as it can produce a substantial impact on accurately estimating human exposure to $PM_{2.5}$ and lead to more reliable assessment of the relationship between $PM_{2.5}$ and disease outcomes overtime.

Compared with the traditional spatial interpolation, an additional time dimension needs to be considered in the spatiotemporal interpolation. There exist few efficient and effective algorithms to interpolate complex spatiotemporal data sets. Some spatiotemporal methods (Appice et al., 2013; Liao et al., 2006) treat space and time separately and reduce the spatiotemporal interpolation problem to a sequence of snapshots of spatial interpolations. Some other spatiotemporal methods (Gräler et al., 2013; Li and Revesz, 2004; Li et al., 2012, 2014; Losser et al., 2014; Pebesma, 2012; Tong et al., 2019a) treat time as another dimension in space and incorporate spatial and temporal dimensions simultaneously. Unfortunately, all these studies did not provide appropriate methods to incorporate time dimension such that the temporal dimension is treated "fairly" compared with the spatial dimension. This problem was referred to as the "time-scale problem" by Li and Revesz (2004), and the scaling ratio is named as *spatiotemporal anisotropy parameter* by Gräler et al. (2013). Only some simple methods were proposed to estimate this parameter. The main reason is the lack of rigorous theory support to figure out the correlation between time dimension and space dimension. Under such circumstances, a black-box approach such as machine learning method is a natural idea and a promising direction to estimate the spatiotemporal anisotropy parameter. Recently, Tong et al. (2019a) addressed this problem by presenting an efficient parallel machine learning method to learn the optimal spatiotemporal anisotropy parameter.

However, the relationship among spatial and temporal dimensions is not just to determine the spatiotemporal anisotropy parameter. Even within the space dimension, there are two subdimensions, for example, longitude and latitude, which might not contribute equally to the air pollutant concentration. Besides, plenty of hidden factors such as meteorology, land use, traffic flow, human activities, etc. also effect the concentration of $PM_{2.5}$. According to the research reports in the literature (Charron and Harrison, 2005; Morgenstern et al., 2007; Oftedal et al., 2008), the $PM_{2.5}$ concentration is influenced by many meteorological and land-use variables, such as precipitation, temperature, humidity, wind speed, road length, distance to nearest road, etc. Aerosol optical density (AOD) or aerosol optical thickness is a measure of the amount of PM suspended in the atmosphere, collected as a part of NASA's earth observing system program (Brokamp et al., 2018). Therefore, satellite remote sensing of aerosols can also be used to assess the surface-level $PM_{2.5}$ concentration at high spatial and temporal resolutions (Al-Saadi et al., 2005).

Most existing spatiotemporal interpolation methods restrict the interpolation models to simple ones, such as IDW and kriging, which usually have explicit and simple mathematical descriptions. On the other hand, the real relationship between $PM_{2.5}$ concentrations and the influential factors is unknown, and can be so complicated that no explicit mathematical model fits for it. Therefore, black-box approaches are preferred in this situation as alternatives to traditional models for input-output mathematical models. Machine learning is the science of getting computers to act without being explicitly programmed. In particular, deep learning can extract high-level, complex abstractions as data representations through a hierarchical learning process (Najafabadi et al., 2015). The hierarchical learning architecture is motivated by the artificial intelligence emulating the deep, layered learning process of the primary sensorial areas of the neocortex in the human brain, which automatically extracts features and abstractions from the underlying data (Arel et al., 2010; Bengio and LeCun, 2007; Bengio et al., 2013). Because the air quality process is inherently complicated, various machine learning and deep learning methods are perfect candidates for black-box approaches to automatically consider the hidden factors and build the model for air pollution data.

We will introduce the state-of-the-art machine learning and deep learning methods for the generation of more accurate estimation of $PM_{2.5}$ concentration on a large geographic scale and over a long-time period.

5.2 Related work

Adopting the machine learning for the $PM_{2.5}$ concentration estimation is relatively a new research line. Gupta and Christopher (2009) used an artificial neural network (ANN) to estimate surface-level $PM_{2.5}$ from Moderate Resolution Imaging Spectroradiometer (MODIS) AOD data and meteorological data in the southeastern United States. Zou et al. (2015) proposed a radial basis function (RBF) neural network (NN) method to estimate $PM_{2.5}$ concentrations in Texas, the United States with various meteorological data and land-use information. Reid et al. (2015) used the generalized boosting model (GBM), containing 11 statistical algorithms[1] and 29 predictor variables, to estimate $PM_{2.5}$ concentrations during the 2008 northern California wildfires. Di et al. (2016) and Kwon (2017) adopted the convolutional neural network (CNN) to extract more useful spatial correlation information around interested point and thus improve the accuracy of $PM_{2.5}$ concentration estimation. In 2017, Hu et al. developed a random forest model incorporating AOD data, meteorological fields, and

---

[1] These algorithms include random forest, bagged trees, GBM, elastic net regression, multivariate adaptive regression splines lasso regression, support vector machines (SVMs), Gaussian processes, generalized linear model, k-nearest neighbors, and generalized additive model.

land-use variables to estimate daily 24 hours averaged ground-level $PM_{2.5}$ concentrations over the conterminous United States in 2011. Later, Chen et al. (2018) applied the random forest model to estimate historical concentrations of $PM_{2.5}$ at daily time scale in China at a national level with MODIS AOD data, meteorological fields, and land-use information. Brokamp et al. (2018) also utilized satellite, meteorologic, atmospheric, and land-use data to train a random forest model capable of accurately predicting daily $PM_{2.5}$ concentrations at a resolution of $1 \times 1\,km$ throughout an urban area encompassing seven counties surrounding Cincinnati, the United States.

In 2017, Fan et al. proposed a spatiotemporal prediction framework for air pollutants levels based on the deep recurrent neural network (DRNN). It is a pure prediction model and cannot be used for the general interpolation purpose. Later in 2018, Qi et al. presented a general and effective approach to solve the interpolation, prediction, and feature analysis of fine-grained air quality in one model. Their approach also employed the RNN as the main ingredient. All previous RNN-based deep learning models only considered the past information for interpolation or prediction. Moreover, they assumed that the current air pollution concentration is only influenced by the past information of the interested point and the current air pollution concentrations of the geographical neighbors. This assumption ignores the temporal correlation among geographical neighbors. On the other hand, when interpolating the spatiotemporal data, the air pollution concentration of the unknown point relies on the close points in both spatial and temporal dimensions. In particular, not only past concentrations but also future concentrations provide valuable information for the estimation of the current concentration.

Most recently, Tong et al. (2019b) developed a novel spatiotemporal interpolation method to predict $PM_{2.5}$ levels based on the bidirectional RNN, which considers not only the past spatiotemporal correlations among the geographical neighbors but also the future information. Their method allows for the generation of more accurate estimation of air pollution on a large geographic scale and over a long-time period. Usually, the memory of past-learned patterns by RNN can fade as time goes on and cause a computation problem—vanishing gradient (Bengio et al., 1994). They explored the long short-term memory (LSTM) (Hochreiter and Schmidhuber, 1997) RNN to tackle this challenge by maintaining an internal state and creating paths where the gradient can flow for long durations. In particular, they employed the bidirectional LSTM DRNN (Bi-LSTM DRNN) to learn the air pollution concentrations. The principle of the Bi-LSTM DRNN is to split the neurons of a regular LSTM DRNN into two directions such that both past and future information can be considered at the same time.

5.3 Commonly used data and data source

There are plenty of hidden factors related to the $PM_{2.5}$ concentration. The input data sets should provide most correlated factors in order to reduce the noise and uncertainty and thus improve the estimation efficiency and accuracy. Many well-recognized organizations are collecting and offering various data related to $PM_{2.5}$. This section introduces several classic sources for commonly used data sets.

$PM_{2.5}$ measurements

There are several sources that provide the daily and even hourly $PM_{2.5}$ concentration data. The US Environmental Protection Agency (EPA) Air Quality System (AQS) (https://www.epa.gov/aqs) is a well-known data source, and provides various air pollution data collected by EPA, state, local, and tribal air pollution control agencies from over thousands of monitors. The Interagency Monitoring of Protected Visual Environment (IMPROVE) (http://vista.cira.colostate.edu/Improve/) is another popular source for different air pollution data.

MODIS AOD data

The satellite-based aerosol optical depth (AOD) measures the light extinction due to aerosol in the whole atmospheric column (Morain and Budge, 2012). Since the aerosols change the way the atmosphere reflects and absorbs visible and infrared light, the AOD measurements have been widely used to estimate $PM_{2.5}$ in various models for its large spatial coverage and repeated daily observations (Streets et al., 2013). The AOD estimates can be generated according to the images retrieved from two MODIS satellites (Terra and Aqua). One recent famous algorithm, named as MAIAC (Lyapustin et al., 2011a, b), is able to generate the satellite-based AOD with a spatial resolution of 1×1 km. The MODIS AOD data for different time periods are available on the Atmosphere Archive and Distribution System of NASA (https://ladsweb.modaps.eosdis.nasa.gov/). Usually, the AOD measurements from Aqua (overpass at 1:30 p.m. local time) and Terra (overpass at 10:30 a.m. local time) are combined to improve the sample size (Hu et al., 2014).

The raw AOD data suffer from missing data problem, which is mainly caused by bright surfaces (e.g., snow coverage) or cloud contamination, especially in winter (Liu et al., 2009). Besides, due to rare events, like forest fires, fireworks, or instrument error, the AOD measurements may contain abnormally large values. Usually the AOD data with values above 1.5 are excluded from modeling, which also creates missing values (Van Donkelaar et al., 2006). Therefore, filling up the missing data is a necessary step in the data preprocessing. Because the spatial distribution of the $PM_{2.5}$ monitor sites is not uniform and the resolution of the ground-level $PM_{2.5}$ data does not match with the MAIAC AOD data. Further preprocessing of either the $PM_{2.5}$ data or the MAIAC AOD data is needed to keep the data's resolution consistent.

Meteorological data

The meteorological data can be obtained from multiple data sources.

1. The EPA AQS (https://www.epa.gov/aqs) not only collects the measurement for various air pollutant concentrations, but also contains meteorological data, descriptive information about each monitoring station (including its geographic location and its operator), and data quality assurance/quality control information.

2. National Climatic Data Center's Integrated Surface Database (ISD) (https://www.ncdc.noaa.gov/isd) provides hourly air temperature, dew point temperature, wind speed, and visibility.
3. North American Regional Reanalysis (NARR) offers various data sources, such as land-surface, ship, radiosonde, pibal, aircraft, satellite, and others, with a spatial resolution of ~32 km and a temporal resolution of 3 hours.
4. North American Land Data Assimilation System (NLDAS) (https://ldas.gsfc.nasa.gov/) also assimilates different kinds of meteorological data at a spatial resolution of ~13 km and a temporal resolution of 1 hour.

Different data sources have different spatial and temporal coverage ranges. Moreover, the data's resolution also varies in both spatial and temporal dimension. Synchronizing the different meteorological data is of great significance.

Land-use data

Land-use data, such like elevation, road density, percentage of urban, etc., serve as proxies for emissions and are used to capture variations at a small a spatial scale (Di et al., 2016). The elevation data at a spatial resolution of ~30 m are available in the National Elevation Dataset (NED) (https://lta.cr.usgs.gov/NED) and the National Map (https://nationalmap.gov/elevation.html). Road network data, including limited access highway, highway, and local roads, can be extracted from ESRI StreetMap USA (Environmental System Research Institute, Inc., Redland, CA). US Census Bureau Topologically Integrated Geographic Recoding and Referencing (TIGER) system (https://www.census.gov/geo/maps-data/data/tiger.html) contains the information about "primary road" and "secondary road" with a 1 × 1 km spatial resolution, from which, the road density (km per km^2) can be calculated as the total length of these two types of roads. The National Land Cover Database (NLCD) (https://www.mrlc.gov/) provides information on forest cover and impervious surface both at the spatial resolution of ~30 m. Greenspace or vegetation coverage can be retrieved as the Normalized Differential Vegetation Index (NDVI), which is available in National Center for Atmospheric Prediction (NCEP) NARR data and MODIS MOD13A2, an NDVI data product.

Population density

The population density data can be obtained from the US Census Bureau. The population information can be weighted by areas of all census tracts intersected within a grid cell.

National Emission Inventory (NEI) emissions

The NEI provides county-wide emissions of various air pollutants, such as PM$_{2.5}$ and PM$_{10}$, and the annual total emissions from point sources (e.g., refineries).

---

[2] Here we need to define *distance* appropriately.

5.4 **Spatiotemporal interpolation**

In numerical analysis, interpolation is a method of constructing or predicting new data points within the range of a discrete set of known data points. Formally, given n sample points x_i and an objective function $Y = Y(x)$, the goal is to predict $Y(x_0)$ at any arbitrary location of interest, from the measured observations $y(x_i)$ of $Y(x)$ at the n sample points x_i. All interpolation algorithms, including inverse distance squared, splines, RBFs, triangulation, estimate the value at a given point as a weighted sum of data values at surrounding points. Almost all these weight functions are decreasing with increasing of separation "distance."[2] Thus nearly all these methods share the same general estimation formula as follows:

$$\hat{y}(x_0) = \sum_{i=1}^{n} \lambda_i \cdot y(x_i), \tag{5.1}$$

where $\{x_1, ..., x_n\}$ are the n sampled points used for the estimation, x_0 is the interested point, $\hat{y}(x_0)$ is the estimated value of the primary variable at x_0, $y(x_i)$ is the observed value at the sampled point x_i, and λ_i is the weight assigned to the sampled point x_i.

Interpolation has become a critical technique in geostatistics. Because geoinformatic data are commonly recorded at scattered or localized sampling locations, it is often necessary to spatially interpolate the data in order to: (a) organize the recording into a regular grid, (b) query the data for estimations at a particular location of interest, or (c) allow for integration of measured data or their derivatives over definite domains of area or space (i.e., calculation of fluxes based on a mass-balance approach [Cambaliza et al., 2014; Mays et al., 2009; Tadić et al., 2015]). Thus, traditional spatial interpolation models have been extensively investigated over years. And interested reader can refer to the excellent surveys on spatial interpolation techniques by Li and Heap (2008, 2011).

For the spatiotemporal interpolation, an additional temporal dimension is added compared with the traditional spatial interpolation. Suppose an area $\mathcal{A}$ can be partitioned into n different grids $\{P_1, ..., P_n\}$ and each grid P_i is regarded as one interested points. Assume there are m covariates or influencing factors. Each grid P_i at a specific time stamp t can be described as a tuple $(x_{i,t}, v_i)$ with $x_{i,t} = (t, x_{i,1}, x_{i,2}, ..., x_{i,m})$, where v_i is the air pollutant concentration, $x_{i,\ell}$ is the ℓth influencing factor. Therefore, the input data set can be denoted as n time series, $\{ts_1, ..., ts_n\}$. Each time series $ts_i = (x_{i,1}, ..., x_{i,T})$ is the sequence of observed data at a grid P_i. Our target is to estimate v for any position in $\mathcal{A}$ at any time.

In the spatial dimension, local air quality is usually influenced by the adjacent areas as air pollutants may disperse or transmit through the atmosphere with the wind. In the temporal dimension, historical states of the air pollutants can affect the current and future states. For instance, the air quality during the last hour will affect the air quality during the next hour. Another example is that the air quality tends to have similarities in the same season over the recent years. In other words, the air quality data in a given area have internal temporal correlation. Therefore, both spatial and temporal correlations (shown in Fig. 5.1) should be taken into account when interpolating air pollutant's concentration spatiotemporally. In addition to all the above influence factors, lots of hidden factors such as meteorology, land use, traffic flow, and human activities can also trigger the change of air quality in both spatial and temporal dimensions and thus affect the concentrations of air pollutants. Due to the inherent complexity of the relationship among various covariants, it is quite challenging to build a perfect mathematical model to estimate the concentrations of air pollutants.

5.5 Machine learning

Machine learning tries to answer the question of whether a computer can "learn." According to Arthur Samuel in 1959 (Samuel, 1959), machine learning gives "computers the ability to learn without being explicitly programmed." Such automated methods of data analysis are urgently needed in current era of big data. We define machine learning is a set of methods that focuses on the automatic representation of the input data and generalization of the learned patterns for use on future unseen data. In other words, a machine learning problem automatically detects or learns patterns from a set of n samples of data $\mathcal{D} = \{(x_i, y_i), i = 1, ..., n\}$, and then tries to predict properties for unknown data. Here, $\mathcal{D}$ is named as *training set* and n is the number

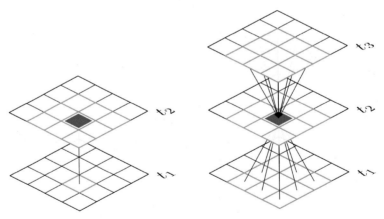

FIG. 5.1

A brief illustration of the spatial and temporal correlations: the *blue solid* cell represents the target point; *blue grids* represent adjacent points of the target point; *red grids* represent correlated points in the temporal dimension, and their influence on the target point is denoted by the *arrows*. The *left subfigure* shows the correlation that was considered by most research work in the literature. The *right subfigure* shows the correlation that is considered by our paper.

of training examples. Each sample is represented by a multidimensional entry x_i and an optional output y_i. Assume each $x_i \in \mathbb{R}^m$ is a m-dimensional vector of numbers. We say each dimension an *attribute* or *feature*.

Machine learning tasks are typically classified into three broad categories (Russell and Norvig, 2003):

- *Supervised learning*: Data usually comes with additional attributes that we want to predict. That is, each sample is (x_i, y_i). The goal is to learn a general function $f : \mathbb{R}^m \to \mathbb{R}$ that maps x_i to y_i.
- *Unsupervised learning*: In this category, data have *no* additional attributes to predict. That is, each sample is x_i. The goal is to discover some interesting structure only according to the input samples. For example, *clustering* is to group similar examples within the data; *density estimation* is to determine the distribution of data within the input space.
- *Reinforcement learning*: This is an interactive learning system, where a player acts in the world so as to maximize his/her rewards-based feedbacks in terms of rewards and punishments. Therefore, reinforcement learning not only depends on the input samples, but also focuses on online performance, which involves finding a balance between exploration (of uncharted territory) and exploitation (of current knowledge) (Williams, 1992).

This section will introduce the commonly used machine learning methods for the spatiotemporal interpolation.

5.5.1 Decision tree learning and random forest

5.5.1.1 Decision tree learning

A *decision tree* is a tree-like graph to facilitate the decision-making process. An example is shown in Fig. 5.2. In a decision tree, an internal node represents a "test" on an attribute (e.g., does the person take exercise in the morning?), each branch represents the outcome of the test, and each leaf node represents a decision after computing all attributes. Each root-to-leaf path shows the decision-making process.

Recall that each entry in the training set in machine learning has the format (x, y) where x is composed of the m features $x_1, ..., x_m$, and the dependent variable, y, is the target variable that is need to be understood, classified, or generalized. *Decision tree learning* is the construction of a decision tree from training set such that this decision tree can be a predictive model for the future decision making, that is, mapping x to appropriate y in the future.

In general, a decision tree can be learned by keeping splitting the training set into subsets recursively based on the tests on different attributes or features. This recursive partitioning process completes when the subset at a node is "homogeneous" enough, for example, an extreme case is that all samples in the subset have the same target value. There are plenty of decision tree learning algorithms. Notable ones include: Iterative Dichotomiser 3 (ID3) (Quinlan, 1986), a successor of ID3 (C4.5)

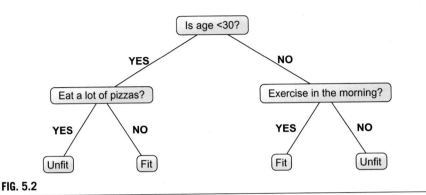

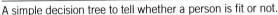

FIG. 5.2

A simple decision tree to tell whether a person is fit or not.

(Quinlan, 2014), Classification And Regression Tree (CART) (Breiman, 2017), CHi-squared Automatic Interaction Detector (CHAID) (Kass, 1980), MARS (Friedman, 1991), and Conditional Inference Trees (Hothorn et al., 2006). Different learning algorithms use different metrics for measuring the homogeneity of the target variable within the subsets. Commonly used metrics include *Gini impurity*, *information gain*, *variance reduction*, etc. All of these algorithms work with the previously mentioned principle of greediness, that is, a variable is selected to partition the sample space in the most "homogeneous" way or maximize some predefined metric. In particular, the algorithms ID3, C4.5, and CART employ the information gain, information gain ratio, and Gini index, respectively.

Most metrics are based on the concept *entropy*, which is borrowed from thermodynamics and is a measure of variability or chaos or randomness. Extended the thermodynamic entropy concept, Shannon (1948) introduced statistical entropy in 1948 to measure the randomness in statistics and the amount of information of a transmitted message in information theory. Suppose X is a discrete random variable by taking the value from a finite set $\mathcal{X}$, and its probability distribution $\{p_X(x)\}_{x \in \mathcal{X}}$. The entropy H_X of the variable X can be defined as:

$$H = -\sum_{x \in \mathcal{X}} p(x) \log_2 p(x), \tag{5.2}$$

where we define by continuity $0 \log_2 0 = 0$. Considering to roll a fair coin, the entropy of this system is $-[0.5 \ln(0.5) + 0.5 \ln(0.5)] = 0.69$, which is the maximum entropy in the system. In other words, if the probable outcomes have same probability of occurrence, there will be the maximum randomness in sample space; if the sample is completely homogeneous, the entropy is zero. Therefore, when partitioning the sample space to grow a decision tree, it is reasonable to select a variable that produces the maximum reduction in system entropy.

Consider a classification decision tree. Let $H(S) = -\sum_{\ell \in \mathcal{L}} p(\ell) \log_2 p(\ell)$, where S is the sample space (or data set), ℓ denotes the class label, and $p(\ell)$ is the proportion of samples labeled by ℓ in S.

- *Information gain*: The information gain after splitting the sample space S by the attribute (or feature) A is

$$IG(S,V) = H(S) - \sum_{a \in A} p(S_a) H(S_a),$$

(5.3)

  where $a \in A$ is a value of the feature V, S_a is the sample subspace formed by the branch with the event $\{A = a\}$, and $p(S_a)$ is the ratio of the number of samples in S_a to the number of instances in S, that is, $p(S_a) = |S_a|/|S|$.

- *Information gain ratio*: The information gain is biased toward multivariate attributes. That is, it tends to split the sample space by the feature which has more values. To avoid this, a more reliable metric *information gain ratio* was introduced to measure the purity or homogeneity. The information gain ratio is calculated as:

$$IGR(S,V) = -IG(S,V) \Big/ \left(\sum_{a \in A} p(S_a) \log p(S_a) \right).$$

(5.4)

- *Gini index or Gini impurity*: Gini index is another commonly used metric when building a decision tree. It measures how often incorrect labeling happens for a randomly chosen element when it is randomly labeled according to the distribution of labels in the subset. An element with label ℓ can be incorrectly labeled with the probability $\sum_{\ell' \in \mathcal{L} - \ell} p(\ell') = 1 - p(\ell)$, where $p(\ell)$ is the fraction of elements labeled with ℓ in the set. Then, the Gini index for the sample space S is defined as:

$$Gini(S) = \sum_{\ell \in \mathcal{L}} p(\ell) \sum_{\ell' \in \mathcal{L} - \ell} p(\ell') = \sum_{\ell \in \mathcal{L}} p(\ell)(1 - p(\ell)) = 1 - \sum_{\ell \in \mathcal{L}} p(\ell)^2.$$

(5.5)

The change of Gini index can be calculated as:

$$\Delta(S,A) = Gini(S) - Gini(S,A),$$

(5.6)

with $Gini(S,A) = \sum_{a \in A} p(S_a) Gini(S_a)$ represents the Gini index of the sample subspace obtained by the partition according to the feature A.

The decision tree learning mirrors human decision making more closely than other approaches (James et al., 2013). Therefore, it is simple to understand and interpret. Compared with a black-box model, for example, ANN, where the explanation for the results is typically difficult to understand, the observed results can be easily explained by Boolean logic in decision tree learning. Besides, the decision tree learning is able to handle both numerical and categorical data (James et al., 2013). Thus, decision tree learning can be adapted to solve both classification and regression problems. Due to its various advantages, the decision tree learning becomes one of the predictive modeling approaches used in statistics, data mining, and machine learning.

5.5.1.2 *Bootstrap aggregating and random forest*

Although decision tree learning have many advantages, it tends to have lower accuracy compared with the other machine learning approaches and it is not very robust to the input, that is, a small change in the training data can result in a big change in the tree, and thus a big change in final predictions (James et al., 2013). Besides, when the decision tree grows very deep the model tends to learn highly irregular patterns and may result in overfitting. The performance of decision tree learning can be improved by the classic ensemble method, bootstrap aggregating. Random forest, a special type of the bootstrap aggregating decision trees, is able to further increase the estimate accuracy and improve the model's the robustness.

Bootstrap aggregating

Bootstrap aggregating, also called *bagging*, is a machine learning ensemble metaalgorithm widely used in statistical classification and regression. It was proposed by Breiman in 1994 to improve the stability and accuracy of base machine learning algorithms and to reduce variance and help to avoid overfitting. Given a standard training set $\mathcal{D}$ of size n, bagging generates k new training sets $\mathcal{D}_\ell$, each of size an, $\alpha \in (0, 1]$, by sampling from $\mathcal{D}$ uniformly and with replacement. By sampling with replacement, some observations may be repeated in each $\mathcal{D}_\ell$. If $\alpha = 1$, then for large n the set $\mathcal{D}_\ell$ is expected to have the fraction $(1 - 1/e)$ ($\approx 63.2\%$) (Aslam et al., 2007) of the unique examples of $\mathcal{D}$, the rest being duplicates. This kind of sample is known as a *bootstrap sample*. The k models are fitted using the above k bootstrap samples and combined by *averaging* the output (for regression) or *voting* (for classification). The bootstrap sampling is able to generate relatively less-correlated training samples and thus fit less correlated decision trees. Therefore, even if some single tree is highly sensitive to noise in its training set, the average of many trees is not. The bootstrap aggregating method generally boosts the performance in the final model greatly, mainly due to the decrease of the model's variance. But as a trade-off, more computation consumption is needed and it also comes at the expense of a little increase of the bias and some loss of interpretability.

Bootstrap

Input: $\mathcal{D} = \{(x_i, y_i), i = 1,\ldots,n\}$, an interested point x, the bag number k, and an $\alpha \in (0, 1]$
Output: Estimate y for x

1. **for** $\ell = 1$ to k **do**
2.   Generate a new training set $\mathcal{D}_\ell$, $|\mathcal{D}_\ell| = \alpha n$, by sampling from D uniformly and with replacement
3.   Learn a model M_ℓ from the $\mathcal{D}$
4.   Denote estimate for the interested point as $M_\ell(x)$
5. **return** $\dfrac{1}{k}\sum_{\ell=1}^{k} M_\ell(x)$

Random forests

Recall that during the construction of a decision tree, usually a feature with the maximum metric value will be selected. Consider the case where several features are very strong correlated with the target output. Then when learning decision trees for the bootstrap samples, these features may still have high metric values and thus be selected with high chances, which could make the resultant trees correlated. In order to de-correlate these trees, random forests inject noise to the feature selection step. More precisely, instead of using the best feature (among all features) for each node split, a subset of features randomly chosen at that node and then a best feature among this subset is use for the split. This process is sometimes called "feature bagging." Typically, when each data entry has m features, $\lfloor \sqrt{m} \rfloor$ features are suggested in each split for classification tasks, and $\lfloor m/3 \rfloor$ is recommended with a minimum node size of five for regression tasks (Friedman et al., 2001).

5.5.2 Support vector machine

SVMs are supervised learning models for both classification and regression tasks. In an SVM model, data points with m features are mapped into a m-dimension Euclidean space. Then, geometric properties can be used to handle the points. For example, a binary classification problem is equal to finding a $(m-1)$-dimensional hyperplane to separate the data points. For nonlinear problems, such as nonlinear classification, SVMs can apply a "kernel trick" by further transforming the feature space such that a linear separation hyperplane can be found in the new space. More formally, an SVM constructs a hyperplane or set of hyperplanes in a high- or infinite-dimensional space, which can be used for classification, regression, or other tasks like outliers detection (Support Vector Machines, n.d.). Vapnik (1998) pointed out in the book "Statistical Learning Theory" that the maximum-margin principle prevents overfitting in high-dimensional input spaces and thus leads to good generalization abilities and accurate predictions.

Linear SVM

Consider a binary classification problem. The input data set is denoted as $\mathcal{D} = \{(x_i, y_i), i = 1, 2, \ldots, n\}$, where $x_i \in \mathbb{R}^m$ is a m-dimensional vector and $y_i \in \{+1, -1\}$ indicates the class to which the point x_i belongs. The target is to learn a hyperplane to separate points with different labels and predict the label for the unknown points in the future.

Assume the training data are linearly separable. In other words, the training data can be discriminated into two classes by a hyperplane. The idea of SVM is to find the separating hyperplane, named as *maximum-margin hyperplane*, that is the farthest apart from the closest training points of each class. It is equal to selecting two parallel separating hyperplanes, also called *margin hyperplanes*, such that the distance between them is as large as possible. The region bounded by these two hyperplanes is named the "margin." The samples, that is, x, on the margin hyperplanes are called the *support vectors*. It is not hard to observe that the maximum-margin hyperplane lies halfway between the two margin hyperplanes (refer to Fig. 5.3).

Any hyperplane can be written as the set of points x satisfying $w \cdot x - b = 0$, where w is the (not necessarily normalized) normal vector to the hyperplane, and $b \in \mathbb{R}$ determines the offset of the hyperplane from the origin along the normal vector w. If the hyperplane passes through the origin, that is, $b = 0$, and w is normalized, the distance from a point x to this hyperplane is the absolute value of the dot product, or inner product, $|w \cdot x|$. In general, the margin hyperplanes can be described by the following equations:

$$w \cdot x - b = 1 \quad \text{and} \quad w \cdot x - b = -1.$$

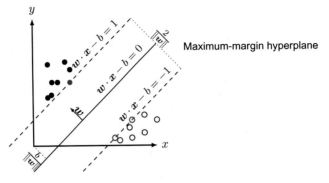

FIG. 5.3

Illustrate the idea of SVM.

The distance between these two hyperplanes is $\dfrac{2}{\|\boldsymbol{w}\|}$. The SVM learns to minimize $\|\boldsymbol{w}\|$ while preventing training data points from falling into the margin. That is, we need to solve the following optimization problem:

$$\text{minimize } \|\boldsymbol{w}\| \tag{5.7}$$
$$\text{subject to } y_i\left(\boldsymbol{w}\cdot\boldsymbol{x}_i - b\right) \geq 1, i = 1,\ldots,n.$$

When the training data are not linearly separable, loss functions are introduced to penalize the misclassification. A common loss function is the hinge loss function, $\max\{0,1-y_i(\boldsymbol{w}\cdot\boldsymbol{x}_i - b)\}$, whose value is proportional to the distance from the margin hyperplane if the data point is on the wrong side of the margin. Then, the optimization problem for the SVM can be formulated as:

$$\text{minimize } \lambda\|\boldsymbol{w}\|^2 + \left[\frac{1}{n}\sum_{i=1}^{n}\max\{0,1-y_i(\boldsymbol{w}\cdot\boldsymbol{x}_i - b)\}\right], \tag{5.8}$$

where the parameter λ determines the trade-off between the maximization of the margin and the permitted training errors. Both Eqs. (5.7), (5.8) can be solved by analyzing the classic Karush-Kuhn-Tucker (KKT) conditions (Kuhn and Tucker, 1951).

Nonlinear SVM

For nonlinear problems, Boser et al. (1992) introduced the so-called "kernel trick" to produce nonlinear decision boundaries. The idea is to project the data points into a higher-dimensional space, where the data points can be linearly separable. The *kernel* is a symmetric semipositive definite function $k(\boldsymbol{x}_i, \boldsymbol{x}_j)$. By substituting the dot products with nonlinear kernel functions, a nonlinear method can be transformed into a linear one. In other words, this substitution allows to learn a linear maximum-margin hyperplane in a transformed feature space. Note that this maximum-margin hyperplane may be nonlinear in the original input space (refer to Fig. 5.4). Some common kernels include:

- polynomial (homogeneous): $k(\boldsymbol{x}_i, \boldsymbol{x}_j) = (\boldsymbol{x}_i.\boldsymbol{x}_j)^d$;
- polynomial (inhomogeneous): $k(\boldsymbol{x}_i, \boldsymbol{x}_j) = (\boldsymbol{x}_i.\boldsymbol{x}_j+1)^d$;

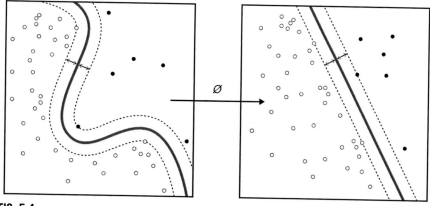

FIG. 5.4

The kernel function transforms the feature space (Wikipedia, n.d.).

- Gaussian RBF: $k(x_i, x_j) = \exp(-\gamma \parallel x_i - x_j \parallel^2)$, for $\gamma > 0$; and
- hyperbolic tangent: $k(x_i, x_j) = \tanh(\kappa x_i \cdot x_j + c)$, for $\kappa > 0$ and $c < 0$.

5.5.3 Neural network

Deep learning, also known as ANN, enables the computer to extract high-level, complex abstractions as data representations through a hierarchical learning process. It can avoid hand-crafted features that are usually expensive to create and require expert knowledge of the field. The hierarchical learning architecture of the deep learning algorithms is motivated by artificial intelligence emulating the deep, layered learning process of the primary sensorial areas of the neocortex in the human brain, which automatically extracts features and abstractions from the underlying data (Arel et al., 2010; Bengio and LeCun, 2007; Bengio et al., 2013).

In the human brain, a neuron (refer to Fig. 5.5) is the basic computational unit, which receives input signals from its dendrites and produces output signals along its (single) axon. Then, the axon branches out and connects via synapses to dendrites of other neurons. The signals that travel along the axons will interact with the dendrites of the other neuron based on the synaptic strength at that synapse. Denote the signal that travel along the axon as x and the synaptic strength as w. In the computational model of a neuron (refer to Fig. 5.5), we simulate the interaction at the synapse by the multiplication $w \cdot x$ and assume all signals carried by the dendrites are summed in the cell body. Besides, we assume that if the final sum is above a certain threshold, the neuron can fire by sending a spike along its axon. The firing rate of the neuron is modeled by an *activation function f*, which represents the frequency of the spikes along the axon. A commonly used activation function is the *sigmoid function* $\sigma(x) = \dfrac{1}{1 + e^{-x}}$, which takes a real-valued input (the signal strength after the sum) and maps it to the range [0, 1]. In the neuron model, the main idea is to learn the synaptic strengths (i.e., the weights w) and to control the strength of influence (i.e., its direction: excitatory (positive weight) or inhibitory (negative weight)) of one neuron on another.

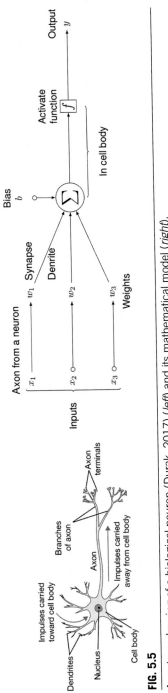

FIG. 5.5

A cartoon drawing of a biological neuron (Durak, 2017) (*left*) and its mathematical model (*right*).

The traditional feedforward NNs are modeled as collections of neurons that are connected as an acyclic graph (i.e., a graph without directed cycles). Usually, NNs are organized as layers of neurons such that neurons between two adjacent layers are connected in some way but neurons within a single layer share no connections. An example is shown in Fig. 5.6. The such NNs can be learned by first initializing weights randomly, then feeding the training points though the network layer by layer (*forward propagation*) and finally propagating back the error and computing gradient descent update on parameters (*back propagation* [BP]).

Consider an NN with L layers. Let d_ℓ denote the number of neurons in layer ℓ, $\ell = 1, 2, \ldots, L$. Suppose $\boldsymbol{W}^{(\ell)} \in \mathbb{R}^{d_\ell \times d_{\ell-1}}$ and $\boldsymbol{b}^{(\ell)} \in \mathbb{R}^{d_\ell}$ are the weight parameter matrix and bias vector, respectively, for the layer ℓ. Denote the output of the layer ℓ by $z^{(\ell)}$, which is also the input of the next layer. Let $f_\ell : \mathbb{R}^{d_{\ell-1}} \to \mathbb{R}^{d_\ell}$ be the set of activation functions between layers $\ell - 1$ and ℓ, each of which is applied to one neuron in layer $\ell - 1$. Then, we have

$$z^\ell = f_\ell(\boldsymbol{W}^{(\ell)} \cdot z^{(\ell-1)} + \boldsymbol{b}^{(\ell)}).$$

By applying the linear transformations and activation functions, the input data can be pushed through the network. Comparing the output and the target output value, a loss function $\mathcal{L}(\cdot)$ can be defined. For instance, the squared loss $\mathcal{L}(\boldsymbol{x}, y) = \frac{1}{2}(z^{(L)} - y)^2$.

We can observe that the loss $\mathcal{L}$ relies on the weight parameter matrix and the bias vector in layer $L - 1$. Then, we can update parameters in the layer $L - 1$ according to the gradient descent update rules. In other words, the parameters are updated in

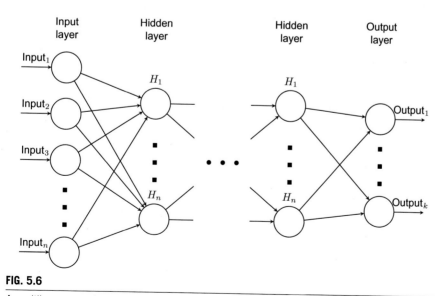

FIG. 5.6

A multilayer neural network with n inputs, at least two hidden layers and one output layer. Each circle is a neuron, which is modeled in Fig. 5.5. Note that there are connections (synapses) between neurons across layers, but not within a layer.

the opposite direction of the gradient of the loss function with respect to the parameters, that is, $\theta = \theta - \eta \cdot \nabla \theta L(\theta)$, where θ is some parameter in $\boldsymbol{W}^{(L-1)}$ and $\boldsymbol{b}^{(\ell)}$, and the *learning rate* η determines the size of the steps we take to reach a (local) minimum. We then can recursively update the parameters for layers $L - 1, L - 2, \ldots, 1$ in a backward propagating manner. There are plenty of gradient descent optimization algorithms, including momentum, Nesterov accelerated gradient, Adagrad, Adadelta, RMSprop, Adam, AdaMax, Nadam, AMSGrad, etc. (Ruder, 2016). These algorithms were designed to optimize different types of NNs.

Since the inception of the NN, it has been successfully applied to image classification (Ciregan et al., 2012), natural language processing (Gers and Schmidhuber, 2001), recommendation systems (Van den Oord et al., 2013), biomedical informatics (Chicco et al., 2014), etc. Typical architecture designs of deep learning include deep convolutional neural networks (CNNs), deep sparse autoencoder (SA), deep recurrent neural networks (RNN), multilayer perceptions (MLP), deep restricted Boltzmann machines (RBM), etc. (Goodfellow et al., 2016). Among them, CNNs are particularly suitable for the learning of discriminative image feature representation, and thus are widely applied in image annotation/tagging problems. DRNN is particularly suitable for time series forecasting and modeling. RNN employs self-connected neurons to implement a cyclic structure in the network, which helps to "memorize" the historical input. In other words, RNN not only considers the current input but also takes into account a trace of previously acquired information via recurrent connections, which allows a direct processing of temporal dependencies.

5.5.4 Convolutional neural network

CNN was inspired by the biological processes in the animal visual cortex (Matsugu et al., 2003). More precisely, the region reacting to stimuli in a cortical neuron is limited in its visual field. The reacting regions, *a.k.a. receptive field*, of different neurons partially overlap so that the entire visual field is covered.

A CNN usually consists of three main types of layers: convolutional (CONV) layer, pooling (POOL) layer, and fully connected (FC) layer. Sometimes, we also explicitly treat the rectified linear unit (ReLU) activation function as a layer. The convolutional layers are the core building blocks of a CNN. In each convolutional layer, convolution operations are applied to the input with the intuition of emulating the response of an individual cortical neuron to visual stimuli. Therefore, an input image is partitioned into partially overlapped regions, each of which simulates the receptive field. Each convolutional operation processes data only for its "receptive field." A pooling layer combines or contracts the outputs of neuron clusters at one layer into a single neuron in the next layer (Ciresan et al., 2011). Pooling layers are periodically inserted in-between successive CONV layers to reduce the size of the representation progressively, which not only makes the input representations (feature dimension) smaller and more manageable but also reduces the amount of parameters and computation in the network. Consequently, high-level features can be extracted from the input image and the overfitting problem can be alleviated at the same time. The commonly used pooling methods are max pooling and average pooling, which take the maximum (and average, respectively) value

Feature map   Feature map   Feature map

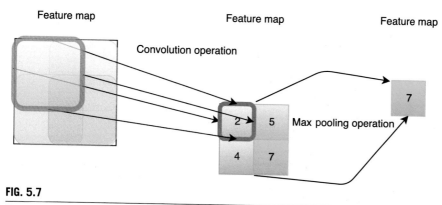

Convolution operation

Max pooling operation

FIG. 5.7

The *left side* shows a convolution operation on the input image or feature map, each cell of which emulates the receptive field. The *right side* operation is an example of the max pooling.

from each of a cluster of neurons at the prior layer. Fig. 5.7 illustrates the convolution and pooling operations. An FC layer connects every neuron to all activations in the previous layer. The purpose of the FC layer is to further increase feature learnability based on the high-level features extracted by the CONV and POOL layers. The ReLU layer applies an element-wise operation (refer to Fig. 5.8) that replaces all negative values by zero. A ReLU layer aims at introducing nonlinearity in the network to better emulate the real-world biological processes.

When building a real CNN, the architecture, including the numbers of different layers and order of different layers, generally depends on the requirement of network capacity and memory cost for a specific application. An simple example with one CONV layer, one POOL layer, and one FC layer is shown in Fig. 5.9.

5.5.5 Recurrent neural network

Suppose the input time series for RNN (refer to Fig. 5.10) is $ts = (x_1, ..., x_T)$, where x_t is the input data at time t. RNN recursively processes input sequence, aiming to learn representations of patterns repeatedly occurred in the past. Therefore, when

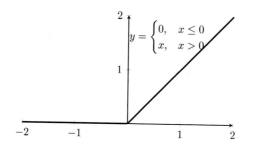

$$y = \begin{cases} 0, & x \le 0 \\ x, & x > 0 \end{cases}$$

FIG. 5.8

A visualization of the ReLU function.

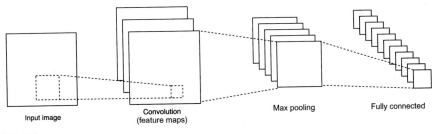

FIG. 5.9

A simple CNN framework with one CONV layer, one POOL layer, and one FC layer.

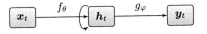

FIG. 5.10

RNN recursively processes input sequence, aiming to learn representations of patterns repeatedly occurred in the past by maintaining its internal hidden state h_t. Here x_t and y_t are the input and output at time t, respectively.

processing the data x_t, RNN maintains its internal hidden state h_t and shares the parameter across all time steps.

$$h_t = f_\theta(x_t, h_{t-1}),$$

where f is the deterministic state transition function and θ is the parameter of f. The output of RNN is computed using the following equation:

$$y_t = g_\varphi(h_t, \ldots, h_1),$$

where function g can be modeled as an NN with weights φ. Finally, stochastic gradient descent and BP (Goodfellow et al., 2016) are used to train the function and find optimal parameters.

However, the memory of the past-learned patterns can fade as time goes on, which causes the computation problem vanishing gradient (Bengio et al., 1994). Long short-term memory (LSTM) (Hochreiter and Schmidhuber, 1997) NN was proposed to tackle this challenge by creating paths where the gradient can flow for long durations. LSTM maintains an internal state c_t, named as *cell state*, that forms the basis for the recurrence, and is passed from time step to the next (Elkaref and Bohnet, 2017). The basic unit (refer to Fig. 5.11) in the hidden layer is a memory block, which consists of three main components:

- memory cells with self-connections memorizing the temporal state;
- a pair of adaptive, multiplicative gating units to control information flow in the block; and
- input gate and output gate controlling the input and output activations into the block.

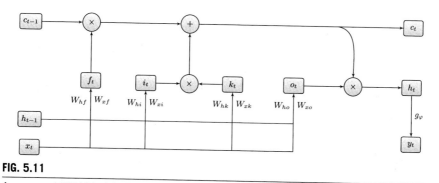

FIG. 5.11

A memory LSTM block/cell (a basic unit in the hidden layer). Blocks are connected recurrently to each other, replacing the usual hidden units h_t of ordinary RNN.

A detailed explanation is as follows. Suppose σ is the activation function like *Sigmoid*, W is the weight matrices and b is the bias vector. The first (leftmost) layer in the LSTM block is the *forget gate layer* f_t, which is to decide what information needs to be thrown away from the cell state. It takes h_{t-1} and x_t as inputs, and outputs a number between 0 and 1 for each number in the cell state c_{t-1}.

$$f_t(x_t, h_{t-1}) = \sigma(W_{xf} \cdot x_t + W_{hf} \cdot h_{t-1} + b_f) = \begin{cases} 1, & \text{if completely remember this,} \\ 0, & \text{if completely forget this.} \end{cases} \quad (5.9)$$

The next step consists of two parts: a input gate layer to decide which value to update and a tanh layer to create a vector of new candidate values, k_t, that could be added to the state. These two layers are combined to get new information or create an update to the state.

$$i_t = \sigma(W_{xi} \cdot x_t + W_{hi} \cdot h_{t-1} + b_i), \quad (5.10)$$

$$k_t = \tanh(W_{xj} \cdot x_t + W_{hj} \cdot h_{t-1} + b_k). \quad (5.11)$$

Then, by multiplying the old state with f_t and adding $i_t \times k_t$, a new state value will be created.

$$c_t = c_{t-1} \times f_t + i_t \times k_t. \quad (5.12)$$

Finally, an output gate layer is applied to control the output. The output is filtered by the cell state.

$$o_t = \sigma(W_{xo} \cdot x_t + W_{ho} \cdot h_{t-1} + b_o), \quad (5.13)$$

$$h_t = \tanh(c_t) \times o_t. \quad (5.14)$$

The classic LSTM is unidirectional as the memory can only flow toward one direction. When interpolating the spatiotemporal data, both past and future air pollution concentrations provide valuable information for the estimation of the current concentration. In order to memorize the past and future information at the same time, bidirectional LSTM splits the neurons of a regular LSTM into two directions: one for positive time direction and the other for negative time direction.

In general, the DRNN can be trained by the algorithm backpropagation through time (BPTT) (Chauvin and Rumelhart, 1995), which is a gradient-based technique and a variant of the classic backpropagation algorithm. BPTT begins by unfolding an RNN in time such that each timestep has one input timestep, one copy of the network, and one output. Given the order dependence of the problem and the internal state from the previous timestep is taken as an input on the subsequent timestep, each timestep may be treated as an additional layer. Then, the backpropagation algorithm is applied to the unfolding network. Errors can be calculated and accumulated for each timestep. The network needs to be rolled back up and the weights are updated.

5.6 Tools

Previous section introduces commonly used machine learning methods that can be applied for spatiotemporal interpolation. This section will introduce practical machine learning tools. Choosing the right tool can be as important as working with the best algorithms. As good tools can automate each step in the applied machine learning process, which will greatly shorten the time from ideas to results.

A good machine learning tool should first have trusted resources such as the effective and efficient implementations of the various algorithms. Then, it should be well maintained, that is, frequent updates to the latest research discovers are available. Finally, good machine learning tools should also provide user-friendly interfaces.

Shogun
Shogun (http://www.shogun-toolbox.org/) is an old open-source machine learning library since 1999. It offers a wide range of unified and efficient machine learning methods. The goal behind its creation is to provide machine learning with transparent and accessible algorithms as well as free machine learning tools to anyone interested in the field. Shogun supports many languages, such as Python, Octave, R, Java/Scala, Lua, C#, Ruby, etc. It also integrates with scientific computing environments of the three popular platforms: Linux/Unix, MacOS, and Windows. However, some report its unfriendly API.

Theano
Theano (http://deeplearning.net/software/theano/) has been powering large-scale computationally intensive scientific investigations since 2007. It is one of the most mature Python deep learning libraries. Theano's main features include tight integration with NumPy, transparent use of graphics processing unit (GPU), efficient symbolic differentiation, speed and stability optimizations, dynamic C code generation, extensive unit-testing, and self-verification.

TensorFlow
TensorFlow (https://www.tensorflow.org/) is an open-source software library developed by the Google Brain Team for high-performance numerical computation, including machine learning and deep NNs. TensorFlow is highly flexible and uses data flow graphs to simulated numerical computations. Such a flexible architecture allows easy deployment of computation across a variety of platforms (central processing units

(CPUs), GPUs, and tensor processing units (TPUs)), and from desktops to clusters of servers to mobile and edge devices. Moreover, TensorFlow allows users to define their own computational architecture, and to write their own higher-level libraries on top of it by using C++ and Python.

Keras

Keras (https://keras.io/) is a high-level NNs API and provides a Python deep learning library. Keras runs on top of TensorFlow, CNTK, or Theano, and it supports both convolutional networks and recurrent networks, as well as combinations of the two. Besides, Keras is able to run seamlessly on CPU and GPU. According to the official site, Keras focuses on four main guiding principles that are user-friendliness, modularity, easy extensibility, and working with Python. It sacrifice the speed for an easier way to express NNs, compared to other libraries. It is still the best choice for any beginner in machine learning due to its simplicity.

Scikit-Learn

Scikit-Learn (http://scikit-learn.org/) is an open-source tool for data mining and data analysis. It is built on NumPy, SciPy, and matplotlib, three popular data science library. Scikit-Learn provides a consistent and friendly API. Its main features include classification, regression, clustering, dimensionality reduction, model selection, and preprocessing.

Deeplearning4j

Eclipse Deeplearning4j (https://deeplearning4j.org/) is the first commercial-grade, open-source, distributed deep learning library written for Java and Scala. Its main features include distributed CPUs and GPUs; Java, Scala, and Python APIs; adapted for microservice architecture; parallel training via iterative reduce; scalable on Hadoop; and GPU support for scaling on AWS. Deeplearning4j has a commercial support arm, Skymind, which bundles Deeplearning4j and other libraries such as TensorFlow and Keras to create a user-friendly and efficient deep learning environment.

Some remote tools

Remote tools are hosted on a server and integrated into the local environment via remote procedure calls. Usually, remote tools are needed for large-scale experiments. Examples of remote tools are Google Prediction API (https://cloud.google.com/prediction/), AWS Machine Learning (https://aws.amazon.com/machine-learning/), and Microsoft Azure Machine Learning Studio (https://azure.microsoft.com/en-us/services/machine-learning-studio/).

References

Al-Saadi, J., Szykman, J., Pierce, R.B., Kittaka, C., Neil, D., Chu, D.A., Remer, L., Gumley, L., Prins, E., Weinstock, L., et al., 2005. Improving national air quality forecasts with satellite aerosol observations. Bull. Am. Meteorol. Soc. 86 (9), 1249–1261.

Appice, A., Ciampi, A., Malerba, D., Guccione, P., 2013. Using trend clusters for spatiotemporal interpolation of missing data in a sensor network. J. Spat. Inf. Sci. 2013, 119–153.

Arel, I., Rose, D.C., Karnowski, T.P., 2010. Deep machine learning—a new frontier in artificial intelligence research [research frontier]. IEEE Comput. Intell. Mag. 5 (4), 13–18.

Aslam, J.A., Popa, R.A., Rivest, R.L., 2007. On estimating the size and confidence of a statistical audit. In: EVT, 7. pp. 8.

Attfield, M.D., Schleiff, P.L., et al., 2012. The diesel exhaust in miners study: a cohort mortality study with emphasis on lung cancer. J. Natl Cancer Inst. 104 (11), 869–883.

Beelen, R., Hoek, G., van den Brandt, P.A., Goldbohm, R.A., Fischer, P., Schouten, L.J., Armstrong, B., Brunekreef, B., 2008. Long-term exposure to traffic-related air pollution and lung cancer risk. Epidemiology 19 (5), 702–710.

Bengio, Y., LeCun, Y., 2007. Scaling learning algorithms towards AI. In: Large-Scale Kernel Machines. 34, The MIT Press, Cambridge, MA, pp. 1–41.

Bengio, Y., Simard, P., Frasconi, P., 1994. Learning long-term dependencies with gradient descent is difficult. IEEE Trans. Neural Netw. 5 (2), 157–166.

Bengio, Y., Courville, A., Vincent, P., 2013. Representation learning: a review and new perspectives. IEEE Trans. Pattern Anal. Mach. Intell. 35 (8), 1798–1828.

Boser, B.E., Guyon, I.M., Vapnik, V.N., 1992. A training algorithm for optimal margin classifiers. In: Proceedings of the Fifth Annual Workshop on Computational Learning Theory, pp. 144–152.

Breiman, L., 1994. Bagging predictor. Department of Statistics, University of California. Technical Report.

Breiman, L., 2017. Classification and Regression Trees. Routledge, New York, NY.

Brokamp, C., Jandarov, R., Hossain, M., Ryan, P., 2018. Predicting daily urban fine particulate matter concentrations using a random forest model. Environ. Sci. Technol. 52 (7), 4173–4179.

Cambaliza, M.O.L., Shepson, P.B., Caulton, D.R., Stirm, B., Samarov, D., Gurney, K.R., Turnbull, J., Davis, K.J., Possolo, A., Karion, A., Sweeney, C., Moser, B., Hendricks, A., Lauvaux, T., Mays, K., Whetstone, J., Huang, J., Razlivanov, I., Miles, N.L., Richardson, S.J., 2014. Assessment of uncertainties of an aircraft-based mass balance approach for quantifying urban greenhouse gas emissions. Atmos. Chem. Phys. 14, 9029–9050.

Charron, A., Harrison, R.M., 2005. Fine ($PM_{2.5}$) and coarse ($PM_{2.5-10}$) particulate matter on a heavily trafficked London highway: sources and processes. Environ. Sci. Technol. 39 (20), 7768–7776.

Chauvin, Y., Rumelhart, D.E., 1995. Backpropagation: Theory, Architectures, and Applications. Psychology Press. London, UK.

Chen, G., Li, S., Knibbs, L.D., Hamm, N., Cao, W., Li, T., Guo, J., Ren, H., Abramson, M.J., Guo, Y., 2018. A machine learning method to estimate $PM_{2.5}$ concentrations across China with remote sensing, meteorological and land use information. Sci. Total Environ. 636, 52–60.

Chicco, D., Sadowski, P., Baldi, P., 2014. Deep autoencoder neural networks for gene ontology annotation predictions. In: Proceedings of the 5th ACM Conference on Bioinformatics, Computational Biology, and Health Informatics, pp. 533–540.

Ciregan, D., Meier, U., Schmidhuber, J., 2012. Multi-column deep neural networks for image classification. In: 2012 IEEE Conference on Computer Vision and Pattern Recognition (CVPR), pp. 3642–3649.

Ciresan, D.C., Meier, U., Masci, J., Gambardella, L.M., Schmidhuber, J., 2011. Flexible, high performance convolutional neural networks for image classification. In: IJCAI Proceedings—International Joint Conference on Artificial Intelligence, 22. pp. 1237.

De Boor, C., 1978. A Practical Guide to Splines. 27 Springer-Verlag, New York, NY. vol.

Di, Q., Kloog, I., Koutrakis, P., Lyapustin, A., Wang, Y., Schwartz, J., 2016. Assessing $PM_{2.5}$ exposures with high spatiotemporal resolution across the continental united states. Environ. Sci. Technol. 50 (9), 4712–4721.

Dockery, D.W., Pope, C.A., et al., 1993. An association between air pollution and mortality in six US cities. N. Engl. J. Med. 329 (24), 1753–1759.

Durak, B.C., 2017. Artificial neural networks. Accessed 27 March 2018 https://wiki.tum.de/display/lfdv/Artificial+Neural+Networks.

Elkaref, M., Bohnet, B., 2017. A simple LSTM model for transition-based dependency parsing. arXiv preprint arXiv:1708.08959.

Fan, J., Li, Q., Hou, J., Feng, X., Karimian, H., Lin, S., 2017. A spatiotemporal prediction framework for air pollution based on deep RNN. In: ISPRS Annals of the Photogrammetry, Remote Sensing and Spatial Information Sciences, 4. p 15.

Friedman, J.H., 1991. Multivariate adaptive regression splines. Ann. Statist, 19 (1), 1–67.

Friedman, J., Hastie, T., Tibshirani, R., 2001. The Elements of Statistical Learning. vol. 1. Springer Series in Statistics, New York, NY.

Gers, F.A., Schmidhuber, E., 2001. LSTM recurrent networks learn simple context-free and context-sensitive languages. IEEE Trans. Neural Netw. 12 (6), 1333–1340.

Ghim, Y.S., Moon, K.C., Lee, S., Kim, Y.P., 2005. Visibility trends in KOREA during the past two decades. J. Air Waste Manage. Assoc. 55, 73–82.

Goodfellow, I., Bengio, Y., Courville, A., 2016. Deep Learning. MIT Press, Cambridge, MA.

Gräler, B., Rehr, M., Gerharz, L.E., Pebesma, E., 2013. Spatio-temporal analysis and interpolation of PM_{10} measurements in Europe for 2009. In: ETC/AM. .

Gupta, P., Christopher, S.A., 2009. Particulate matter air quality assessment using integrated surface, satellite, and meteorological products: 2. A neural network approach. J. Geophys. Res. 114, https://doi.org/10.1029/2008JD011497. D20205.

Hochreiter, S., Schmidhuber, J., 1997. Long short-term memory. Neural Comput. 9 (8), 1735–1780.

Hong, Y.C., Lee, J.T., Kim, H., Ha, E.H., Schwartz, J., Christiani, D.C., 2002. Effects of air pollutants on acute stroke mortality. Environ. Health Perspect. 110, 187.

Hothorn, T., Hornik, K., Zeileis, A., 2006. Unbiased recursive partitioning: a conditional inference framework. J. Comput. Graph. Stat. 15 (3), 651–674.

Hu, X., Waller, L.A., Lyapustin, A., Wang, Y., Al-Hamdan, M.Z., Crosson, W.L., Estes Jr., M.G., Estes, S.M., Quattrochi, D.A., Puttaswamy, S.J., Liu, Y., 2014. Estimating ground-level $PM_{2.5}$ concentrations in the southeastern united states using MAIAC AOD retrievals and a two-stage model. Remote Sens. Environ. 140, 220–232.

Hu, X., Belle, J.H., Meng, X., Wildani, A., Waller, L.A., Strickland, M.J., Liu, Y., 2017. Estimating $PM_{2.5}$ concentrations in the conterminous United States using the random forest approach. Environ. Sci. Technol. 51 (12), 6936–6944.

Hystad, P., Demers, P.A., Johnson, K.C., Carpiano, R.M., Brauer, M., 2013. Long-term residential exposure to air pollution and lung cancer risk. Epidemiology 24 (5), 762–772.

International Agency for Research on Cancer, 2013. Outdoor air pollution a leading environmental cause of cancer deaths. https://www.iarc.fr/en/media-centre/iarcnews/pdf/pr221_E.pdf.

James, G., Witten, D., Hastie, T., Tibshirani, R., 2013. An Introduction to Statistical Learning. vol. 112 Springer, New York.

Kass, G.V., 1980. An exploratory technique for investigating large quantities of categorical data. Appl. Stat. 29 (2), 119–127.

Krige, D.G., 1951. A statistical approach to some mine valuation and allied problems on the Witwatersrand, Doctoral dissertation. University of the Witwatersrand.

Kuhn, H.W., Tucker, A.W., 1951. Nonlinear programming. In: Proceedings of the Second Berkeley Symposium on Mathematical Statistics and Probability. University of California Press, Oakland, CA, pp. 481–492.

Kwon, B.J., 2017. Deep Convolutional Neural Networks for Estimating PM2.5 Concentration Levels (Ph.D. thesis). DGIST.

Laden, F., Neas, L.M., Dockery, D.W., Schwartz, J., 2000. Association of fine particulate matter from different sources with daily mortality in six U.S. cities. Environ. Health Perspect. 108, 941–947.

Li, J., Heap, A.D., 2008. A Review of Spatial Interpolation Methods for Environmental Scientists. vol. 137 Geoscience Australia, Canberra.

Li, J., Heap, A.D., 2011. A review of comparative studies of spatial interpolation methods in environmental sciences: performance and impact factors. Ecol. Inf. 6, 228–241.

Li, L., Revesz, P., 2004. Interpolation methods for spatio-temporal geographic data. Comput. Environ. Urban Syst. 28, 201–227.

Li, L., Tian, J., Zhang, X., Holt, J.B., Piltner, R., 2012. Estimating population exposure to fine particulate matter in the Conterminous US using shape function-based spatiotemporal interpolation method: a county level analysis. GSTF Int. J. Comput. 1, 24–30.

Li, L., Losser, T., Yorke, C., Piltner, R., 2014. Fast inverse distance weighting-based spatiotemporal interpolation: a web-based application of interpolating daily fine particulate matter PM$_{2.5}$ in the contiguous US using parallel programming and K-D tree. Int. J. Environ. Res. Public Health 11, 9101–9141.

Liao, D., Peuquet, D.J., Duan, Y., Whitsel, E.A., Dou, J., Smith, R.L., Lin, H.M., Chen, J.C., Heiss, G., 2006. GIS approaches for the estimation of residential-level ambient PM concentrations. Environ. Health Perspect, 114 (9), 1374–1380.

Liu, Y., Paciorek, C.J., Koutrakis, P., 2009. Estimating regional spatial and temporal variability of PM$_{2.5}$ concentrations using satellite data, meteorology, and land use information. Environ. Health Perspect. 117 (6), 886.

Losser, T., Li, L., Piltner, R., 2014. A spatiotemporal interpolation method using radial basis functions for geospatiotemporal big data. In: COM.geo, pp. 17–24.

Lyapustin, A., Martonchik, J., Wang, Y., Laszlo, I., Korkin, S., 2011. Multiangle implementation of atmospheric correction (MAIAC): 1. Radiative transfer basis and look-up tables. J. Geophys. Res. 116, https://doi.org/10.1029/2010JD014985. D03210.

Lyapustin, A., Wang, Y., Laszlo, I., Kahn, R., Korkin, S., Remer, L., Levy, R., Reid, J.S., 2011. Multiangle implementation of atmospheric correction (MAIAC): 2. Aerosol algorithm. J. Geophys. Res. 116, https://doi.org/10.1029/2010JD014986. D03211.

Matsugu, M., Mori, K., Mitari, Y., Kaneda, Y., 2003. Subject independent facial expression recognition with robust face detection using a convolutional neural network. Neural Netw. 16 (5–6), 555–559.

Mays, K.L., Shepson, P.B., Stirm, B.H., Karion, A., Sweeney, C., Gurney, K.R., 2009. Aircraft-based measurements of the carbon footprint of indianapolis. Environ. Sci. Technol. 43, 7816–7823.

Morain, S.A., Budge, A.M., 2012. Environmental Tracking for Public Health Surveillance. CRC Press, Boca Raton, FL.

Morgenstern, V., Zutavern, A., Cyrys, J., Brockow, I., Gehring, U., Koletzko, S., Bauer, C.P., Reinhardt, D., Wichmann, H.E., Heinrich, J., 2007. Respiratory health and individual estimated exposure to traffic-related air pollutants in a cohort of young children. Occup. Environ. Med. 64 (1), 8–16.

Næss, Ø., Nafstad, P., Aamodt, G., Claussen, B., Rosland, P., 2006. Relation between concentration of air pollution and cause-specific mortality: four-year exposures to nitrogen dioxide and particulate matter pollutants in 470 neighborhoods in Oslo, Norway. Am. J. Epidemiol. 165 (4), 435–443.

Najafabadi, M.M., Villanustre, F., Khoshgoftaar, T.M., Seliya, N., Wald, R., Muharemagic, E., 2015. Deep learning applications and challenges in big data analytics. J. Big Data 2 (1), 1.

Oftedal, B., Walker, S.E., Gram, F., McInnes, H., Nafstad, P., 2008. Modelling long-term averages of local ambient air pollution in Oslo, Norway: evaluation of nitrogen dioxide, PM10 and $PM_{2.5}$. Int. J. Environ. Pollut. 36 (1–3), 110–126.

Pebesma, E., 2012. spacetime: spatiotemporal data in R. J. Stat. Softw. 51, 1–30.

Pope III, C.A., Burnett, R.T., Thun, M.J., Calle, E.E., Krewski, D., Ito, K., Thurston, G.D., 2002. Lung cancer, cardiopulmonary mortality, and long-term exposure to fine particulate air pollution. JAMA 287 (9), 1132–1141.

Puett, R.C., Hart, J.E., Yanosky, J.D., Spiegelman, D., Wang, M., Fisher, J.A., Hong, B., Laden, F., 2014. Particulate matter air pollution exposure, distance to road, and incident lung cancer in the nurses' health study cohort. Environ. Health Perspect. 122 (9), 926.

Qi, Z., Wang, T., Song, G., Hu, W., Li, X., Zhang, Z.M., 2018. Deep air learning: interpolation, prediction, and feature analysis of fine-grained air quality. IEEE Trans. Knowl. Data Eng. 30 (12), 2285–2297.

Quinlan, J.R., 1986. Induction of decision trees. Mach. Learn. 1 (1), 81–106.

Quinlan, J.R., 2014. C4.5: Programs for Machine Learning. Morgan Kaufmann Publishers, San Mateo, CA.

Raaschou-Nielsen, O., Andersen, Z.J., et al., 2013. Air pollution and lung cancer incidence in 17 European cohorts: prospective analyses from the European Study of Cohorts for Air Pollution Effects (ESCAPE). Lancet Oncol. 14 (9), 813–822.

Reid, C.E., Jerrett, M., Petersen, M.L., Pfister, G.G., Morefield, P.E., Tager, I.B., Raffuse, S.M., Balmes, J.R., 2015. Spatiotemporal prediction of fine particulate matter during the 2008 Northern California Wildfires using machine learning. Environ. Sci. Technol. 49 (6), 3887–3896.

Ruder, S., 2016. An overview of gradient descent optimization algorithms. arXiv preprint arXiv:1609.04747.

Russell, S.J., Norvig, P., 2003. Artificial Intelligence: A Modern Approach, second ed. Pearson Education. Prentice-Hall, Upper Saddle River, NJ.

Samuel, A.L., 1959. Some studies in machine learning using the game of checkers. IBM J. Res. Dev. 3 (3), 210–229.

Shannon, C.E., 1948. A note on the concept of entropy. Bell Syst. Tech. J. 27 (3), 379–423.

Shepard, D., 1968. A two-dimensional interpolation function for irregularly-spaced data. In: Proceedings of the 23rd ACM National Conference, pp. 517–524.

Sloane, C.S., Watson, J., Chow, J., Pritchett, L., Richards, L.W., 1991. Size-segregated fine particle measurements by chemical species and their impact on visibility impairment in Denver. Atmos. Environ. A Gen. Top. 25, 1013–1024.

Streets, D.G., Canty, T., Carmichael, G.R., de Foy, B., Dickerson, R.R., Duncan, B.N., Edwards, D.P., Haynes, J.A., Henze, D.K., Houyoux, M.R., et al., 2013. Emissions estimation from satellite retrievals: a review of current capability. Atmos. Environ. 77, 1011–1042.

Support Vector Machines, Available from: https://web.archive.org/web/20171108151644/, http://scikit-learn.org/stable/modules/svm.html (Accessed 4 June 2018).

Tadić, J.M., Ilić, V., Biraud, S., 2015. Examination of geostatistical and machine-learning techniques as interpolators in anisotropic atmospheric environments. Atmos. Environ. 111, 28–38.

Tong, W., Li, L., Zhou, X., Franklin, J., 2019a. Efficient spatiotemporal interpolation with spark machine learning. Earth Sci. Inform. 12 (1), 87–96. https://doi.org/10.1007/s12145-018-0364-4.

Tong, W., Li, L., Zhou, X., Hamilton, A., Zhang, K., 2019b. Deep learning PM2.5 concentrations with bidirectional LSTM RNN. Air Qual. Atmos. Health 12 (4), 411–423. https://doi.org/10.1007/s11869-018-0647-4.

Van den Oord, A.Dieleman, S., Schrauwen, B., 2013. Deep content-based music recommendation. In: Advances in Neural Information Processing Systems, pp. 2643–2651.

Van Donkelaar, A., Martin, R.V., Park, R.J., 2006. Estimating ground-level $PM_{2.5}$ using aerosol optical depth determined from satellite remote sensing. J. Geophys. Res. 111, https://doi.org/10.1029/2005JD006996. D21201.

Vapnik, V., 1998. Statistical Learning Theory. Wiley, New York, NY.

Wikipedia Support Vector Machine n.d. Accessed 11 June 2018 https://www.wikiwand.com/en/Support_vector_machine.

Williams, R.J., 1992. Simple statistical gradient-following algorithms for connectionist reinforcement learning. In: Machine Learning, pp. 229–256.

Zanobetti, A., Schwartz, J., 2009. The effect of fine and coarse particulate air pollution on mortality: a national analysis. Environ. Health Perspect. 117, 898–903.

Zou, B., Wang, M., Wan, N., Wilson, J.G., Fang, X., Tang, Y., 2015. Spatial modeling of $PM_{2.5}$ concentrations with a multifactoral radial basis function neural network. Environ. Sci. Pollut. Res. 22 (14), 10395–10404.

Zurflueh, E.G., 1967. Applications of two-dimensional linear wavelength filtering. Geophysics 32, 1015–1035.

Integrate machine learning and geostatistics for high-resolution mapping of ground-level PM$_{2.5}$ concentrations

Ying Liu[a,b], Guofeng Cao[a,b], Naizhuo Zhao[b]

[a]*Department of Geosciences, Texas Tech University, Lubbock, TX, United States*
[b]*Center for Geospatial Technology, Texas Tech University, Lubbock, TX, United States*

6.1 Introduction

Fine particulate matter with aerosol dynamic diameters equal to or less than 2.5 micrometers (PM$_{2.5}$) has been identified as one of the three leading risk factors for human health (Lim et al., 2013; Pope et al., 2002). Exposure to PM$_{2.5}$ is estimated to cause 3.2 million premature deaths every year globally (Lim et al., 2013). To study and mitigate the adverse effects of PM$_{2.5}$ exposure on public health, accurately measuring ground-level PM$_{2.5}$ concentrations is of essential importance (Liu et al., 2018; van Donkelaar et al., 2015b; Di et al., 2016).

The ground-based monitoring systems can provide accurate PM$_{2.5}$ measurements. However, no country in the world has yet established a monitoring network with a satisfying population coverage (Liu et al., 2018). Even in the United States (US), the relatively developed PM$_{2.5}$ monitoring network with approximately 2500 monitoring stations leave many people living in suburban and rural areas unmonitored (Liu et al., 2018). To monitor the ground-level PM$_{2.5}$ concentrations at large geographical scales, remote sensing has proven to be a useful tool. A close relationship was found between the ground-level PM$_{2.5}$ concentration and satellite-observed aerosol optical depth (AOD) (Engel-Cox et al., 2004), which can be exploited to improve the PM$_{2.5}$ concentration mapping for large geographical areas. This relationship can be quantified by numerical chemical transport models (CTM), e.g., typically the GEOS (Goddard Earth Observing System)-Chem model, and then applied to the remotely sensed AOD imagery for PM$_{2.5}$ mapping (Liu et al., 2004; van Donkelaar et al., 2006). Recently, the Atmospheric Composition Analysis Group (ACAG) at Dalhousie University released CTM-derived ground-level PM$_{2.5}$

concentration datasets at 50 km and 10 km spatial resolutions with a combined use of Moderate-Resolution Imaging Spectroradiometer (MODIS) and Multi-angle Imaging SpectroRadiometer (MISR) AOD images in 2014 and 2016, respectively (de Sherbinin et al., 2014; van Donkelaar et al., 2015a). These two datasets have global coverage for more than ten-year period, but the relatively coarse spatial (50 km or 10 km) and temporal resolutions (annual) limit their applications in practical studies at finer geographic scales (e.g., neighborhood scales).

In addition to the CTM-based numerical models, the relationship between the AOD and other meteorological and geographical variables can also be quantified statistically. Statistical methods can be first used to model the relationship between the in situ measurements of $PM_{2.5}$ concentrations and related variables, and then make estimation at the unknown locations. Land use regression (LUR) models represent statistical methods to combine land use information with ground measurements in $PM_{2.5}$ concentrations mapping (Hoek et al., 2008). Moore et al. (2007) utilized a LUR model to estimate the distribution of annual average ground-level $PM_{2.5}$ in Los Angeles, California, with the adoption of variables (i.e., traffic intensity, population, land use types, digital elevation, and distance to the waterbodies) related to $PM_{2.5}$ concentrations at a 2.3-km resolution and achieved a higher agreement with the in situ $PM_{2.5}$ concentrations than that of the CTM-derived $PM_{2.5}$ concentration dataset at the 10-km resolution produced by ACAG. Compared with the CTM-derived numerical methods, statistical methods are more straightforward, less computationally demanding, and can produce datasets with flexible spatial and temporal resolution (Dons et al., 2014). Wu et al. (2015) used a LUR model to generate a set of hourly 816 m × 816 m $PM_{2.5}$ concentration images for Beijing, China. LUR models are essentially standard regression models applied to geospatial data. Shortcomings of the LUR models are apparent; the most notable is the failure to consider the spatial nonstationarity, complex nonlinear relationship with predictors, and complex spatial autocorrelation exhibited in geospatial data (Haining and Haining, 2003; Hoek et al., 2008). More sophisticated spatial statistical methods, such as geographically weighted regression (GWR), have been used to mitigate these issues. van Donkelaar et al. (2016) adopted a geographically weighted regression model to downscale the CTM-derived $PM_{2.5}$ images to 1 km × 1 km with an apparent accuracy improvement for the global area. However, the GWR model is essentially linear, and cannot effectively model the nonlinear effects of different predict variables on $PM_{2.5}$ concentrations. The yearly temporal resolution of the new $PM_{2.5}$ dataset (van Donkelaar et al., 2016) makes it incompetent to study seasonal trends in $PM_{2.5}$ concentrations or short-term influences of $PM_{2.5}$ exposure on human health.

Random forests (RF) is a commonly used machine learning method to capture the complex nonlinear relationships between the response variable and related predictors (Breiman, 2001). In geostatistics literature, regression kriging (RK) represents a practical approach to integrate the linear regression and conventional kriging method for geostatistical mapping while accounting for spatial dependence (via kriging) of response variable and linear relationship with predict variables (via linear regression) and. Recently, a random forests-based regression kriging (RFRK) was proposed to replace the linear regression component in RK with random forests to take into account the complex nonlinear relationships between response and predict variables (Hengl et al., 2015). The RFRK has shown performance advantages in mapping applications, including mapping $PM_{2.5}$ concentrations (Liu et al., 2018). The previous study of Liu et al. (2018) adopted the RFRK method to refine a CTM-derived dataset of $PM_{2.5}$ concentrations (with 10-km resolution) into a dataset with finer spatial resolution (1 km). Despite the improvements in spatial resolution and accuracy, the reliance on CTM derived dataset in Liu et al. (2018) limits the temporal resolution of results (at yearly). To address this issue, this study extends the work of Liu et al. (2018) and applied the RFRK directly to in situ $PM_{2.5}$ measurements and closely related meteorological variables (i.e., total precipitation, mean temperature, average dew-point temperature, and vapor pressure deficit) and geographic variables (e.g., anthropogenic or socioeconomic development factors) to improve the ground-level $PM_{2.5}$ concentration mapping. The removal of the reliance on CTM makes it possible to generate $PM_{2.5}$ concentration dataset with high temporal and spatial resolution.

Therefore, the major objective of this study is to produce a monthly time-series $PM_{2.5}$ concentration dataset at a 500-m spatial resolution for the contiguous US using the RFRK statistical approach. To fulfill the objective, we first discuss the selection of predict variables that are closely related to the distribution of $PM_{2.5}$ concentrations. Then, we describe the adopted RFRK approach that integrates RF and RK to produce monthly $PM_{2.5}$ concentration maps. Next, we evaluate the accuracy of our monthly $PM_{2.5}$ dataset and analyze seasonal patterns of $PM_{2.5}$ concentrations over the United States. To showcase the advantages of the RFRK approach, we compare the accuracy of our results with those of the traditional LUR model. Finally, we discuss the importance of the predict variables in estimating $PM_{2.5}$ concentrations and advantages of the machine learning-based geostatistical approach.

6.2 Data and preprocessing

In this study, we selected seven variables to consider effects of meteorology, human activities, emission sources, and terrain on distribution of ground-level $PM_{2.5}$ concentrations.

6.2.1 Meteorological variables

The monthly precipitation, air temperature, dew-point temperature, and vapor pressure at a 4 km × 4 km spatial resolution for January 2014 to December 2014 are obtained from the Parameter-elevation Regressions on Independent Slopes Model (PRISM) Climate Group of Oregon State University (available from: http://prism. oregonstate.edu. Accessed 14 October 2018). These gridded meteorological parameter datasets are produced by the PRISM, which imitates thought process of expert climatologists drawing climate maps and considers the influence of elevation and slope orientation on the meteorological factors (Daly et al., 2008). We resample the monthly datasets from 4 km to 500 m to make the spatial resolution of all the covariates in the RF model consistent. The resampling is accomplished by the bilinear approach in which the value of a pixel in a resampled image is determined by the average of distance-weighted values of the four nearest cells in the original image (Chang, 2006).

6.2.2 Geographical variables

Previous studies have demonstrated that anthropogenic activities significantly influence ground-level $PM_{2.5}$ concentrations (Braniš et al., 2005; Pérez et al., 2010; Bao et al., 2016). Brightness of nighttime lights extracted from the Defense Meteorological Satellite Program's Operational Linescan System (DMSP-OLS) stable lights annual composites is a good proxy of population density, traffic flow, and economic productivity (Chen and Nordhaus, 2011; Levin and Duke, 2012; Tian et al., 2013; Zhao et al., 2011) and has been used to estimate $PM_{2.5}$ concentrations (Li et al., 2017; Liu et al., 2018). In this study, we employed the new-generation nighttime lights image products, i.e., the VIIRS-DNB monthly image composites, to replace the old DMSP-OLS stable lights annual composites. The VIIRS-DNB monthly image composites are obtained from the National Oceanic and Atmospheric Administration's (NOAA) National Centers for Environmental Information (NCEI) (available from https://ngdc.noaa.gov/eog/viirs/download_dnb_composites.html. Accessed 14 October 2018). Besides a finer spatial resolution (500 m vs. 1 km), VIIRS-DNB monthly image products have a larger quantization range than DMSP-OLS annual image products (14 bit vs. 6 bit) so that saturated pixels in the VIIRS-DNB image products are nonexistent (Zhao et al., 2017). Thus, brightness of nighttime lights extracted from the VIIRS-DNB image products can more accurately capture variations of $PM_{2.5}$ concentrations across urban areas where saturated pixels prevalently exist in the DMSP-OLS image products.

Ground-level $PM_{2.5}$ concentrations also have significant correlations with land cover/land use (LCLU) types as LCLU types potentially reflect emission sources of $PM_{2.5}$ (Liu et al., 2009; Nowak et al., 2006; Hoek et al., 2008; Wu et al., 2012). Common LCLU data are categorical and thus tend to generate prominent uncertainties with other numerical spatial data in a geostatistical model (Cao et al., 2011a, b). Thus, in this study, we use normalized difference vegetation index (NDVI) as a numerical proxy of LCLU types (Liu et al., 2018). The NDVI data are extracted from the 16-day MODIS (MOD13A1) vegetation index products at a 500 m × 500 m spatial resolution (Didan et al., 2015). It needs to be particularly emphasized that all of the 16-day NDVI images are aggregated to one image by Eq. (6.2):

$$NDVI_{combined} = Maximum\left[NDVI_1, NDVI_2, ..., NDVI_n\right] \quad (6.1)$$

where $NDVI_1, NDVI_2, ..., NDVI_n$ are the NDVI values for the same pixel from the multitemporal MOD13A1 images collected between April and October in 2014. NDVI values for estimating any months' $PM_{2.5}$ concentrations are always from the combined NDVI image. In each individual 16-day MOD13A1 image, a considerable proportion of pixels are contaminated by clouds and flagged as "no data."

Thus, if we combine the 16-day DNVI images to 12 images for the 12 months of 2014, a large number of pixels still cannot be valued in the individual monthly images. More importantly, in wintertime, croplands or grasslands in northern US are very likely to be covered by snow or observed as bare soil, while the maximal NDVI from summertime multitemporal images can best identify the LCLU type of a pixel.

The digital elevation model (DEM) data at a 500-m spatial resolution are acquired from the Consortium for Spatial Information (Jarvis et al., 2008). Elevation is used as a variable in the RF model for considering constraints of terrain on dispersion, diffusion, and deposition of $PM_{2.5}$ pollutants (Beaver et al., 2010).

6.2.3 Insitu $PM_{2.5}$ measurements

In Liu et al. (2018) study, in situ $PM_{2.5}$ concentrations that were used to train the RF model were all collected from the US Environmental Protection Agency's (EPA) monitoring stations. In the current study, totally 12,860 (10,882 + 1978) daily average $PM_{2.5}$ concentration records are retrieved from the EPA (EPA, 2016) and the Interagency Monitoring of Protected Visual Environments (IMPROVE) (IMPROVE, 2017). These $PM_{2.5}$ concentrations are measured by 964 and 166 monitoring stations in the networks of EPA and IMPROVE, respectively. Most EPA monitoring stations are located in the populated urban areas. $PM_{2.5}$ measurements from the IMPROVE are essential supplementations to improve training of the RF model because the IMPROVE's monitoring sites are mostly located in national parks where the environment is much less affected by anthropogenic activities (see Fig. 6.1 for the locations of the monitoring stations). $PM_{2.5}$ concentrations measured at the same stations and within the same month are averaged and then are treated as reference data for training and evaluating the model in this study.

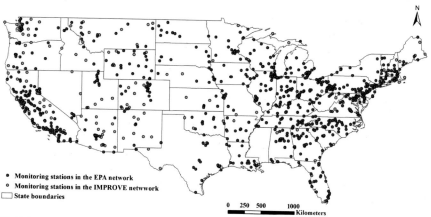

FIG. 6.1

The distribution of $PM_{2.5}$ monitoring stations in the networks of EPA and IMPROVE.

6.3 Method

6.3.1 The random forests-based regression kriging model

Let the $PM_{2.5}$ concentration ($\hat{p}$) at a location (s_0) be the sum of a deterministic and a stochastic components. Then, the RK model is written as:

$$\hat{p}(s_0) = \hat{m}(s_0) + \hat{\varepsilon}(s_0) = X_0^T \cdot \hat{\beta} + \lambda^T \cdot (z - X \cdot \hat{\beta}) \qquad (6.2)$$

where $\hat{m}(s_0)$ is the fitted deterministic part and $\hat{\varepsilon}(s_0)$ represents the spatially correlated residuals that can be modeled with kriging. X_0 is a vector of covariates at the prediction location s_0, $\hat{\beta}$ is the vector of estimated regression coefficients, λ is a vector of kriging weights used to interpolate the residuals, z is a vector of observations at sampling locations, and X is a matrix of predictors at the sampling locations. If $\hat{\beta}$ is estimated by a linear regression function, then Eq. (6.2) can be called a linear regression kriging model, which is the most widely used type of regression kriging in previous geographic studies (Hengl et al., 2015). In reality, the relationships between $PM_{2.5}$ concentration and its related variables are typically nonlinear (Liu et al., 2018). Thus, the idea of RFRK is to replace the linear regression term in the RK model with a nonlinear machine learning method, random forests, to better capture the correlations between $PM_{2.5}$ concentration and each covariate. The deterministic part is predicted by RF and thus Eq. (6.2) can be revised as:

$$\hat{p}(s_0) = f_{RF}(NTL_0, NDV_0, ELE_0, TEM_0, DEW_0, PRE_0, AIR_0, \hat{\beta}) + \hat{\varepsilon}(s_0) \qquad (6.3)$$

where *NTL, NDV, ELE, TEM, DEW, PRE, AIR* denote brightness of nighttime lights, NDVI, elevation, mean temperature, dew-point temperature, precipitation, and air pressure, respectively. As in Eq. (6.2), $\hat{\varepsilon}(s_0)$ represents the spatially correlated residuals that can be modeled with kriging to account for complex spatial autocorrelation. The *NTL, NDV, ELE, TEM, DEW, PRE, AIR* values of each monitoring station are extracted from the corresponding imagery datasets and then put into the RF model to establish the relationships between $PM_{2.5}$ concentration and covariates. The "randomForest" (Liaw and Wiener, 2002) and "gstat" (Pebesma et al., 2016) packages are used for training the RF model and interpolating the residuals in R programming environment, respectively (R Core Team, 2017). The monthly average $PM_{2.5}$ concentration dataset for the contiguous US during January 2014 to December 2014 is produced at a spatial resolution of 500 m × 500 m (henceforth referred to as RFRK-$PM_{2.5}$ concentration dataset).

6.3.2 Land use regression model

Land use regression (LUR) model is originally designed for estimating the traffic-related ambient air pollutants and has been successfully utilized in mapping the ground-level $PM_{2.5}$ concentrations at fine spatial resolution (Eeftens et al., 2012; Henderson et al., 2007; Wu et al., 2015). In the LUR model, the relationship between in situ PM$_{2.5}$ measurements and the related environmental predictors are established by the least-square based multivariate regression methods (e.g., linear regression, spline function, logarithmic algorithm). Then the relationship can be applied to locations without ground-based monitoring stations. Based on the same predictors, we apply the LUR model to produce another series of ground-level PM$_{2.5}$ concentration dataset at the same spatial resolution (henceforth referred to as LUR-PM$_{2.5}$ concentration dataset) and compare its accuracy with the RFRK-PM$_{2.5}$ dataset. Thus, the PM$_{2.5}$ concentration at a location s_0 and estimated by the LUR model, $PM(s)_{LUR}$, can be expressed by Eq. (6.4):

$$PM(s)_{LUR} = \beta_0 + f\left(x_1(s)\right) + f\left(x_2(s)\right) + \ldots + f\left(x_n(s)\right) \tag{6.4}$$

where $f(\bullet)$ is either a linear or a spline function with a smoothing term of three, $x = [x_1, x_2, \ldots, x_n]$ denotes the independent variables, and β_0 is a constant term.

6.3.3 Accuracy assessment

A 10-fold cross-validation method is adopted to quantitatively assess accuracy of the RFRK-PM$_{2.5}$ and the LUR-PM$_{2.5}$ concentration datasets. In each fold of the validation, 90% of the in situ PM$_{2.5}$ measurements are selected to compose a training dataset for the RFRK and the LUR models, and the remaining 10% of the PM$_{2.5}$ data are used for test. Absolute mean error (MAE), root mean squared error (RMSE), and R^2 of the regression function of the estimated PM$_{2.5}$ concentration on the in situ PM$_{2.5}$ concentration are used to gauge and compare accuracy of the estimations from the two models.

We quantify the importance of each predictor on modeling PM$_{2.5}$ concentrations by the percentage of increased mean square error (%IncMSE). The %IncMSE indicates the increased mean square error of a variable as a result of the variable being randomly shuffled while leaving the other predictors unchanged. Thus, a higher %IncMSE of a variable, more important the independent variable for the PM$_{2.5}$ estimation. The partial dependence curves show the relationships between PM$_{2.5}$ concentrations and the predictors. The variable importance and partial dependence curves are extracted by season (i.e., spring for March, April, and May; summer for June, July, and August; fall for September, October, and November; winter for December, January, and February).

6.4 Results
6.4.1 Model performance

Fig. 6.2 illustrates the differences of the average RMSEs, MAEs, and R^{2}s of the 10-fold cross-validations for each month between the RF and the LUR models. Except October, the RMSEs and MAEs of the RF model are apparently lower than those of the LUR model. For the October, the RMSE and MAE of the LUR model are slightly

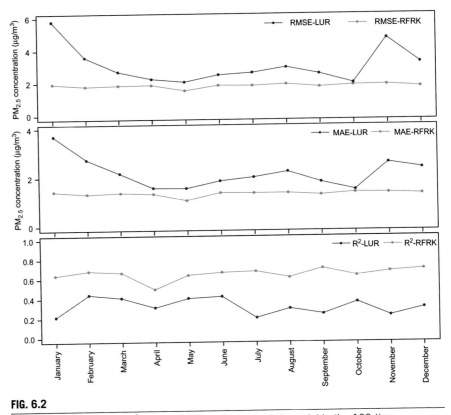

FIG. 6.2

The RMSEs, MAEs, and R^2s of the RF mode and the LUR model in the 100-time crossvalidation during each month in 2014.

higher than those of the RF model. Compared with the LUR-PM$_{2.5}$ concentration dataset, accuracy of the RFRK-PM$_{2.5}$ concentration maps is more stable, demonstrating that the RMSEs and MAEs of the RF model maintain 2.0 μg/m^3 (±0.05 μg/m^3) and 1.7 μg/m^3 (±0.1 μg/m^3) while those of the LUR model vary greatly across the 12 months (see Fig. 6.2). Specifically, the LUR-PM$_{2.5}$ concentration dataset shows much larger errors in winter season. The better performance of the RF model can also be shown by the larger fitting R^2 values for all of the 12 months. Except for April, the R^2 values of the RF model are all higher than 0.6 while those of the LUR model are nearly all smaller than 0.4.

6.4.2 Spatiotemporal variation in RFRK-PM$_{2.5}$ concentrations

Fig. 6.3 displays the RFRK-derived monthly average PM$_{2.5}$ concentration images for 12 monthly of 2014 at the 500 m × 500 m spatial resolution. For any one given month, the PM$_{2.5}$ concentrations are apparently higher in the Eastern part than in the

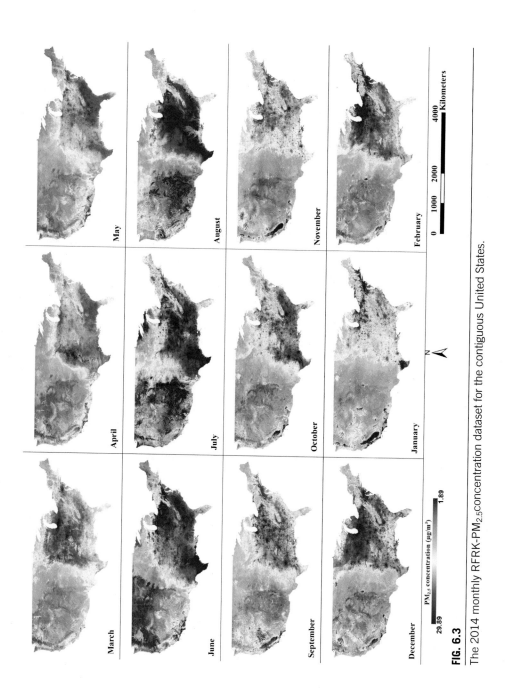

FIG. 6.3

The 2014 monthly RFRK-PM$_{2.5}$concentration dataset for the contiguous United States.

Western part. The large population densities in the Eastern region of the contiguous US lead to more emissions $PM_{2.5}$ from commercial and industrial activities and therefore associate with high $PM_{2.5}$ levels. California Valley is an exception in the West region. The California Valley, especially in the winter season (i.e., November, December, and January), has apparently higher $PM_{2.5}$ concentrations than its surrounding regions. The closed terrain of the California Valley greatly constrains diffusion and dispersion of air pollutants and so leads to high $PM_{2.5}$ concentrations. Furthermore, the dominant meteorological conditions of California valley in winter are cool and moist with low wind speed, which promotes the formations of the secondary $PM_{2.5}$ components (Motallebi et al., 2003).

Although similar spatial pattern of $PM_{2.5}$ concentrations is found during the whole year, variations still can be seen across different months. In Fig. 6.3, bluer represents higher $PM_{2.5}$ concentration while browner donates lower $PM_{2.5}$ level. More and darker blue areas appear in the three months (i.e., June, July, and August) during the summer season, yet more and darker browner areas occur on the RFRK-$PM_{2.5}$ images of the three months (i.e., November, December, and January) during the winter season. To quantitatively compare the $PM_{2.5}$ concentrations across different seasons, the average $PM_{2.5}$ concentrations for the 12 months are extracted from the RFRK-$PM_{2.5}$ dataset (Fig. 6.4). The average $PM_{2.5}$ concentration in summer is 7.33 $\mu g/m^3$, which is 55% higher than the mean of winter (4.74 $\mu g/m^3$). Many previous studies indicate that high (low) temperature-related meteorological condition is more likely to associate with high (low) $PM_{2.5}$ concentration (Liu et al., 2017; Zhao et al., 2018). People were more sensitive and vulnerable in hot weather. Thus, the modifying effects of season or temperature on ground-level $PM_{2.5}$ concentration should be further considered when investigating the associations between $PM_{2.5}$ exposure and its health outcomes.

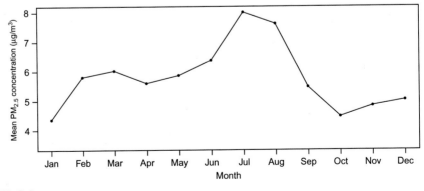

FIG. 6.4

The monthly average $PM_{2.5}$ concentrations extracted from the RFRK-$PM_{2.5}$ concentration dataset.

6.5 **Discussion**

The %IncMSEs for all the predictors are calculated by seasons to quantify the variable importance in training the RF trend. In any one of the four seasons, brightness of NTL always has higher %IncMSE than most of the other variables. Since brightness of NTL is a good proxy of population and the number of anthropogenic activities (Chen and Nordhaus, 2011; Levin and Duke, 2012; Tian et al., 2013; Zhao et al., 2011), the large %IncMSE indicates large influence of human activities on ground-level $PM_{2.5}$ concentrations. In wintertime, dramatic increases in wood and coal combustion for heating make the impacts of human activities on $PM_{2.5}$ concentrations larger (Gehrig and Buchmann, 2003; Gorin et al., 2006; Zheng et al., 2009; Gu et al., 2011), resulting in remarkably large %IncMSE (67.28%) of brightness of NTL in winter. Besides brightness of NTL, elevation also has considerably large %IncMSEs in the RF models. In the contiguous US population and the number of human activities, and consequently the amount of $PM_{2.5}$ pollutants, are decreased with the rise in elevation (Cohen and Small, 1998). Thus, elevation can indirectly influence ground-level $PM_{2.5}$ concentrations besides the direct impacts on dispersion, diffusion, and deposition of $PM_{2.5}$ pollutants (Beaver et al., 2010).

In the contiguous US, $PM_{2.5}$ concentrations in winter and summer are generally higher than those in spring and fall (Zhao et al., 2018). Different from the high $PM_{2.5}$ concentrations in wintertime that are mainly derived from anthropogenic activities (i.e., wood and coal combustion for heating), the high $PM_{2.5}$ concentrations in summertime are primarily generated by natural factors. The large air humidity in conjunction with high air temperature is likely to lead to high $PM_{2.5}$ concentrations by trapping more particulate matter and other pollutants (Greene et al., 1999; Dixon et al., 2016). In the seven predict variables, the four meteorological factors have relatively small %IncMSE, yet dew-point temperature has an exceptionally large %IncMSE (i.e., 41.54%) in summer (see Table 6.1).

In addition to the variable importance, the response of the estimated $PM_{2.5}$ concentration on each predictor is clearly nonlinear and varies across seasons (see Fig. 6.5 ear). As mentioned before, brightness of NTL is a good proxy for

Table 6.1 The percentage of IncMSE for the predictors in the RF model during spring, summer, fall, and winter

	Spring	Summer	Fall	Winter
Brightness of NTL	56.30%	34.26%	22.59%	67.28%
NDVI	41.51%	20.81%	15.07%	33.84%
Elevation	66.61%	44.61%	11.51%	42.88%
Temperature	40.81%	27.65%	9.46%	30.20%
Dew-point temperature	33.72%	41.54%	10.24%	31.40%
Precipitation	46.12%	32.34%	8.41%	27.90%
Vapor pressure deficit	39.29%	17.83%	11.38%	29.40%

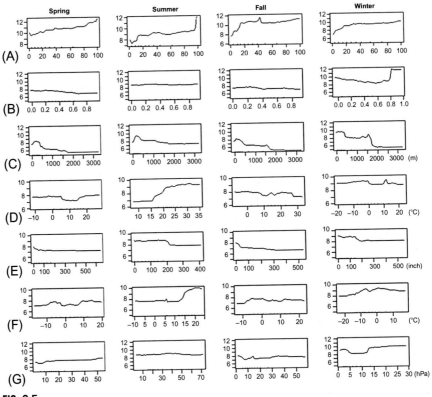

FIG. 6.5

The partial dependence curves of RF-derived PM$_{2.5}$ concentrations on the predictors by seasons: (A) for brightness of nighttime lights, (B) for NDVI, (C) for digital elevation, (D) for temperature, (E) for precipitation, (F) for dew-point temperature, and (G) for vapor pressure deficit.

anthropogenic activities and so brighter regions produce more PM$_{2.5}$ emissions and should be more polluted. Fig. 6.5A illustrates the positive effect of the brightness of NTL on the estimated PM$_{2.5}$ concentrations during the whole year, especially in regions with brightness of NTL less than 40 in spring and 20 in other seasons. The estimated PM$_{2.5}$ concentration increases the fastest with the increase in the brightness of NTL in fall. The vegetation index, NDVI, is widely used as an indicator for land cover and land use types. Higher NDVI means more vegetation coverage and so associates with less socioeconomic activities and lower PM$_{2.5}$ concentrations. Hence, the partial dependence curves in Fig. 6.5B display such negative effects of NDVI on the estimated PM$_{2.5}$ concentrations. Such negative effects present apparently in winter with the NDVI less than 0.75 while are weak in spring, summer, and fall. In winter, with an extremely high NDVI (i.e., >0.75), the estimated PM$_{2.5}$ concentration sharply increases with the rise of NDVI. Wood combustion for heating in the houses close to forest may be the reason leading to the dramatic increase in PM$_{2.5}$ concentrations in wintertime. Terrain feature can greatly affect the transmission of PM$_{2.5}$

components, including their dispersion, diffusion, and deposition. Low-altitude regions surrounded by highland constrain the dispersion and diffusion of air flow and facilitate the deposition of air pollutants and consequently associate with high $PM_{2.5}$ concentration. Fig. 6.5C clearly shows such negative effect occurs in the areas with the elevation lower than 1500 m. With the elevation reaching out of 1500 m, the $PM_{2.5}$ concentration becomes stable or even no change with the elevation due to the existence of small human activities and $PM_{2.5}$ emissions.

Among the meteorological variables, the impact of precipitation on the estimated $PM_{2.5}$ concentrations is constant across the four seasons. The partial dependence plots in Fig. 6.5E suggest a weakly negative trend between $PM_{2.5}$ levels and the precipitation. A relatively strong effect of rainfall on $PM_{2.5}$ concentrations occurs in winter with the precipitation less than 200 inches. The influences of the other three meteorological variables (i.e., temperature, dew-point temperature, and vapor pressure) are more complicated. First, the estimated $PM_{2.5}$ concentration has an apparent positive response on air temperature in summer when it is higher than 15°C (Fig. 6.5D), while the $PM_{2.5}$ concentration steadily decreases with the increase of air temperature in fall. Situations become more complex in spring and winter. In spring, a slightly declining trend appears on the partial dependence curve with the air temperature lower than 15°C in spring, but it turns to an upward trend after that. The shape of the winter partial dependence curve of air temperature is completely converse to that in spring. In winter, the partial dependence curve of air temperature changes from rising to falling at the frozen point. Second, seasonal effects on the partial dependence curve of dew-point temperature can be found in Fig. 6.5F. Similar patterns can be detected from the curves in spring and fall. The estimated $PM_{2.5}$ concentration slightly increases with the dew-point temperature less than −5°C, then starts to decrease until 8°C, and finally becomes stable. In the cold winter, the dew-point temperature is positively correlated with the estimated $PM_{2.5}$ concentration before the frozen point and presents a negative effect afterwards. The effect of dew-point temperature on $PM_{2.5}$ concentrations in summer is relatively simple. As the dew-point temperature is higher than 15°C, a strong positive effect appears on the estimated $PM_{2.5}$ levels. Finally, the partial dependence plots of vapor pressure indicate low pressure has relatively more influence on the estimations of $PM_{2.5}$ concentrations (Fig. 6.5G). The $PM_{2.5}$ concentrations are negatively related with air pressure in fall (i.e., <8 hPa) and winter (i.e., <5 hPa). The effects of vapor pressure on $PM_{2.5}$ levels are extremely weak in spring and summer.

6.6 Conclusion

This study highlights the potential of combining machine learning and geostatistical methods on mapping concentrations of air pollutants. We remove the reliance on the CTM-derived dataset in previous study (Liu et al., 2018) but directly use precipitation, air temperature, dew-point temperature, and vapor pressure in the RFRK to estimate the impacts of meteorological factors on ground-level $PM_{2.5}$ concentrations. Compared with the commonly used LUR model, the RFRK method can simultaneously consider the nonlinear relationship with predictor variables (with RF) and the complex spatial effects (with kriging) in a practical and effective manner. Without the reliance on the numerical CTM that are usually computationally demanding, the adopted RFRK approach greatly improves computing efficiency and is able to produce accurate $PM_{2.5}$ concentrations at fine spatiotemporal resolutions with consumer-grade computers. Additionally, we also show the effectiveness of the selected geographical variables in ground-level $PM_{2.5}$ concentration mapping, particularly the brightness of NTL extracted from the VIIRS-DNB monthly image composites as a comprehensive indicator to take account into the influence of human activities on ground-level $PM_{2.5}$ concentrations. In the future, we plan to exploit the capabilities of deep learning methods (e.g., recurrent neural network) to replace the present random forests in the hybrid model to produce more accurate $PM_{2.5}$ concentration maps at finer spatiotemporal resolutions.

References

Bao, C., Chai, P., Lin, H., Zhang, Z., Ye, Z., Gu, M., Lu, H., Shen, P., Jin, M., Wang, J., Chen, K., 2016. Association of PM2.5 pollution with the pattern of human activity: a case study of a developed city in eastern China. J. Air Waste Manage. Assoc. 66 (12), 1202–1213.

Beaver, S., Palazoglu, A., Singh, A., Soong, S.-T., Tanrikulu, S., 2010. Identification of weather patterns impacting 24-h average fine particulate matter pollution. Atmos. Environ. 44 (14), 1761–1771.

Braniš, M., Řezáčová, P., Domasová, M., 2005. The effect of outdoor air and indoor human activity on mass concentrations of PM10, PM2.5, and PM1 in a classroom. Environ. Res. 99 (2), 143–149.

Breiman, L., 2001. Random forests. Mach. Learn. 45 (1), 5–32.

Cao, G., Kyriakidis, P.C., Goodchild, M.F., 2011a. A multinomial logistic mixed model for the prediction of categorical spatial data. Int. J. Geogr. Inf. Sci. 25 (12), 2071–2086.

Cao, G., Kyriakidis, P.C., Goodchild, M.F., 2011b. Combining spatial transition probabilities for stochastic simulation of categorical fields. Int. J. Geogr. Inf. Sci. 25 (11), 1773–1791.

Chang, K.T., 2006. Introduction to Geographic Information Systems. McGraw-Hill Higher Education. McGraw-Hill Higher Education, Boston117–122.

Chen, X., Nordhaus, W.D., 2011. Using luminosity data as a proxy for economic statistics. Proc. Natl. Acad. Sci. 108 (21), 8589–8594.

Cohen, J.E., Small, C., 1998. Hypsographic demography: the distribution of human population by altitude. PNAS 95 (24), 14009–14014.

Daly, C., Halbleib, M., Smith, J.I., Gibson, W.P., Doggett, M.K., Taylor, G.H., Pasteris, P.P., 2008. Physiographically sensitive mapping of climatological temperature and precipitation across the conterminous United States. Int. J. Climatol. 28 (15), 2031–2064.

Di, Q., Kloog, I., Koutrakis, P., Lyapustin, A., Wang, Y., Schwartz, J., 2016. Assessing PM2.5 exposures with high spatiotemporal resolution across the continental United States. Environ. Sci. Technol. 50 (9), 4712–4721.

Didan, K., Munoz, A.B., Solano, R., Huete, A., 2015. MODIS Vegetation Index User's Guide (MOD13 Series)., Vegetation Index and Phenology Lab, The University of Arizona. 1–38.

Dixon, P.G., Allen, M., Gosling, S.N., Hondula, D.M., Ingole, V., Lucas, R., Vanos, J., 2016. Perspectives on the synoptic climate classification and its role in interdisciplinary research. Geogr. Compass 10, 147–164.

van Donkelaar, A., Martin, R.V., Spurr, R.J., Burnett, R.T., 2015a. High-resolution satellite-derived PM2. 5 from optimal estimation and geographically weighted regression over North America. Environ. Sci. Technol. 49 (17), 10482–10491.

van Donkelaar, A., Martin, R.V., Brauer, M., Boys, B.L., 2015. Use of satellite observations for long-term exposure assessment of global concentrations of fine particulate matter. Environ. Health Perspect. 123 (2), 135–143.

van Donkelaar, A., Martin, R.V., Brauer, M., Hsu, N.C., Kahn, R.A., Levy, R.C., Winker, D.M., 2016. Global estimates of fine particulate matter using a combined geophysical-statistical method with information from satellites, models, and monitors. Environ. Sci. Technol. 50 (7), 3762–3772.

van Donkelaar, A., Martin, R.V., Park, R.J., 2006. Estimating ground-level PM2.5 using aerosol optical depth determined from satellite remote sensing. J. Geophys. Res. Atmos. 111 (D21).

Dons, E., Van Poppel, M., Panis, L.I., De Prins, S., Berghmans, P., Koppen, G., Matheeussen, C., 2014. Land use regression models as a tool for short, medium and long term exposure to traffic related air pollution. Sci. Total Environ. 476, 378–386.

Eeftens, M., Beelen, R., de Hoogh, K., Bellander, T., Cesaroni, G., Cirach, M., Dimakopoulou, K., 2012. Development of land use regression models for PM2.5, PM2.5 absorbance, PM10 and PMcoarse in 20 European study areas; results of the ESCAPE project. Environ. Sci. Technol. 46 (20), 11195–11205.

Engel-Cox, J.A., Hoff, R.M., Haymet, A.D.J., 2004. Recommendations on the use of satellite remote-sensing data for urban air quality. J. Air Waste Manage. Assoc. 54 (11), 1360–1371.

Environmental Protection Agency (EPA), 2016. Air Quality Measurements. http://aqsdr1.epa.gov/aqsweb/aqstmp/airdata/download_files.html. [Accessed October 26, 2018].

Gehrig, R., Buchmann, B., 2003. Characterising seasonal variations and spatial distribution of ambient PM10 and PM2.5 concentrations based on long-term Swiss monitoring data. Atmos. Environ. 37, 2571–2580.

Gorin, C.A., Collett, J.L., Herckes, P., 2006. Wood smoke contribution to winter aerosol in Fresno, CA. J. Air Waste Manage. Assoc. 56, 1584–1590.

Greene, J.S., Kalkstein, L.S., Ye, H., Smoyer, K., 1999. Relationships between synoptic climatology and atmospheric pollution at 4 US cities. Theor. Appl. Climatol. 62, 163–174.

Gu, J., Bai, Z., Li, W., Wu, L., Liu, A., Dong, H., Xie, Y., 2011. Chemical composition of $PM_{2.5}$ during winter in Tianjin, China. Particuology 9 (3), 215–221.

Haining, R.P., Haining, R., 2003. Spatial Data Analysis: Theory and Practice. Cambridge University Press.

Henderson, S.B., Beckerman, B., Jerrett, M., Brauer, M., 2007. Application of land use regression to estimate long-term concentrations of traffic-related nitrogen oxides and fine particulate matter. Environ. Sci. Technol. 41 (7), 2422–2428.

Hengl, T., Heuvelink, G.B., Kempen, B., Leenaars, J.G., Walsh, M.G., Shepherd, K.D., Sila, A., et al., 2015. Mapping soil properties of Africa at 250 m resolution: random forests significantly improve current predictions. PLoS ONE 10 (6), e0125814.

Hoek, G., Beelen, R., De Hoogh, K., Vienneau, D., Gulliver, J., Fischer, P., Briggs, D., 2008. A review of land-use regression models to assess spatial variation of outdoor air pollution. Atmos. Environ. 42 (33), 7561–7578.

Interagency Monitoring of Protected Visual Environment (IMPROVE), 2017. IMPROVE Data. Available from, http://vista.cira.colostate.edu/Improve/improve-data/. [Accessed 26 October, 2018].

Jarvis, A., Reuter, H.I., Nelson, A., Guevara, E., 2008. Hole-filled SRTM for the globe Version 4 CGIAR-CSI SRTM 90m Database. http://srtm.csi.cgiar.org. [Accessed 26 October, 2018].

Levin, N., Duke, Y., 2012. High spatial resolution night-time light images for demographic and socio-economic studies. Remote Sens. Environ. 119, 1–10.

Li, T., Shen, H., Yuan, Q., Zhang, X., Zhang, L., 2017. Estimating ground-level PM2.5 by fusing satellite and station observations: a geo-intelligent deep learning approach. Geophys. Res. Lett. 44 (23), 11985–11993.

Liaw, A., Wiener, M., 2002. Classification and regression by randomForest. R News 2 (3), 18–22.

Lim, S.S., Vos, T., Flaxman, A.D., Danaei, G., Shibuya, K., Adair-Rohani, H., Aryee, M., 2013. A comparative risk assessment of burden of disease and injury attributable to 67 risk factors and risk factor clusters in 21 regions, 1990–2010: a systematic analysis for the Global Burden of Disease Study 2010. Lancet 380 (9859), 2224–2260.

Liu, Y., Cao, G., Zhao, N., Mulligan, K., Ye, X., 2018. Improve ground-level $PM_{2.5}$ concentration mapping using a random forests-based geostatistical approach. Environ. Pollut. 235, 272–282.

Liu, Y., Paciorek, C.J., Koutrakis, P., 2009. Estimating regional spatial and temporal variability of PM2. 5 concentrations using satellite data, meteorology, and land use information. Environ. Health Perspect. 117 (6), 886.

Liu, Y., Park, R.J., Jacob, D.J., Li, Q., Kilaru, V., Sarnat, J.A., 2004. Mapping annual mean ground level $PM_{2.5}$ concentrations using Multiangle Imaging Spectroradiometer aerosol optical thickness over the contiguous United States. J. Geophys. Res. Atmos. 109 (D22).

Liu, Y., Zhao, N., Vanos, J.K., Cao, G., 2017. Effects of synoptic weather on ground-level PM2. 5 concentrations in the United States. Atmos. Environ. 148, 297–305.

Moore, D.K., Jerrett, M., Mack, W.J., Künzli, N., 2007. A land use regression model for predicting ambient fine particulate matter across Los Angeles, CA. J. Environ. Monit. 9 (3), 246–252.

Motallebi, N., Tran, H., Croes, B.E., Larsen, L.C., 2003. Day-of-week patterns of particulate matter and its chemical components at selected sites in California. J. Air Waste Manage. Assoc. 53 (7), 876–888.

Nowak, D.J., Crane, D.E., Stevens, J.C., 2006. Air pollution removal by urban trees and shrubs in the United States. Urban For. Urban Green. 4 (3-4), 115–123.

Pebesma, E., Graeler, B., Pebesma, M.E., 2016. Package 'gstat'.

Pérez, N., Pey, J., Cusack, M., Reche, C., Querol, X., Alastuey, A., Viana, M., 2010. Variability of particle number, black carbon, and PM10, PM2. 5, and PM1 levels and speciation: influence of road traffic emissions on urban air quality. Aerosol Sci. Technol. 44 (7), 487–499.

Pope III, C.A., Burnett, R.T., Thun, M.J., Calle, E.E., Krewski, D., Ito, K., Thurston, G.D., 2002. Lung cancer, cardiopulmonary mortality, and long-term exposure to fine particulate air pollution. JAMA 287 (9), 1132–1141.

R Core Team, 2017. R: a Language and Environment for Statistical Computing. R Foundation for Statistical Computing, Vienna, Austria. [Online]. Available: https://www.R-project.org/.

de Sherbinin, A., Levy, M.A., Zell, E., Weber, S., Jaiteh, M., 2014. Using satellite data to develop environmental indicators. Environ. Res. Lett. 9 (8), 084013.

Tian, J., Zhao, N., Samson, E.L., Wang, S., 2013. Brightness of nighttime lights as a proxy for freight traffic: a case study of China. IEEE J. Sel. Top. Appl. Earth Observ. Remote Sens. 7 (1), 206–212.

Wu, J., Li, J., Peng, J., Li, W., Xu, G., Dong, C., 2015. Applying land use regression model to estimate spatial variation of PM2. 5 in Beijing, China. Environ. Sci. Pollut. Res. 22 (9), 7045–7061.

Wu, S., Mickley, L.J., Kaplan, J.O., Jacob, D.J., 2012. Impacts of changes in land use and land cover on atmospheric chemistry and air quality over the 21st century. Atmos. Chem. Phys. 12 (3), 1597–1609.

Zhao, N., Currit, N., Samson, E., 2011. Net primary production and gross domestic product in China derived from satellite imagery. Ecol. Econ. 70 (5), 921–928.

Zhao, N., Liu, Y., Cao, G., Samson, E.L., Zhang, J., 2017. Forecasting China's GDP at the pixel level using nighttime lights time series and population images. GISci. Remote Sens. 54 (3), 407–425.

Zhao, N., Liu, Y., Vanos, J.K., Cao, G., 2018. Day-of-week and seasonal patterns of PM2.5 concentrations over the United States: time-series analyses using the Prophet procedure. Atmos. Environ. 192, 116–127.

Zheng, L., Porter, E.N., Sjodin, A., Needham, L.L., Lee, S., Russell, A.G., Mulholland, J.A., 2009. Characterization of $PM_{2.5}$-bound polycyclic aromatic hydrocarbons in Atlanta—seasonal variations at urban, suburban, and rural ambient air monitoring sites. Atmos. Environ. 43, 4187–4193.

Further reading

NASA, 2018. MODIS/Terra Vegetation Indices 16-Day L3 Global 500m SIN Grid V006. https://lpdaac.usgs.gov/dataset_discovery/modis/modis_products_table/mod13a1_v006. [Accessed June 26, 2018].

Spatiotemporal interpolation methods for air pollution

7

Lixin Li[a], Weitian Tong[b], Reinhard Piltner[c]

[a]*Department of Computer Science, Georgia Southern University, Statesboro, GA, United States*
[b]*Department of Computer Science, Eastern Michigan University, Ypsilanti, MI, United States*
[c]*Department of Mathematical Sciences, Georgia Southern University, Statesboro, GA,*
United States

7.1 Introduction

Spatial interpolation methods have been well developed to estimate values at unknown locations based upon values that are spatially sampled in Geographic Information System (GIS). These methods assume a stronger correlation among points that are closer than those farther apart. They are characterized as either deterministic or stochastic depending on whether statistical properties are utilized. Deterministic interpolation methods determine an unknown value using mathematical functions with predefined parameters such as distances in inverse distance weighting (IDW) (Robichaud and Ménard, 2014; Shepard, 1968) and areas or volumes in shape function (SF)-based methods (Li and Revesz, 2004; Zienkiewics and Taylor, 2000). Many previous studies have used deterministic interpolation methods, such as radial basis functions (RBFs) (Franke and Schaback, 1998), spline (de Boor, 2001), natural neighbor (Sibson, 1981), and trend surfaces (Zurflueh, 1967). Stochastic interpolation methods such as Kriging (Krige, 1966) investigate the spatial autocorrelation and give estimates of model errors. Stochastic interpolation methods have been used to handle areal interpolation uncertainty (Geddes et al., 2013), model-data fusion (sometimes called *analysis*) (Blond and Vautard, 2004; Pagowski et al., 2010), and optimal interpolation (Robichaud et al., 2016).

Although spatial interpolation methods have been widely adopted in various GIS applications, many critical problems remain unsolved. One of them is that traditional spatial interpolation methods tend to treat space and time separately when interpolation needs to be conducted in a continuous space-time domain. The primary strategy identified from the literature is to reduce spatiotemporal interpolation problems to a sequence of snapshots of spatial interpolations (Liao et al., 2006). In order to interpolate at an unsampled time instance, temporal interpolation can then be conducted based on the spatial interpolation results at each location (Borak and Jasinski, 2009).

Integrating space and time simultaneously is shown to yield better interpolation results than treating them separately for certain typical GIS applications (Li et al., 2012). Unfortunately, there are relatively fewer models for spatiotemporal interpolation compared with spatial interpolation, especially in the application of air pollution over a large geographic area. The first exception is a study that investigated the Kriging-based spatiotemporal interpolation approaches for daily mean PM_{10} concentrations (Gräler et al., 2013). The methods used included separate daily variogram estimates, temporally evolving variograms, the metric model, the separable covariance model, and the product-sum model, and are combined with multiple linear regression. These methods were applied to daily mean rural background PM_{10} concentrations across Europe for the year 2005. The second exception is IDW-based spatiotemporal interpolation methods in Li et al. (2014) and Tong et al. (2019b) that extended the traditional spatial IDW to interpolate daily $PM_{2.5}$ concentrations at the centroids of census block groups and counties across the contiguous United States. In these studies, various IDW-based spatiotemporal interpolation methods with different parameter configurations were evaluated and compared by cross-validation. Parallel programming techniques and an advanced data structure, named k-d tree, were adapted to address the computational challenges. The third exception is an SF-based spatiotemporal interpolation method in Li et al. (2016b) that extends the popular SFs in engineering applications such as finite element algorithms. This study compared the SF-based method with the IDW-based methods in Li et al. (2014) using the same $PM_{2.5}$ data and combined interpolation results with population data to estimate the population exposure to $PM_{2.5}$ in the contiguous United States. Furthermore, Zou et al. (2011) reviewed some air pollution exposure assessment methods utilized in epidemiological studies and the use of GIS for resolving problems with spatiotemporal attributes. In summary, in the era of big data, there is a need to develop and evaluate spatiotemporal methods that produce good interpolation results with computational efficiency to handle the increasing amount of air pollution data over a large geographic area.

Some recent research on spatiotemporal data interpolation and/or health effects include Xu et al. (2014), Amato and Vecchia (2018), Mei et al. (2016, 2017), Yang et al. (2017), Dunea et al. (2016), Susanto et al. (2016), Liang et al. (2017), Delikhoon et al. (2018), Bruno et al. (2016), Li et al. (2016a), Chen et al. (2017), Singh and Toshniwal (2019), Nyhan et al. (2016), Lassman et al. (2017), Safaie et al. (2017), and Wang et al. (2019). It is worth noting that there is a new line of research that applies machine learning to solve spatiotemporal interpolation problems (Brokamp et al., 2018; Chen et al., 2018; Fan et al., 2017; Gupta and Christopher, 2009; Hu et al., 2017; Qi et al., 2018; Reid et al., 2015; Tong et al., 2019a; Wang and Song, 2018; Zou et al., 2015).

In this chapter, we will focus on deterministic spatiotemporal interpolation methods based on SFs, IDW, and RBFs. First, we introduce two SF-based spatiotemporal interpolation methods in Section 7.2 and show how to implement the algorithms. In Chapter 8, we will show how to apply and compare these specific SF-based interpolation methods on a set of real-time air pollution data over a large geographic area. Second, we introduce IDW-based spatiotemporal interpolation methods in Section 7.3. Third, we introduce RBF-based spatiotemporal interpolation methods in Section 7.4.

7.2 SF-based spatiotemporal interpolation
7.2.1 Extension approach for SF-based spatiotemporal interpolation

In order to integrate space and time simultaneously, we developed an "extension approach" to conduct the SF-based spatiotemporal interpolation. This approach treats time as another dimension in space and therefore, extending the spatiotemporal interpolation problem into a higher-dimensional spatial interpolation problem (Li, 2003). Some applications using the extension approach can be found in Li and Revesz (2004) and Li et al. (2006, 2008, 2011, 2012).

Using the extension approach of SF-based interpolation method, we treat time as the imaginary third dimension z in space. Therefore, SF-based spatiotemporal interpolation method for two-dimensional (2D) space and one-dimensional (1D) time problems can be summarized as

$$w(x,y,t) = N_1(x,y,t)w_1 + N_2(x,y,t)w_2 \\ + N_3(x,y,t)w_3 + N_4(x,y,t)w_4 \tag{7.1}$$

where N_1, N_2, N_3, and N_4 are the following SFs.

$$N_1(x,y,t) = \frac{V_1}{V}, N_2(x,y,t) = \frac{V_2}{V} \tag{7.2}$$
$$N_3(x,y,t) = \frac{V_3}{V}, N_4(x,y,t) = \frac{V_4}{V}$$

V_1, V_2, V_3, and V_4 are the volumes of the four subtetrahedra $ww_2w_3w_4$, $w_1ww_3w_4$, $w_1w_2ww_4$, and $w_1w_2w_3w$, respectively; and V is the volume of the outside tetrahedron $w_1w_2w_3w_4$ as shown in Fig. 7.1.
Alternatively, the SFs can be computed as

$$N_1(x,y,t) = \frac{a_1 + b_1x + c_1y + d_1t}{6V}$$
$$N_2(x,y,t) = \frac{a_2 + b_2x + c_2y + d_2t}{6V} \tag{7.3}$$
$$N_3(x,y,t) = \frac{a_3 + b_3x + c_3y + d_3t}{6V}$$
$$N_4(x,y,t) = \frac{a_4 + b_4x + c_4y + d_4t}{6V}$$

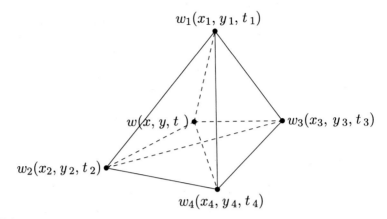

$w_1(x_1, y_1, t_1)$

$w(x, y, t)$ $w_3(x_3, y_3, t_3)$

$w_2(x_2, y_2, t_2)$

$w_4(x_4, y_4, t_4)$

FIG. 7.1

A tetrahedral element. Computing 3D shape functions by tetrahedral volume divisions. w_1, w_2, w_3, and w_4 are measured values, while the value w at the location (x, y, t) is unknown and needs to be interpolated.

The volume V can be computed using the corner coordinates (x_i, y_i, t_i) $(i = 1, 2, 3, 4)$ in the determinant of a matrix as

$$V = \frac{1}{6} \det \begin{bmatrix} 1 & x_1 & y_1 & t_1 \\ 1 & x_2 & y_2 & t_2 \\ 1 & x_3 & y_3 & t_3 \\ 1 & x_4 & y_4 & t_4 \end{bmatrix} \tag{7.4}$$

By expanding the other relevant determinants into their cofactors, we have

$$a_1 = \det \begin{bmatrix} x_2 & y_2 & t_2 \\ x_3 & y_3 & t_3 \\ x_4 & y_4 & t_4 \end{bmatrix} \quad b_1 = -\det \begin{bmatrix} 1 & y_2 & t_2 \\ 1 & y_3 & t_3 \\ 1 & y_4 & t_4 \end{bmatrix}$$

$$c_1 = -\det \begin{bmatrix} x_2 & 1 & t_2 \\ x_3 & 1 & t_3 \\ x_4 & 1 & t_4 \end{bmatrix} \quad d_1 = -\det \begin{bmatrix} x_2 & y_2 & 1 \\ x_3 & y_3 & 1 \\ x_4 & y_4 & 1 \end{bmatrix}$$

with the other constants defined by cyclic interchange of the subscripts in the order of 4, 1, 2, 3 (Zienkiewics and Taylor, 2000).

The extension method is based on three-dimensional (3D) Delaunay triangulation. During implementation, we first generate a 3D Delaunay triangulation from a set of input points with (x, y, t) coordinates. The triangulation process generates an index for the containing tetrahedron in the Delaunay mesh, as well as barycentric coordinates associated with the query point within its containing tetrahedron. Barycentric coordinates are equivalent to the N_1, N_2, N_3, and N_4 coefficients in Eqs. (7.1), (7.2).

Table 7.1 Algorithm 1 using the SF-based extension method: *interpolate (A, B, dtMesh)*

Input:	*A* — a matrix representing the **known** data in the form [*t x y w*], where *t* is the time and *x* and *y* are the location coordinates, and *w* is the known measured value;
	B — a matrix of **query** points in the form [*t x y*], where *t* is the time and *x* and *y* are the location coordinates;
	dtMesh (optional) — the triangulation used to perform the interpolation;
Output:	*res* — a matrix containing the query point coordinates and the interpolated value at each query point in the form [*t x y w*].
1.	Check format of input arguments;
	// if the *dtMesh* is not provided, calculate one
2.	**if** *dtMesh* is null
	dtMesh = calculate *Delaunay Triangulation* using *t, x, y* of *A*;
	end
	// *pl* is the coordinates of the simplex containing the query point
	// *bc* are the barycentric coordinates
3.	[*pl, bc*] = *dtMesh.pointLocation*(B);
4.	*triVals* = extract *w* from *A* at each corner identified in *pl*;
	//calculate the interpolated values at each query point
5.	*Vq* = *bc* ·*triVals*;
6.	Set *Vq* to *NaN* if no containing tetrahedron was found;
7.	*res* = append *Vq* to *B*.

The interpolated value at the query point is then calculated as the dot product of the value at each corner of the containing tetrahedron with the barycentric coordinates. Error handling is incorporated to handle an edge case where query points are not located within the convex hull of the Delaunay mesh. The algorithm of the extension method is presented in Table 7.1.

7.2.2 Reduction approach for SF-based spatiotemporal interpolation

Alternatively, we developed a "reduction approach." This approach reduces the spatiotemporal interpolation problem to a regular spatial interpolation case using two steps. First, we interpolate (using any 1D interpolation in time) the value of interest overtime at each sample point. Second, by substituting the desired time instant into some regular 2D spatial interpolation functions, we can get spatiotemporal interpolation results (Li, 2003).

Assume the value at node i at time t_1 is w_{i1}, and at time t_2 the value is w_{i2}. The value at the node i at any time between t_1 and t_2 can be approximated using a 1D time SF in the following way as

$$w_i(t) = \frac{t_2 - t}{t_2 - t_1} w_{i1} + \frac{t - t_1}{t_2 - t_1} w_{i2} \qquad (7.5)$$

The reduction method is based on 2D Delaunay triangulation. We first generate a 2D Delaunay triangulation using a set of monitoring site locations with (x, y) coordinates. Then, we need to find the containing triangle of a query point. The SF-based interpolation result at a query point (x, y) located inside the triangle can be obtained by using the measurement values w_1, w_2, and w_3 at the three corner vertices as below (Li and Revesz, 2004):

$$w(x,y) = N_1(x,y)w_1 + N_2(x,y)w_2 + N_3(x,y)w_3 \qquad (7.6)$$

where N_1, N_2, and N_3 are the following linear SFs (Li and Revesz, 2002):

$$N_1(x,y) = \frac{A_1}{A}, N_2(x,y) = \frac{A_2}{A}, N_3(x,y) = \frac{A_3}{A} \qquad (7.7)$$

A_1, A_2, and A_3 are the areas of the three subtriangles ww_2w_3, w_1ww_3, and w_1w_2w, respectively; and A is the area of the outside triangle $w_1w_2w_3$.

Using Eqs. (7.5), (7.6), the interpolation function for any point constraint to a triangular element with corner vertices at any time between t_1 and t_2 can be expressed as follows (Li and Revesz, 2002):

$$N_1(x,y) = \frac{A_1}{A}, N_2(x,y) = \frac{A_2}{A}, N_3(x,y) = \frac{A_3}{A} \qquad (7.8)$$

Since only 2D coordinate data are used in the reduction method, in order to deal with data at missing times, a linear time interpolation derived from Eq. (7.5) is used in Eq. (7.8).

Similar to the extension method, query points must be within the convex hull of the triangulation when using the reduction method. Since the triangulation in 2D geographic space only addresses the location coordinates, an additional restriction applies to the reduction method in the time domain. A query point must request a time that is between known measurements at each corner of the containing 2D triangle. If this is not the case, then values at the containing triangle corners can be extrapolated in the time dimension (prior to interpolating the value of the query point in the 2D coordinate dimensions using SFs). The algorithm of the reduction method is presented in Table 7.2.

7.3 IDW-based spatiotemporal interpolation

Similar as SF-based interpolation methods, IDW interpolation assumes that each measured point has a local influence that diminishes with distance. Instead of forming triangular or tetrehedral meshes as in SF-based interpolation methods, IDW needs to find nearest neighbors. Points at a shorter distance are given high weights, whereas points at a far distance are given small weights.

Table 7.2 Algorithm 2 using the SF-based reduction method: *interpolate2D (A, B, Da, T)*

Input:	*A*—a matrix representing the **known** data in the form [*t x y w*], where *t* is the time and *x* and *y* are the location coordinates, and *w* is the known measured value;
	B—a matrix of **query** points in the form [*t x y*], where *t* is the time and *x* and *y* are the location coordinates;
	dtMesh (optional)—the triangulation used to perform the interpolation;
	Da (optional)—a matrix containing the time interpolated values for all locations in *A*;
	T (optional)—a vector containing all times in *A*;
Output:	*res*—a matrix containing the query point coordinates and the interpolated value at each query point in the form [*t x y w*]
	dtMesh—the triangulation used to perform the interpolation;
	Da—a matrix containing the time interpolated values for all locations in *A*;
	T—a vector containing all distinct times in *A*.
1.	Check format of input arguments;
	// if the *dtMesh*, *Da*, *T* are not provided, calculate them
2.	**if** *dtMesh*, *Da*, or *T* is null
	dtMesh = calculate *Delaunay Triangulation* using *t*, *x*, *y* of *A*;
	Da = calculate time-interpolated value at the Cartesian product of all locations in *A* with all times in *T*
	T = collect unique times in *A*;
	end
	// *pl* is the coordinates of the simplex containing the query point
	// *bc* are the barycentric coordinates
3.	[*pl*, *bc*] = *dtMesh.pointLocation*(B);
4.	*triVals* = extract *w* from *A* at each corner identified in *pl*;
	//calculate the interpolated values at each query point
5.	*Vq* = *bc* ·*triVals*;
6.	Set *Vq* to *NaN* if no containing tetrahedron was found;
7.	*res* = append *Vq* to *B*.

According to Johnston et al. (2001), the general formula of IDW interpolation is the following:

$$w(x,y) = \sum_{i=1}^{N} \lambda_i w_i, \lambda_i = \frac{\left(\dfrac{1}{d_i}\right)^p}{\sum_{k=1}^{N}\left(\dfrac{1}{d_k}\right)^p} \tag{7.9}$$

where $w(x, y)$ is the predicted value at location (x, y), N is the number of nearest known points surrounding (x, y), λ_i are the weights assigned to each known point value w_i at location (x_i, y_i), d_i are the Euclidean distances between each (x_i, y_i) and (x, y), and p is the exponent, which influences the weighting of w_i on w.

IDW-based reduction and extension approaches to 2D problem have been introduced in Li and Revesz (2004).

7.3.1 Extension approach for IDW-based spatiotemporal interpolation

Since this method treats time as a third dimension, the formula of IDW-based spatiotemporal interpolation using the extension approach is

$$w(x, y, t) = \sum_{i=1}^{N} \lambda_i w_i, \lambda_i = \frac{\left(\dfrac{1}{d_i}\right)^p}{\sum_{k=1}^{N}\left(\dfrac{1}{d_k}\right)^p} \tag{7.10}$$

with $d_i = \sqrt{(x_i - x)^2 + (y_i - y)^2 + (t_i - t)^2}$.

7.3.2 Reduction approach for IDW-based spatiotemporal interpolation

Assume we are interested in the value of the unsampled point at location (x, y) and time t. The IDW-based reduction method first finds the nearest neighbors of each unsampled point and calculates the corresponding weights λ_i using 2D Euclidean distance $d_i = \sqrt{(x_i - x)^2 + (y_i - y)^2}$. Then, it calculates for each neighbor the value at time t by some temporal interpolation method. If we use 1D time SF (Eq. 7.5) for the temporal interpolation, the formula of IDW-based reduction method can be expressed as

$$w(x, y, t) = \sum_{i=1}^{N} \lambda_i w_i(t), \lambda_i = \frac{\left(\dfrac{1}{d_i}\right)^p}{\sum_{k=1}^{N}\left(\dfrac{1}{d_k}\right)^p} \tag{7.11}$$

where

$$w_i(t) = \frac{t_{i2} - t}{t_{i2} - t_{i1}} w_{i1} + \frac{t - t_{i1}}{t_{i2} - t_{i1}} w_{i2} \tag{7.12}$$

Each neighbor may have different beginning and ending times t_{i1} and t_{i2} in Eq. (7.12) if each points are sampled at different times.

7.4 **RBF-based spatiotemporal interpolation**

In the recent decades, the use of RBFs has gained popularity for interpolating data and for approximating solutions of partial differential equations (Buhmann, 2003; Fasshauer, 2007; Franke and Schaback, 1998; Kansa, 1990; Liu and Gu, 2005; Piltner, 2019; Wendland, 2005; Wu, 1995). RBFs are also useful in computer graphics and medical imaging, RBFs showed some advantages in modeling complicated surfaces (Morse et al., 2001; Schaback, 1995).

In the context of spatiotemporal air pollution modeling, Losser et al. (2014) explored how to design and implement RBF-based spatiotemporal interpolation methods to assess the trend of daily $PM_{2.5}$ concentrations for the contiguous United States. In this study, time values are calculated with the help of a factor under the assumption that spatial and temporal dimensions are equally important when interpolating a continuous changing phenomenon in the space-time domain. Various RBF-based spatiotemporal interpolation methods were evaluated by leave-one-out cross-validation. Using various meteorological data and land-use information, Zou et al. (2015) presented a neural network based on RBF to estimate $PM_{2.5}$ concentrations in Texas. Most recently, Piltner (2018, 2019) used RBFs satisfying the 2D biharmonic equation $\Delta\Delta w = 0$ and a quadruharmonic equation $\Delta\Delta\Delta\Delta w = 0$ (see Table 7.5). The motivation for the use of biharmonic and polyharmonic solution functions is to avoid oscillating behavior of functions.

By using RBFs, it became possible to deal with higher-dimensional problems in a similar way as dealing with 2D and 3D problems. After choosing N points in a domain under consideration, we can easily construct N linearly independent basis functions by using distances measured from the chosen points. In 2D, the distance r from a point (x_q, y_q) is used: $r = \sqrt{(x-x_q)^2 + (y-y_q)^2}$. For a 3D problem involving spatial coordinates x, y, and z, the distance $r = \sqrt{(x-x_q)^2 + (y-y_q)^2 + (z-z_q)^2}$ is used. In the case of an N-dimensional problem with points $\mathbf{p} = (p_1, p_2, ..., p_N)$ and $\mathbf{q} = (q_1, q_2, ..., q_N)$, the Euclidian distance is $r = \|\mathbf{p}-\mathbf{q}\| = \sqrt{(\mathbf{p}-\mathbf{q})\cdot(\mathbf{p}-\mathbf{q})}$.

If we want to interpolate time-dependent 2D spatial data, we can use the following coordinate for the RBFs:

$$r = \sqrt{(x-x_q)^2 + (y-y_q)^2 + c^2(t-t_q)^2} \tag{7.13}$$

The parameter c has the unit [distance/time]. Depending on the given data with spatial and time units, an appropriate numerical value for the velocity parameter has to be chosen. This could involve a few numerical experiments to find a suitable value for c.

For an interpolation task, we first choose an RBF $\varphi(r)$ from a catalog of RBFs and then add some low-order polynomial functions in P to get the interpolation function (Fasshauer, 2007; Morse et al., 2001):

$$w\left(\mathbf{x_p}\right) = \sum_{q=1}^{N} \lambda_q \, \varphi\left(\|\, \mathbf{x_p} - \mathbf{x_q}\,\|\right) + P(\mathbf{x_p}) \tag{7.14}$$

For a 3D problem, the polynomial part can be chosen as $P = a + bx + cy + dz$.

Some of the most commonly used RBFs with global support are listed in Table 7.3. In order to obtain a banded matrix for the linear system of equations, RBFs with local support can be used. In Table 7.4, RBFs for compactly supported cases derived by Wendland (2005) are listed.

For 2D spatial interpolation problems, Piltner suggested in Piltner (2019) the use of compactly supported RBFs which satisfy biharmonic or polyharmonic differential equations (see Table 7.5). From the bending of elastic plates governed by a biharmonic partial differential equation, we know that the bending functions have a nice looking smooth behavior.

Table 7.3 Global RBFs.

Gaussian	$\varphi(r) = e^{-cr^2}$
Multiquadric	$\varphi(r) = \sqrt{r^2 + c^2}$
Inverse multiquadric	$\varphi(r) = 1/\sqrt{r^2 + c^2}$
Thin plate spline	$\varphi(r) = r^2 \log r$
Polyharmonic	$\varphi(r) = r^{2n} \log r$

Table 7.4 Selected Wendland RBFs for $0 \le r \le 1$.

$\varphi(r)$	Continuity
$\varphi(r) = (1-r)_+^2$	C^0
$\varphi(r) = (1-r)_+^4 (4r + 1)$	C^2
$\varphi(r) = (1-r)_+^6 (35r^2 + 18r + 3)$	C^4
$\varphi(r) = (1-r)_+^8 (32r^3 + 25r^2 + 8r + 1)$	C^6

Table 7.5 Compactly supported 2D Trefftz RBFs (see Piltner, 2019) with C^1 continuity $\left(\text{support radius} = a, \ R = \dfrac{r}{a}\right)$.

RBF	PDE
$\Phi(R) = 2R^2 \ln(R) + (1 - R^2)$	$\Delta\Delta\Phi = 0$
$w = \dfrac{1}{288} R^6 \ln(R) - \dfrac{1}{216} R^6 + \dfrac{1}{64} R^4$ $-\dfrac{1747}{1728} R^2 + 1 + \dfrac{857}{432} R^2 \ln(R)$	$\Delta\Delta\Delta\Delta w = 0$ and $\Delta\Delta w = \dfrac{1}{a^4} \Phi(R)$

For each data point $\mathbf{x} = \mathbf{x_i}$ (where $i = 1, \ldots, N$), we prescribe the function value for w as

$$w(\mathbf{x_i}) = \sum_j^N \lambda_j \varphi(\| \mathbf{x_i} - \mathbf{x_j} \|) + P(\mathbf{x_i}) = w_i \tag{7.15}$$

In this way, we generate N linear equations. For the example of a 3D spatial interpolation problem where the polynomial part involves four additional unknown parameters a, b, c, and d, we have to add four constraint equations to obtain the following system of $(N + 4)$ equations:

$$\lambda_1 \varphi(\| \mathbf{x_i} - \mathbf{x_1} \|) + \cdots + \lambda_N \varphi(\| \mathbf{x_i} - \mathbf{x_N} \|) + a + bx_i + cy_i + dz_i = w_i \tag{7.16}$$

$$\sum_j^N \lambda_j = 0$$

$$\sum_j^N \lambda_j x_j = 0$$

$$\sum_j^N \lambda_j y_j = 0$$

$$\sum_j^N \lambda_j z_j = 0$$

In matrix notation, the system of equations can be written as

$$\begin{bmatrix} \varphi_{11} & \cdots & \varphi_{1N} & 1 & x_1 & y_1 & z_1 \\ \vdots & \ddots & \vdots & \vdots & \vdots & \vdots & \vdots \\ \varphi_{N1} & \cdots & \varphi_{NN} & 1 & x_N & y_N & z_N \\ 1 & \cdots & 1 & 0 & 0 & 0 & 0 \\ x_1 & \cdots & x_N & 0 & 0 & 0 & 0 \\ y_1 & \cdots & y_N & 0 & 0 & 0 & 0 \\ z_1 & \cdots & z_N & 0 & 0 & 0 & 0 \end{bmatrix} \begin{bmatrix} \lambda_1 \\ \vdots \\ \lambda_N \\ a \\ b \\ c \\ d \end{bmatrix} = \begin{bmatrix} w_1 \\ \vdots \\ w_N \\ 0 \\ 0 \\ 0 \\ 0 \end{bmatrix} \tag{7.17}$$

where

$$\varphi_{ij} = \varphi(\| \mathbf{x_i} - \mathbf{x_j} \|) \tag{7.18}$$

The coefficient matrix in the linear system of equations is symmetric.

References

Amato, U., Vecchia, B.D., 2018. On Shepard-Gupta-type operators. J. Inequal. Appl. 2018 (1), 232. https://doi.org/10.1186/s13660-018-1823-7.
Blond, N., Vautard, R., 2004. Three-dimensional ozone analyses and their use for short-term ozone forecasts. J. Geophys. Res. Atmos. 109 (D17). https://doi.org/10.1029/2004JD004515.

Borak, J.S., Jasinski, M.F., 2009. Effective interpolation of incomplete satellite-derived leaf-area index time series for the continental united states. Agric. For. Meteorol. 149 (2), 320–332.

Brokamp, C., Jandarov, R., Hossain, M., Ryan, P., 2018. Predicting daily urban fine particulate matter concentrations using a random forest model. Environ. Sci. Technol. 52 (7), 4173–4179.

Bruno, F., Cameletti, M., Franco-Villoria, M., Greco, F., Ignaccolo, R., Ippoliti, L., Valentini, P., Ventrucci, M., 2016. A survey on ecological regression for health hazard associated with air pollution. Spat. Statist. 18, 276–299. https://doi.org/10.1016/j.spasta.2016.05.003.

Buhmann, M.D., 2003. Radial Basis Functions: Theory and Implementations. Cambridge University Press, Cambridge, MA.

Chen, Y., Liu, X., Li, X., Liu, X., Yao, Y., Hu, G., Xu, X., Pei, F., 2017. Delineating urban functional areas with building-level social media data: a dynamic time warping (DTW) distance based k-medoids method. Landsc. Urban Plan. 160, 48–60. https://doi.org/10.1016/j.landurbplan.2016.12.001.

Chen, G., Li, S., Knibbs, L.D., Hamm, N., Cao, W., Li, T., Guo, J., Ren, H., Abramson, M.J., Guo, Y., 2018. A machine learning method to estimate $PM_{2.5}$ concentrations across China with remote sensing, meteorological and land use information. Sci. Total Environ. 636, 52–60.

de Boor, C., 2001. A Practical Guide to Splines. vol. 27 Springer, New York, NY.

Delikhoon, M., Fazlzadeh, M., Sorooshian, A., Baghani, A.N., Golaki, M., Ashournejad, Q., Barkhordari, A., 2018. Characteristics and health effects of formaldehyde and acetaldehyde in an urban area in Iran. Environ. Pollut. 242, 938–951. https://doi.org/10.1016/j.envpol.2018.07.037.

Dunea, D., Iordache, S., Pohoata, A., 2016. Fine particulate matter in urban environments: a trigger of respiratory symptoms in sensitive children. Int. J. Environ. Res. Public Health 13 (12). https://doi.org/10.3390/ijerph13121246.

Fan, J., Li, Q., Hou, J., Feng, X., Karimian, H., Lin, S., 2017. A spatiotemporal prediction framework for air pollution based on deep RNN. ISPRS Ann. J. Photogramm. Remote Sens. Spat. Inf. Sci. 4, 15.

Fasshauer, G.E., 2007. Meshfree approximation methods with MATLAB. Interdisciplinary Mathematical Sciences, World Scientific, Singapore.

Franke, C., Schaback, R., 1998. Solving partial differential equations by collocation using radial basis functions. Appl. Math. Comput. 93, 73–82.

Geddes, A., Elston, D.A., Hodgson, M.E.A., Birnie, R.V., 2013. Stochastic model-based methods for handling uncertainty in areal interpolation. Int. J. Geogr. Inf. Sci. 27 (4), 785–803. https://doi.org/10.1080/13658816.2012.722636.

Gräler, B., Rehr, M., Gerharz, L., Pebesma, E., 2013. Spatio-temporal analysis and interpolation of PM_{10} measurements in Europe for 2009, pp. 1–29. ETC/ACM Technical Paper.

Gupta, P., Christopher, S.A., 2009. Particulate matter air quality assessment using integrated surface, satellite, and meteorological products: 2. A neural network approach. J. Geophys. Res. Atmos. 114 (D20), 1–14.

Hu, X., Belle, J.H., Meng, X., Wildani, A., Waller, L.A., Strickland, M.J., Liu, Y., 2017. Estimating $PM_{2.5}$ concentrations in the conterminous United States using the random forest approach. Environ. Sci. Technol. 51 (12), 6936–6944.

Johnston, K., Ver Hoef, J.M.V., Krivoruchko, K., Lucas, N., 2001. Using ArcGIS Geostatistical Analyst. ESRI Press, Redlands, CA.

Kansa, E.J., 1990. Multiquadrics-A scattered data approximation scheme with applications to computational fluid-dynamics-I surface approximations and partial derivative estimates. Comput. Math. Appl. 66 (8–9), 127–145. https://doi.org/10.1016/0898-1221(90)90270-T.

Krige, D.G., 1966. Two dimensional weighted moving average trend surfaces for ore evaluation. J. Soc. Afr. Inst. Min. Metall. 66, 13–38.

Lassman, W., Ford, B., Gan, R.W., Pfister, G., Magzamen, S., Fischer, E.V., Pierce, J.R., 2017. Spatial and temporal estimates of population exposure to wildfire smoke during the Washington State 2012 wildfire season using blended model, satellite, and in situ data. GeoHealth 1 (3), 106–121. https://doi.org/10.1002/2017GH000049.

Li, L., 2003, May. Spatiotemporal Interpolation Methods in GIS (Ph.D. thesis), University of Nebraska-Lincoln, Lincoln, Nebraska.

Li, L., Revesz, P., 2002. A comparison of spatio-temporal interpolation methods. In: Proceedings of the Second International Conference on GIScience 2002. Lecture Notes in Computer Science, vol. 2478. Springer, pp. 145–160.

Li, L., Revesz, P., 2004. Interpolation methods for spatio-temporal geographic data. J. Comput. Environ. Urban Syst. 28 (3), 201–227.

Li, L., Zhang, X., Piltner, R., 2006. A spatiotemporal database for ozone in the conterminous U.S. In: Proceedings of the Thirteenth International Symposium on Temporal Representation and Reasoning. IEEE, Budapest, Hungary, pp. 168–176.

Li, L., Zhang, X., Piltner, R., 2008. An application of the shape function based spatiotemporal interpolation method on ozone and population exposure in the contiguous U.S. J. Environ. Inf. 12 (2), 120–128.

Li, L., Zhang, X., Holt, J., Tian, J., Piltner, R., 2011. Spatiotemporal interpolation methods for air pollution exposure. In: Proceedings of the Ninth Symposium on Abstraction, Reformulation, and Approximation. AAAI (Association for the Advancement of Artificial Intelligence), Parador de Cardona, Spain, pp. 75–81.

Li, L., Zhang, X., Holt, J.B., Tian, J., Piltner, R., 2012. Estimating population exposure to fine particulate matter in the conterminous U.S. using shape function-based spatiotemporal interpolation method: a county level analysis. GSTF Int. J. Comput. 1, 24–30.

Li, L., Losser, T., Yorke, C., Piltner, R., 2014. Fast inverse distance weighting-based spatiotemporal interpolation: a web-based application of interpolating daily fine particulate matter PM$_{2.5}$ in the contiguous U.S. using parallel programming and k-d tree. Int. J. Environ. Res. Public Health 11 (9), 9101–9141.

Li, H., Fan, H., Mao, F., 2016. A visualization approach to air pollution data exploration—a case study of air quality index (PM2.5) in Beijing, China. Atmosphere 7 (3). https://doi.org/10.3390/atmos7030035.

Li, L., Zhou, X., Kalo, M., Piltner, R., 2016. Spatiotemporal interpolation methods for the application of estimating population exposure to fine particulate matter in the contiguous U.S. and a real-time web application. Int. J. Environ. Res. Public Health 13 (8), 749. https://doi.org/10.3390/ijerph13080749. 20 p.

Liang, F., Gao, M., Xiao, Q., Carmichael, G.R., Pan, X., Liu, Y., 2017. Evaluation of a data fusion approach to estimate daily PM2.5 levels in North China. Environ. Res. 158, 54–60. https://doi.org/10.1016/j.envres.2017.06.001.

Liao, D., Peuquet, D.J., Duan, Y., Whitsel, E.A., Dou, J., Smith, R.L., Lin, H.-M., Chen, J.-C., Heiss, G., 2006. GIS approaches for the estimation of residential-level ambient PM concentrations. Environ. Health Perspect. 114 (9), 1374–1380.

Liu, G.-R., Gu, Y.-T., 2005. An Introduction to Meshfree Methods and Their Programming. Springer, New York, NY.

Losser, T., Li, L., Piltner, R., 2014. A spatiotemporal interpolation method using radial basis functions for geospatiotemporal big data. In: Proceedings of the 5th International Conference on Computing for Geospatial Research and Application. IEEE, Washington, D.C, pp. 17–24.

Mei, G., Xu, N., Xu, L., 2016. Improving GPU-accelerated adaptive IDW interpolation algorithm using fast kNN search. SpringerPlus 5 (1), 1389. https://doi.org/10.1186/s40064-016-3035-2.

Mei, G., Xu, L., Xu, N., 2017. Accelerating adaptive inverse distance weighting interpolation algorithm on a graphics processing unit. R. Soc. Open Sci. 4 (9), 170436. https://doi.org/10.1098/rsos.170436.

Morse, B.S., Yoo, T.S., Rheingans, P., Chen, D.T., Subramanian, K.R., 2001. Interpolating implicit surfaces from scattered surface data using compactly supported radial basis functions. In: SMI'01: Proceedings of the International Conference on Shape Modeling and Applications, pp. 89–98.

Nyhan, M., Grauwin, S., Britter, R., Misstear, B., McNabola, A., Laden, F., Barrett, S.R.H., Ratti, C., 2016. Exposure track-the impact of mobile-device-based mobility patterns on quantifying population exposure to air pollution. Environ. Sci. Technol. 50 (17), 9671–9681. https://doi.org/10.1021/acs.est.6b02385.

Pagowski, M., Grell, G.A., McKeen, S.A., Peckham, S.E., Devenyi, D., 2010. Three-dimensional variational data assimilation of ozone and fine particulate matter observations: some results using the weather research and forecasting—chemistry model and grid-point statistical interpolation. Q. J. R. Meteorol. Soc. 136 (653), 2013–2024. https://doi.org/10.1002/qj.700.

Piltner, R., 2018. Exploring new options for data interpolation with radial basis functions. In: MOBIMEDIA'18 Proceedings of the 11th EAI International Conference on Mobile Multimedia Communications, Workshop on Environmental Health and Air Pollution. EAI, Qingdao, People's Republic of China, pp. 228–231.

Piltner, R., 2019. Some remarks on Trefftz type approximations. Eng. Anal. Bound. Elem. 101, 102–112. https://doi.org/10.1016/j.enganabound.2018.12.010.

Qi, Z., Wang, T., Song, G., Hu, W., Li, X., Zhang, Z.M., 2018. Deep air learning: interpolation, prediction, and feature analysis of fine-grained air quality. IEEE Trans. Knowl. Data Eng 30 (12), 2285–2297.

Reid, C.E., Jerrett, M., Petersen, M.L., Pfister, G.G., Morefield, P.E., Tager, I.B., Raffuse, S.M., Balmes, J.R., 2015. Spatiotemporal prediction of fine particulate matter during the 2008 Northern California wildfires using machine learning. Environ. Sci. Technol. 49 (6), 3887–3896.

Robichaud, A., Ménard, R., 2014. Multi-year objective analyses of warm season ground-level ozone and $PM_{2.5}$ over North America using real-time observations and Canadian operational air quality models. Atmos. Chem. Phys. 14, 1769–1800.

Robichaud, A., Ménard, R., Zaïtseva, Y., Anselmo, D., 2016. Multi-pollutant surface objective analyses and mapping of air quality health index over North America. Air Qual. Atmos. Health, 1–17. https://doi.org/10.1007/s11869-015-0385-9.

Safaie, A., Dang, C., Qiu, H., Radha, H., Phanikumar, M.S., 2017. Manifold methods for assimilating geophysical and meteorological data in earth system models and their components. J. Hydrol. 544, 383–396. https://doi.org/10.1016/j.jhydrol.2016.11.009.

Schaback, R., 1995. Creating surfaces from scattered data using radial basis functions. In: Mathematical Methods for Curves and Surfaces. Vanderbilt University Press, Nashville, TN, pp. 477–496.

Shepard, D., 1968. A two-dimensional interpolation function for irregularly spaced data. In: Proceedings of the 23rd National Conference ACM. ACM, New York, NY, pp. 517–524.

Sibson, R., 1981. A brief description of natural neighbor interpolation. In: Barnett, V. (Ed.), Interpreting Multivariate Data. John Wiley & Sons, Inc., New York, NY, pp. 21–36.

Singh, B., Toshniwal, D., 2019. MOWM: multiple overlapping window method for RBF based missing value prediction on big data. Expert Syst. Appl. 122, 303–318. https://doi.org/10.1016/j.eswa.2018.12.060.

Susanto, F., De Souza, P., He, J., 2016. Spatiotemporal interpolation for environmental modelling. Sensors 16 (8). https://doi.org/10.3390/s16081245.

Tong, W., Li, L., Zhou, X., Hamilton, A., Zhang, K., 2019. Deep learning air pollution with bidirectional LSTM RNN. Air Qual. Atmos. Health 12 (4), 411–423.

Tong, W., Li, L., Zhou, X., Franklin, J., 2019. Efficient spatiotemporal interpolation with Spark machine learning. Earth Sci. Inf. 12 (1), 87–96.

Wang, J., Song, G., 2018. A deep spatial-temporal ensemble model for air quality prediction. Neurocomputing 314, 198–206.

Wang, J., Zhang, J., Feng, Y., 2019. Characterizing the spatial variability of soil particle size distribution in an underground coal mining area: an approach combining multi-fractal theory and geostatistics. CATENA 176, 94–103. https://doi.org/10.1016/j.catena.2019.01.011.

Wendland, H., 2005. Scattered Data Approximation. vol. 17 Cambridge University Press, Cambridge.

Wu, Z., 1995. Compactly supported positive definite radial functions. Adv. Comput. Math. 4 (1), 283–292. https://doi.org/10.1007/BF03177517.

Xu, M., Guo, Y., Zhang, Y., Westerdahl, D., Mo, Y., Liang, F., Pan, X., 2014. Spatiotemporal analysis of particulate air pollution and ischemic heart disease mortality in Beijing, China. Environ. Health 13 (1), 109. https://doi.org/10.1186/1476-069X-13-109.

Yang, W., Wang, G., Bi, C., 2017. Analysis of long-range transport effects on PM2.5 during a short severe haze in Beijing, China. Aerosol Air Qual. Res. 17 (6), 1610–1622. https://doi.org/10.4209/aaqr.2016.06.0220.

Zienkiewics, O.C., Taylor, R.L., 2000. Finite Element Method, vol. 1, The Basis. Butterworth-Heinemann, London.

Zou, B., Wilson, J.G., Zhan, F.B., Zeng, Y., 2011. Air pollution exposure assessment methods utilized in epidemiological studies. J. Environ. Monit. 11, 475–490.

Zou, B., Wang, M., Wan, N., Wilson, J.G., Fang, X., Tang, Y., 2015. Spatial modeling of $PM_{2.5}$ concentrations with a multifactoral radial basis function neural network. Environ. Sci. Pollut. Res. 22 (14), 10395–10404.

Zurflueh, E.G., 1967. Applications of two-dimensional linear wavelength filtering. Geophysics 32, 1015–1035.

Sensing air quality: Spatiotemporal interpolation and visualization of real-time air pollution data for the contiguous United States

Marc Kalo[a], Xiaolu Zhou[b], Lixin Li[a], Weitian Tong[c], Reinhard Piltner[d]

[a]*Department of Computer Science, Georgia Southern University, Statesboro, GA, United States*
[b]*Department of Geography, Texas Christian University, Fort Worth, TX, United States*
[c]*Department of Computer Science, Eastern Michigan University, Ypsilanti, MI, United States*
[d]*Department of Mathematical Sciences, Georgia Southern University, Statesboro, GA, United States*

8.1 Introduction

Air pollution is a major environment-related health threat and a risk factor for both acute and chronic respiratory diseases (Seaton et al., 1995). Outdoor air pollution is a major contributor to these unhealthy living environments (Künzli et al., 2000). Particle pollution, known as particulate matter (PM), is composed of microscopic solids or liquid droplets that are so small that can get deep into the lungs and cause serious health problems (Zanobetti and Schwartz, 2009). It has been shown that those with cardiovascular or respiratory conditions and the youth and elderly are the most susceptible to the adverse effects of PM (Laden et al., 2000; Pope et al., 2004). Gaseous pollutants such as ozone can also cause many health issues (Girardot et al., 2006; Triche et al., 2006). Therefore, the effects of particle pollutants and gaseous pollutants pose a serious threat to heath (Maynard, 2015).

Estimating air pollution concentrations smoothly across geographic regions is an important problem. There has been general consensus that society would benefit from being better engaged and educated about the complex relationship between air quality and health condition. In order to better monitor pollution distribution and control its influence on public health, it is important to accurately estimate air pollution concentrations at any location and time. For example, Li et al. (2016) estimate population exposure to $PM_{2.5}$ (fine PM) in the contiguous United States for the year 2009 based on the spatiotemporally interpolated daily $PM_{2.5}$ concentrations at the centroids of census block groups.

Spatial interpolation methods, such as inverse distance weighting (IDW), spline, and kriging, have been well developed in geographic information systems (GIS) to estimate values at unknown locations based upon values that are spatially sampled (Li and Heap, 2011; Zou et al., 2016). However, unlike spatial data in traditional GIS applications such as digital elevation model, air pollution data vary in both temporal and spatial dimensions. This calls for appropriate theories and methods to achieve better interpolation of the observed spatiotemporal data. Although many GIS researchers treat space and time separately (Liao et al., 2006), integrating space and time simultaneously is proven to yield better interpolation results (Li et al., 2012). Furthermore, the advance of information and communications technologies enables modern sensors to monitor variables (such as air pollution concentrations) at an increasing temporal resolution, resulting in rich spatiotemporal data sets. There is a need to develop interpolation approaches that can handle big data and support real-time analysis.

This chapter contributes to the literature in two aspects. The first contribution is to develop efficient spatiotemporal interpolation methods and search for optimal parameters to achieve good accuracy. The increasing amount of air pollution data requires efficient methods to handle the demanding computational tasks. Several interpolation methods have been explored in previous studies. Although many GIS applications provide interpolation tools, most current interpolation methods only apply to spatial data. Air pollution data not only have spatial attributes, but also change with time. When interpolating across space and time, the choice of the time scale versus the distance scale is an important issue that has affects the accuracy of interpolation. We demonstrate the computational power and improved performance of our methods, which outperform the previous work in terms of speed and accuracy.

The second contribution of this chapter is to increase public awareness of variations in the air quality by designing and implementing a real-time system to visualize and query pollution data using the proposed interpolation method through a web application with a friendly user interface. Visualization has been an effective tool for not only conveying information in vast amounts of data but also assisting data-driven analysis and decision making (Keim et al., 2013). However, few studies explored air pollutant visualization (Elbir, 2004; Qu et al., 2007), especially at a large geographic scale and in real time. We develop a real-time system in this study that provides a fast and real-time visualization for researchers to drill down into the vast amount of pollutant data, and for policy makers to quickly respond to abnormal air pollution incidences.

8.2 Spatiotemporal interpolation

Spatial interpolation methods have been well developed to estimate values at unknown locations based upon values that are spatially sampled in GIS. These methods assume a stronger correlation among points that are closer than those farther apart. They are characterized as either deterministic or stochastic depending on whether statistical properties are utilized. Deterministic interpolation methods determine an unknown value using mathematical functions with predefined parameters such as distances in IDW (Robichaud and Ménard, 2014; Shepard, 1968) and areas or volumes in shape function (SF)-based methods (Li and Revesz, 2004; Zienkiewics and Taylor, 2000). Many previous studies have used deterministic interpolation methods, such as are radial basis functions (Franke and Schaback, 1998), spline (de Boor, 2001), natural neighbor (Sibson, 1981), and trend surfaces (Zurflueh, 1967). Stochastic interpolation methods such as kriging (Krige, 1966) investigate the spatial autocorrelation and give estimates of model errors. Stochastic interpolation methods have been used to handle areal interpolation uncertainty (Geddes et al., 2013), model-data fusion (sometimes called *analysis*) (Blond and Vautard, 2004; Pagowski et al., 2010), and optimal interpolation (Robichaud et al., 2016).

Although spatial interpolation methods have been widely adopted in various GIS applications, many critical problems remain unsolved. One of them is that traditional spatial interpolation methods tend to treat space and time separately when interpolation needs to be conducted in a continuous space-time domain. The primary strategy identified from the literature is to reduce spatiotemporal interpolation problems to a sequence of snapshots of spatial interpolations (Liao et al., 2006). In order to interpolate at an unsampled time instance, temporal interpolation can then be conducted based on the spatial interpolation results at each location (Borak and Jasinski, 2009).

Integrating space and time simultaneously is shown to yield better interpolation results than treating them separately for certain typical GIS applications (Li et al., 2012). Unfortunately, there are relatively fewer models for spatiotemporal interpolation compared with spatial interpolation, especially in the application of air pollution over a large geographic area. The first exception is a study that investigated the kriging-based spatiotemporal interpolation approaches for daily mean PM_{10} concentrations (Gräler et al., 2013). The methods used included separate daily variogram estimates, temporally evolving variograms, the metric model, the separable covariance model, and the product-sum model, and are combined with multiple linear regression. These methods were applied to daily mean rural background PM_{10} concentrations across Europe for the year 2005. The second exception is IDW-based spatiotemporal interpolation methods in Li et al. (2014) that extended the traditional spatial IDW to interpolate daily $PM_{2.5}$ concentrations at the centroids of census block groups and counties across the contiguous United States for the year 2009. In this study, various IDW-based spatiotemporal interpolation methods with different parameter configurations were evaluated and compared by cross-validation. Parallel programming techniques and an advanced data structure, named *k-d* tree, were adapted to address the computational challenges. The third exception is an SF-based spatiotemporal interpolation method in Li et al. (2016) that extends the popular SFs in engineering applications such as finite element algorithms. This study compared the SF-based method with the IDW-based methods in Li et al. (2014) using the same $PM_{2.5}$ data and combined interpolation results with population data to estimate the population exposure to $PM_{2.5}$ in the contiguous United States. Furthermore, Zou et al. (2011) reviewed some air pollution exposure assessment methods utilized in epidemiological studies and the use of GIS for resolving problems with spatiotemporal attributes. In summary, in the era of big data, there is a need to develop and evaluate spatiotemporal methods that produce good interpolation results with computational efficiency to handle the increasing amount of air pollution data over a large geographic area.

Since the comparison result of the SF- and IDW-based spatiotemporal interpolation methods using the $PM_{2.5}$ data shows that a SF-based method outperforms the IDW-based method even in the IDW-based method's best scenarios (Li et al., 2016), we investigate more SF-based spatiotemporal interpolation methods and compare them in this chapter. In the rest of this section, we introduce two SF-based spatiotemporal interpolation approaches and show how we implement them.

8.2.1 Extension approach for SF-based spatiotemporal interpolation

In order to integrate space and time simultaneously, we developed an "extension approach" to conduct the SF-based spatiotemporal interpolation. This approach treats time as another dimension in space and therefore, extending the spatiotemporal interpolation problem into a higher-dimensional spatial interpolation problem (Li, 2003).

Using the extension approach of SF-based interpolation method, we treat time as the imaginary third dimension z in space. Therefore, SF-based spatiotemporal interpolation method for two-dimensional (2D) space and one-dimensional (1D) time problems can be summarized as:

$$w(x,y,t) = N_1(x,y,t)w_1 + N_2(x,y,t)w_2 + N_3(x,y,t)w_3 + N_4(x,y,t)w_4, \tag{8.1}$$

where N_1, N_2, N_3, and N_4 are the following SFs.

$$N_1(x,y,t) = \frac{V_1}{V}, \quad N_2(x,y,t) = \frac{V_2}{V}, \tag{8.2}$$

$$N_3(x,y,t) = \frac{V_3}{V}, \quad N_4(x,y,t) = \frac{V_4}{V}.$$

V_1, V_2, V_3, and V_4 are the volumes of the four subtetrahedra $ww_2w_3w_4$, $w_1ww_3w_4$, $w_1w_2ww_4$, and $w_1w_2w_3w$, respectively; and V is the volume of the outside tetrahedron $w_1w_2w_3w_4$ as shown in Fig. 8.1.

Alternatively, the SFs can be computed as:

$$N_1(x,y,t) = \frac{a_1 + b_1x + c_1y + d_1t}{6V}, \tag{8.3}$$

$$N_2(x,y,t) = \frac{a_2 + b_2x + c_2y + d_2t}{6V},$$

$$N_3(x,y,t) = \frac{a_3 + b_3x + c_3y + d_3t}{6V},$$

$$N_4(x,y,t) = \frac{a_4 + b_4x + c_4y + d_4t}{6V}.$$

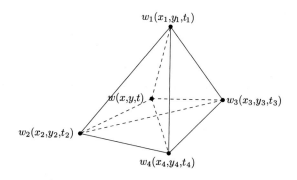

FIG. 8.1

A tetrahedral element. Computing 3D SFs by tetrahedral volume divisions. w_1, w_2, w_3, and w_4 are measured values, while the value w at the location (x, y, t) is unknown and needs to be interpolated.

The volume $\mathcal{V}$ can be computed using the corner coordinates (x_i, y_i, t_i) $(i = 1, 2, 3, 4)$ in the determinant of a matrix as:

$$\mathcal{V} = \frac{1}{6} \det \begin{bmatrix} 1 & x_1 & y_1 & t_1 \\ 1 & x_2 & y_2 & t_2 \\ 1 & x_3 & y_3 & t_3 \\ 1 & x_4 & y_4 & t_4 \end{bmatrix}. \tag{8.4}$$

By expanding the other relevant determinants into their cofactors, we have

$$a_1 = \det \begin{bmatrix} x_2 & y_2 & t_2 \\ x_3 & y_3 & t_3 \\ x_4 & y_4 & t_4 \end{bmatrix}$$

$$b_1 = -\det \begin{bmatrix} 1 & y_2 & t_2 \\ 1 & y_3 & t_3 \\ 1 & y_4 & t_4 \end{bmatrix}$$

$$c_1 = -\det \begin{bmatrix} x_2 & 1 & t_2 \\ x_3 & 1 & t_3 \\ x_4 & 1 & t_4 \end{bmatrix}$$

$$d_1 = -\det \begin{bmatrix} x_2 & y_2 & 1 \\ x_3 & y_3 & 1 \\ x_4 & y_4 & 1 \end{bmatrix}$$

with the other constants defined by cyclic interchange of the subscripts in the order of 4, 1, 2, 3 (Zienkiewics and Taylor, 2000).

The extension method is based on three-dimensional (3D) Delaunay triangulation. During implementation, we first generate a 3D Delaunay triangulation from a set of input points with (x, y, t) coordinates. The triangulation process generates an

Table 8.1 Algorithm using the SF-based extension method: *interpolate(A, B, dtMesh)*.

Input:	A—a matrix representing the **known** data in the form $\begin{bmatrix} t & x & y & w \end{bmatrix}$, where t is the time, x and y are the location coordinates, and w is the known measured value;
	B—a matrix of **query** points in the form $\begin{bmatrix} t & x & y \end{bmatrix}$, where t is the time, x and y are the location coordinates;
	dtMesh (optional)—the triangulation used to perform the interpolation;
Output:	*res*—a matrix containing the query point coordinates and the interpolated value at each query point in the form $\begin{bmatrix} t & x & y & w \end{bmatrix}$.

1. Check format of input arguments;
2. // if the *dtMesh* is not provided, calculate one
3. **if** *dtMesh* is null
4. *dtMesh* = calculate *Delaunay triangulation* using t, x, y of A;
5. **end**
6. // *pl* are the coordinates of the simplex containing the query point
7. // *bc* are the barycentric coordinates
8. [*pl*, *bc*] = *dtMesh.pointLocation*(B);
9. *triV als* = extract w from A at each corner identified in *pl*;
10. //calculate the interpolated values at each query point
11. $Vq = bc \cdot triV\,als$;
12. Set *Vq* to *NaN* if no containing tetrahedron was found;
13. *res* = append *Vq* to *B*.

index for the containing tetrahedron in the Delaunay mesh, as well as barycentric coordinates associated with the query point within its containing tetrahedron. Barycentric coordinates are equivalent to the N_1, N_2, N_3, and N_4 coefficients in equations (Figs. 8.1 and 8.2).

The interpolated value at the query point is then calculated as the dot product of the value at each corner of the containing tetrahedron with the barycentric coordinates. Error handling is incorporated to handle an edge case where query points are not located within the convex hull of the Delaunay mesh. The algorithm of the extension method is presented in Table 8.1.

8.2.2 Reduction approach for SF-based spatiotemporal interpolation

Alternatively, we developed a "reduction approach." This approach reduces the spatiotemporal interpolation problem to a regular spatial interpolation case using two steps. First, we interpolate (using any 1D interpolation in time) the value of interest overtime at each sample point. Second, by substituting the desired time instant into some regular 2D spatial interpolation functions, we can get spatiotemporal interpolation results (Li, 2003).

Assume the value at node i at time t_1 is w_{i1}, and at time t_2 the value is w_{i2}. The value at the node i at any time between t_1 and t_2 can be approximated using a 1D time SF in the following way as:

$$w_i(t) = \frac{t_2 - t}{t_2 - t_1} w_{i1} + \frac{t - t_1}{t_2 - t_1} w_{i2}. \tag{8.5}$$

The reduction method is based on the 2D Delaunay triangulation. We first generate a 2D Delaunay triangulation using a set of monitoring site locations with (x, y) coordinates. Then, we need to find the containing triangle of a query point. The SF-based interpolation result at a query point (x, y) located inside the triangle can be obtained by using the measurement values w_1, w_2, and w_3 at the three corner vertices as follows (Li and Revesz, 2004):

$$w(x, y) = N_1(x, y)w_1 + N_2(x, y)w_2 + N_3(x, y)w_3 \tag{8.6}$$

where N_1, N_2, and N_3 are the following linear SFs (Li and Revesz, 2002):

$$N_1(x, y) = \frac{A_1}{A}, \quad N_2(x, y) = \frac{A_2}{A}, \quad N_3(x, y) = \frac{A_3}{A}. \tag{8.7}$$

A_1, A_2, and A_3 are the areas of the three subtriangles ww_2w_3, w_1ww_3, and w_1w_2w, respectively; and A is the area of the outside triangle $w_1w_2w_3$.

Using Eqs. (8.5), (8.6), the interpolation function for any point constraint to a triangular element with corner vertices at any time between t_1 and t_2 can be expressed as follows (Li and Revesz, 2002):

$$
\begin{aligned}
w(x, y, t) = {}& N_1(x, y)\left[\frac{t_2 - t}{t_2 - t_1} w_{11} + \frac{t - t_1}{t_2 - t_1} w_{12}\right] \\
& + N_2(x, y)\left[\frac{t_2 - t}{t_2 - t_1} w_{21} + \frac{t - t_1}{t_2 - t_1} w_{22}\right] \\
& + N_3(x, y)\left[\frac{t_2 - t}{t_2 - t_1} w_{31} + \frac{t - t_1}{t_2 - t_1} w_{32}\right] \\
= {}& \frac{t_2 - t}{t_2 - t_1}[N_1(x, y)w_{11} + N_2(x, y)w_{21} + N_3(x, y)w_{31}] \\
& + \frac{t - t_1}{t_2 - t_1}[N_1(x, y)w_{12} + N_2(x, y)w_{22} + N_3(x, y)w_{32}].
\end{aligned}
\tag{8.8}
$$

Since only 2D coordinate data are used in the reduction method, in order to deal with data at missing times, a linear time interpolation derived from Eq. (8.5) is used in Eq. (8.8).

Similar to the extension method, query points must be within the convex hull of the triangulation when using the reduction method. Since the triangulation in 2D geographic space only addresses the location coordinates, an additional restriction applies to the reduction method in the time domain. A query point must request a time that is between known measurements at each corner of the containing 2D triangle.

Table 8.2 Algorithm using the SF-based reduction method: *interpolate2D(A, B, Da, T)*.

Input:	A—a matrix representing the **known** data in the form $\begin{bmatrix} t & x & y & w \end{bmatrix}$, where t is the time, x and y are the location coordinates, and w is the known measured value;
	B—a matrix of **query** points in the form $\begin{bmatrix} t & x & y \end{bmatrix}$, where t is the time, x and y are the location coordinates;
	dtMesh (optional)—the triangulation used to perform the interpolation;
	Da (optional)—a matrix containing the time-interpolated values for all locations in A;
	T (optional)—a vector containing all times in A;
Output:	*res*—a matrix containing the query point coordinates and the interpolated value at each query point in the form $\begin{bmatrix} t & x & y & w \end{bmatrix}$
	dtMesh—the triangulation used to perform the interpolation;
	Da—a matrix containing the time-interpolated values for all locations in A;
	T—a vector containing all distinct times in A.

1. Check format of input arguments;
2. // if the *dtMesh, Da, T* are not provided, calculate them
3. **if** *dtMesh, Da*, or T is null
4. *dtMesh* = calculate *Delaunay triangulation* using t, x, y of A;
5. *Da* = calculate time-interpolated value at the Cartesian product
6. of all locations in A with all times in T
7. T = collect unique times in A;
8. **end**
9. // *pl* are the coordinates of the simplex containing the query point
10. // *bc* are the barycentric coordinates
11. [*pl, bc*] = *dtMesh.pointLocation*(B);
12. *triVals* = extract w from A at each corner identified in *pl*;
13. //calculate the interpolated values at each query point
14. $Vq = bc \cdot triVals$;
15. Set Vq to *NaN* if no containing tetrahedron was found;
16. *res* = append Vq to B.

If this is not the case, then values at the containing triangle corners can be extrapolated in the time dimension (prior to interpolating the value of the query point in the 2D coordinate dimensions using SFs). The algorithm of the reduction method is presented in Table 8.2.

8.3 Experimental data and cross-validation for evaluating spatiotemporal interpolation methods

8.3.1 Experimental data

The experimental data used to evaluate and compare spatiotemporal interpolation methods were air pollution data obtained from US Environmental

Protection Agency (EPA). In order to directly compare SF-based spatiotemporal interpolation methods with previous work on IDW-based methods (Li et al., 2014, 2016), we used the same experimental data as the previous research that consists of three data sets. The first data set contains 146,125 $PM_{2.5}$ measurements from the year 2009. $PM_{2.5}$ refers to fine particles with a mean aerodynamic diameter less than or equal to 2.5 μm. The second data set contains centroid coordinates of 3109 counties in the contiguous United States. The third data set contains the centroid coordinates of 207,630 census block groups in the contiguous United States.

8.3.2 Cross-validation

Both leave-one-out cross-validation (LOOCV) and k-fold cross-validation were used in this study. Error statistics were calculated using each cross-validation methodology for both extension and reduction methods. In order to explore the performances of algorithms, we calculated six error statistics: mean absolute error (MAE), mean squared error (MSE), root mean squared error (RMSE), mean absolute relative error (MARE), mean squared relative error (MSRE), and root mean squared relative error (RMSRE). They are defined in formula (8.9), where N is the number of observations, I_i is the interpolated value, and O_i is the original measurement value.

$$\text{MAE} = \frac{\sum_{i=1}^{N} |I_i - O_i|}{N}, \tag{8.9}$$

$$\text{MSE} = \frac{\sum_{i=1}^{N} (I_i - O_i)^2}{N},$$

$$\text{RMSE} = \sqrt{\frac{\sum_{i=1}^{N} (I_i - O_i)^2}{N}},$$

$$\text{MARE} = \frac{\sum_{i=1}^{N} \frac{|I_i - O_i|}{O_i}}{N},$$

$$\text{MSRE} = \frac{\sum_{i=1}^{N} \frac{(I_i - O_i)^2}{O_i}}{N},$$

$$\text{RMSRE} = \sqrt{\frac{\sum_{i=1}^{N} \frac{(I_i - O_i)^2}{O_i}}{N}}.$$

Earlier works have shown that the k-fold cross-validation method would overestimate the error due to the small number of points available in the sample data sets (Li et al., 2014). Randomly excluding points that would otherwise be close (in the time and space dimensions) to the query point greatly affect the performance outcome of interpolation algorithms using this data set. With this limitation in mind, we decided to explore a method of selecting locations such that they are evenly distributed across the possible locations in the data set.

The objective is to create a stratification of data (the folds) such that geographic regions are equally represented by the data in each fold. Individual locations should not be present in more than onefold, but each region should be represented in all folds. A measure of spatial dispersion is used to maximize the spread of each fold and make sure that the locations are not too clustered in any fold. If the points are too clustered in a fold, we run the risk of points in the test folds not being contained within the convex hull of the triangular mesh generated by the sample data folds.

A measure of spatial dispersion, the average nearest neighbor (ANN) distance, was utilized as described in Clark and Evans (1954). The ANN is a measure of the spacing of individuals in a population. It is calculated by finding the distance from each individual to its nearest neighbor and then calculating the mean distance value. A greater ANN distance indicates that a population is better dispersed.

In order to get a baseline measurement of spatial dispersion, the following method was used. In all, 10-folds were generated using the $PM_{2.5}$ data set by partitioning the data based on the random locations. The ANN distance in each fold was then calculated treating the latitude and longitude measurements as coordinates and calculating the Euclidian distance between locations. The average of the 10 ANN distances within each fold was then calculated. Several experiments where tried with random partitions of the data. In each case, the ANN score was approximately 1.2.

To create a stratification of the locations in the $PM_{2.5}$ data with a better spatial dispersion, the following method was used:

1. Find the number of unique locations in the data set.
2. Set k to be the floor of the number of unique locations divided by the number of folds.
3. Use k-means++ as described by Arthur and Vassilvitskii (2007) to find k centroids.
4. Since each centroid could have a different number of points assigned to it with k-means++, use the Hungarian linear assignment algorithm as described by Munkres (1957) to assign k points to each centroid by minimizing the distance of each point to each centroid.
5. Assign each of k points associated with each centroid at random to each of the k-folds.
6. Since the floor function is used in step 2, a centroid could have $k + 1$ points associated with it. Assign any leftover points to random folds.

This method was run several times on the $PM_{2.5}$ data. Each trial had an ANN distance of approximately 1.6, which was much better dispersed than a random partition of locations (which had an ANN of 1.2). The stratification created with this method is referred in this chapter as "location aware" k-fold cross-validation as opposed to the naïve "random" k-fold cross-validation.

We compared the model accuracy and computational performance using different cross-validation and stratification methods, which will be presented in Section 8.5.1.

8.4 Real-time air pollution visualization

In order to visualize the real-time pollution distribution, an interpolation method needs to be used to create good approximations of the data at unmeasured times and locations. Since a real-time approach is desired, the SF-based reduction method is chosen because it provides great computational performance with reasonable error performance as compared to other methods as shown in Section 8.5.1.

8.4.1 Real-time air pollution data source

The US EPA and several other agencies developed the AirNow system to provide the public with easy access to nationwide air quality information. This program collects pollution measurement data from thousands of monitoring sites across the United States. A summarized index of pollution data known as the Air Quality Index (AQI) is made available to the public through the AirNow program. The AirNow system provides capabilities for visualizing the spatial distribution of real-time air pollution concentrations and constructing temporal profile of air pollution concentrations at any measurement location. Currently, the AirNow system provides pollutant data at measurement sites and created a web service to visualize AQI for certain zip codes. However, this system only covers a discrete set of locations. As a result, many rural areas cannot be queried in the system.

Customizing and presenting AirNow data to the public in more effective ways could help raise awareness of the environmental pollution problem and its public health effects. In this study, we develop a real-time air pollution system to visualize the spatial distribution and construct temporal profile of air pollution at any location. This allows air pollution to be estimated in the continuous space-time domain so that it can be correlated to more fine-grained demographic measurements for public health applications.

8.4.2 Visualization for existing data

The first objective of the visualization approach explored in this chapter is to convey information derived from existing pollution data measured at monitoring sites in an intuitive way. For example, an existing set of $PM_{2.5}$ data measurements can be shown on a map by rendering a marker at the geographic location where the measurement is taken. This allows a researcher to see spatial patterns in measurements such as a cluster of measurement sites and their proximity to major cities or natural terrain. Several dimensions of information need to be conveyed, such as locations of the monitoring sites, air pollution values measured at different times, missing values at certain times, and the changing trend at each monitoring site. We use several visualization techniques to achieve the goal.

First, the color of the map markers is adjusted to show different classifications of the data. We color each map marker to show whether it is measured or interpolated in the time dimension. Using the SF-based reduction method, we first start with a set of known measurements from monitoring sites. If a monitoring site contains a gap in its data at a particular time, we do temporal interpolation using Eq. (8.5) to resolve the missing information. A real measurement is colored blue and an interpolated measurement is colored red.

Second, we dynamically change the size of the map marker based on the magnitude of the measurement. Using a circle as a map marker, the radius of the circle can be adjusted to be larger or smaller based on the relative size of the measurement. This provides an intuitive indication of patterns and outliers in large data sets without overwhelming a user with raw data and numbers.

Third, *trends* are integrated into the visualization. Using a map as the main method of visualization allows a researcher to explore patterns that are evident across the 2D geographic area. However, data that are overlaid on a map intuitively represent only one time instance. To allow a researcher to jump between analysis in both the space and time dimensions, *trends* are integrated into the visualization. A typical pattern followed by a researcher can be displayed on a map at a particular point in time. After noticing an outlier at a particular measurement site in the geographic dimension, the researcher could click on the map marker to bring up the information tag. The researcher could then click on the hyperlink in the information tag to navigate to historical measurements at that particular site. This would allow the researcher to see if the outlier in the geographic dimension also represents an outlier in the time dimension.

The result of these visualization techniques will be presented in Section 8.5.2.1.

8.4.3 Visualization for interpolated data

The second objective is to visualize the interpolated values across the space. In order to achieve this objective, we use heatmaps to represent the pollution intensity. Most heatmaps employ a density-based approach. In other words, the geographic areas where there are more measurement sites always show a more intense color even if the actual measured values at the location are relatively low. Conversely, a relatively high measurement might show as a less intense color if it was the only measurement in the area. In the application of air pollution, we need a value-based heatmap that ignores the density of measurements in a given area.

To address this issue, we developed an approach to use the Google Maps density-based heatmap to simulate a value-based heatmap. Rather than adding the actual measurement data to the heatmap rendering directly, data were added to the heatmap in a regular grid with equal spacing. In this way, the data points are at a constant density and will render across the entire viewport in a uniform manner. The regularly spaced points can then be weighted with a normalized value representing the pollution data measurement at that location. We use the 2D SF-based interpolation method in Eq. (8.6) to "fill" gaps between known measurement sites by providing a good approximation of the data at times and locations where a measurement is not known.

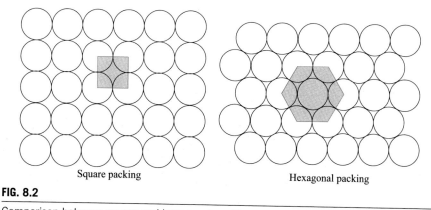

Square packing   Hexagonal packing

FIG. 8.2

Comparison between square and hexagonal packing.

In order to create the smoothest possible visualization from circular heatmap elements, circles should overlap in the most efficient way possible. In geometry, "circle packing" is the study of the arrangement of circles (of equal or varying sizes) on a given surface such that no overlapping occurs and such that all circles touch one another (Conway and Sloane, 1999). The packing density of an arrangement is the proportion of the surface covered by the circles.

If using the square lattice, the packing density of nonoverlapping circles is 78.54%. In theory, the largest proportion of a surface that can be covered by non-overlapping circles is about 90.7%. This coverage is obtained using an arrangement of circles with centers at the points of a hexagonal lattice. The first claim of a proof of this was made by Axel Thue in 1892. This proof was considered somewhat incomplete, and it is generally believed that the first complete proof was produced in 1940 by Làszlò Fejes Tòth. A history of this specialized field of geometry is covered by Conway and Sloane (1999). To create a hexagonal lattice, we adjusted the location of each query point so that every other row of query points is offset on the x-axis by half the distance between circle centers as shown in Fig. 8.2.

The result of the process and visualization effect of heatmaps and hexagonal packing will be presented in Section 8.5.2.2.

8.4.4 Web application

A third objective is to design and develop a web application to support real-time visualization. In order to monitor and visualize the spatial variation of various pollution measurements in real time, the application uses a three-tiered architecture as shown in Fig. 8.3. The AirNow government website service provides hourly updates of pollution measurement data from sites across North America. The middle-tier server periodically downloads pollution data from the AirNow service and stores it in the database backend. The server also provides the pollution data to the front-end client and answers point queries by performing SF-based reduction interpolation in real time. The client application incorporates the visualization techniques presented in Sections 8.4.2 and 8.4.3.

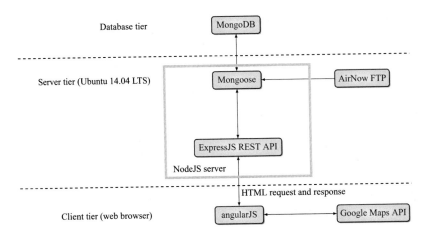

FIG. 8.3

The client, server, and database tiers and technologies used to implement the web application.

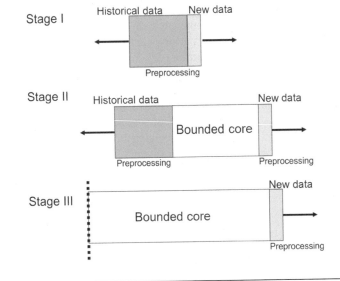

FIG. 8.4

Core of data exists on the timeline. The *pink area* is new data that were just downloaded representing current measurements. The *blue area* is historical data available from the AirNow server.

    To realize the real-time support, a "bounded core" of data is maintained in the backend database. For a newly deployed server, there are two tasks: to download the available historical data from the AirNow and to fetch the new data for the most recent hour. Fig. 8.4 shows a graphical representation of the pollution data as it is downloaded and processed on the server. At the beginning, the core will be set as empty. For historical data (the blue chunk in Fig. 8.4), in order to avoid overloading the server to process all the historical data at once, only data from the most recent 5 h away from the bounded core are processed. The process continues until all the available historical data set from the AirNow server are processed and stored in the bounded core. For the data in the most recent 1 h (the pink chunk in Fig. 8.4), this process will continue run as long as our server functions.

The result of the web application will be presented in Section 8.5.2.3.

8.5 **Result**
8.5.1 **Model evaluation**

Fig. 8.5 shows the results of plotting error values across time scales using the extension method and the random 10-fold stratification as described in Section 8.3.2. The generated time scales are plotted on the x-axis and the error is plotted on the y-axis. Fig. 8.6 shows the results using the location aware stratification. The same experiment was run using the reduction method. It was expected that the error curve would be flat across the entire range of time scales since the reduction method has been shown to be time-scale invariant. This was indeed the case, so the graphs for the reduction method are not presented here.

Table 8.3 presents the value of each error statistic using the random 10-fold cross-validation on the reduction method with any time scale and the extension method with the best and worst time scales. Table 8.4 presents the value of each error statistic using the location aware 10-fold cross-validation on the same set of methods. It is interesting to note that with the MSE and RMSE metrics, the reduction method performs better than the worst case of the extension method, but not as good as the best case of the extension method (where a good time scale is selected). Using the location aware stratification, the relative and percentage error metrics with the reduction method actually seem to perform better than the extension method even with the best choice of time scale.

In addition, we also compared our results with a number of previous works with interpolation algorithms (Li et al., 2014). It is interesting to note that with a good choice of time scale, the extension method has the lowest MARE of any previous works, where a time scale of 0.2705 results in a MARE of 0.35878 (Table 8.5).

When calculating large data sets or selecting an algorithm for real-time visualizations, the efficiency of the algorithm is important. The performance of the SF-based extension and reduction methods are compared to the results of the IDW algorithm in Li et al. (2014). The performance was compared by calculating the interpolated values of $PM_{2.5}$ measurements at the centroids of the census block groups in the contiguous United States for every day in 2009. Since there are 365 days and 207,630 census block group centroid, this results in the need for 74,746,800 interpolated query points. The results of performance tests in CPU time consumption are shown in Table 8.6, as well as the reported number of threads and processor type running the tests. It is obvious from Table 8.6 that

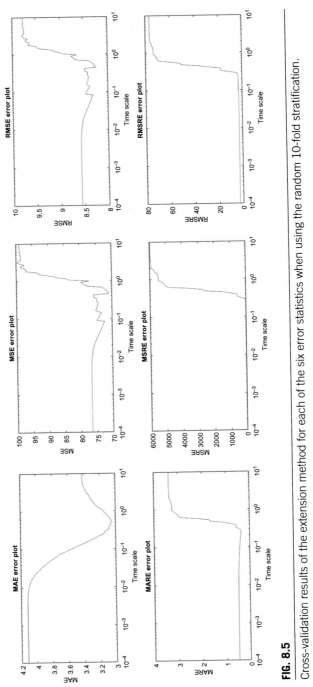

FIG. 8.5

Cross-validation results of the extension method for each of the six error statistics when using the random 10-fold stratification.

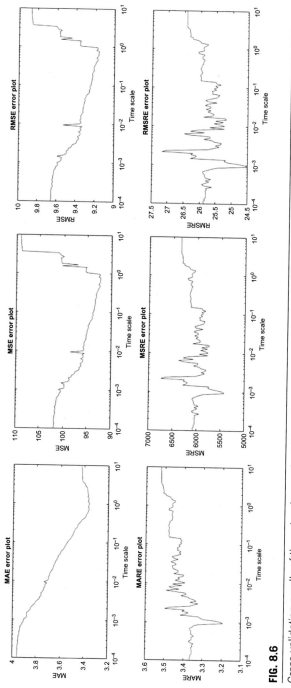

FIG. 8.6

Cross-validation results of the extension method for each of the six error statistics when using the location aware 10-fold stratification.

Table 8.3 Cross-validation results using the random 10-fold stratification.

Error statistics	Reduction method (time-scale invariant)	Extension method			
		Error best case	Time-scale best case	Error worst case	Time-scale worst case
MAE	4.1271	3.0525	0.54159	4.01	<0.02
MSE	76.6	71.019	0.43976	98	>2.5
RMSE	8.5662	8.2451	0.43976	9.71	>2.5
MARE	0.47904	0.36752	0.2705	3.37	>2.5
MSRE	1.9645	35.065	0.036123	5944	>2.5
RMSRE	1.3481	3.0499	0.036123	76.5	>2.5

Table 8.4 Cross-validation results using the location aware 10-fold stratification.

Error statistics	Reduction method (time-scale invariant)	Extension method			
		Error best case	Time-scale best case	Error worst case	Time-scale worst case
MAE	3.78	3.3636	0.75774	3.9602	<10^4
MSE	98.1	92.254	0.75774	109.35	>5
RMSE	9.48	9.1593	0.75774	9.8896	>5
MARE	2.98	3.3154	0.001408	3.5246	>5
MSRE	4556.7	5623.8	0.0089497	6384.8	>5
RMSRE	22.01	24.939	0.0089497	26.502	>5

both SF-based extension and reduction algorithms in this chapter are much more efficient in computational time than the IDW-based extension algorithm in Li et al. (2014).

8.5.2 Real-time air pollution visualization result

8.5.2.1 Visualization result for existing data

This section presents the visualizations for the existing data measured at monitoring sites. Fig. 8.7A shows map markers representing $PM_{2.5}$ measurement data overlaid on a Google Maps application. In order to provide the ability to drill down and see the numbers behind the data, the circles can be clicked to pop up an information tag that shows the source of the data, the magnitude of the measurement, and the measurement units. Fig. 8.7B shows an information tag for a map marker. Given the importance of the Delaunay triangulation to the SF-based methods of interpolation, the ability to see the resultant triangulation for a given data set could help to provide

Table 8.5 MARE and RMSE error statistics for different interpolation algorithms with various choices of time scale.

Algorithm using SF-based method	Cross-validation	Error statistic	Time-scale	Error value
Reduction	LOOCV	MARE	Invariant	0.4703
Reduction	LOOCV	RMSE	Invariant	1.387441
Extension	LOOCV	MARE	0.1086	0.4183
Extension	LOOCV	RMSE	0.1086	5.981279
Extension	LOOCV	MARE	0.2705	0.35878
Extension	LOOCV	RMSE	0.2705	6.1633
Reduction	10-fold (location aware)	MARE	Invariant	2.98
Reduction	10-fold (location aware)	RMSE	Invariant	22.01
Extension	10-fold (location aware)	MARE	0.001408	3.3154
Extension	10-fold (location aware)	RMSE	0.0089497	24.939
Reduction	10-fold (random)	MARE	Invariant	0.47904
Reduction	10-fold (random)	RMSE	Invariant	1.3481
Extension	10-fold (random)	MARE	0.2705	0.36752
Extension	10-fold (random)	RMSE	0.036123	3.0499

Table 8.6 Performance of spatiotemporal interpolation algorithms on the 2009 $PM_{2.5}$ data set using the census block group centroids.

Algorithm	Processor	Threads	CPU time (s)
SF-based extension	AMD Phenom X6	1	638
SF-based extension	Intel I5-6600K	1	355
SF-based extension	Intel I5-6600K	4	113.7
SF-based reduction	AMD Phenom X6	1	182
SF-based reduction	Intel I5-6600K	1	78.5
SF-based reduction	Intel I5-6600K	4	25.3
IDW (Li et al., 2014)	Intel I7-3630 QM	8	2756

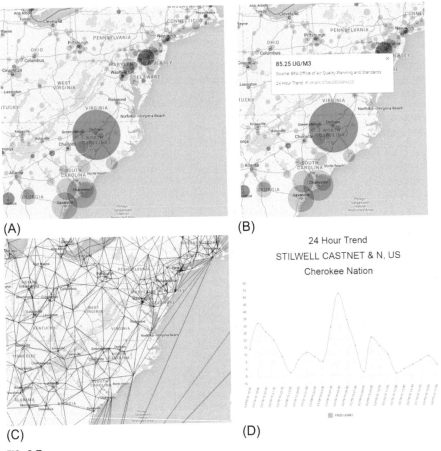

(A) (B)

(C) (D)

FIG. 8.7

(A) Overlay of map markers representing PM$_{2.5}$ data that were collected at measurement sites along the east coast of the United States on March 17, 2016 at 8:00 GMT. (B) Information tag displayed after clicking a map marker. (C) Triangulation generated for PM$_{2.5}$ data that were collected at measurement sites along the east coast of the United States on March 17, 2016 at 8:00 GMT. (D) A trend showing PM$_{2.5}$ measurements for a 24-h period at a given station.

further insight into a data set. Therefore, the visualization approach incorporates the ability to see the triangulation that results from using the reduction method. Fig. 8.7C shows a triangulation overlaid on a section of a map. Fig. 8.7D shows a trend showing PM$_{2.5}$ measurements for a 24-h period.

8.5.2.2 Visualization result for interpolated data

To visualize the interpolated pollutant values, we developed value-based heatmap. Fig. 8.8A shows a regular grid of heatmap points overlaid on a map. Higher

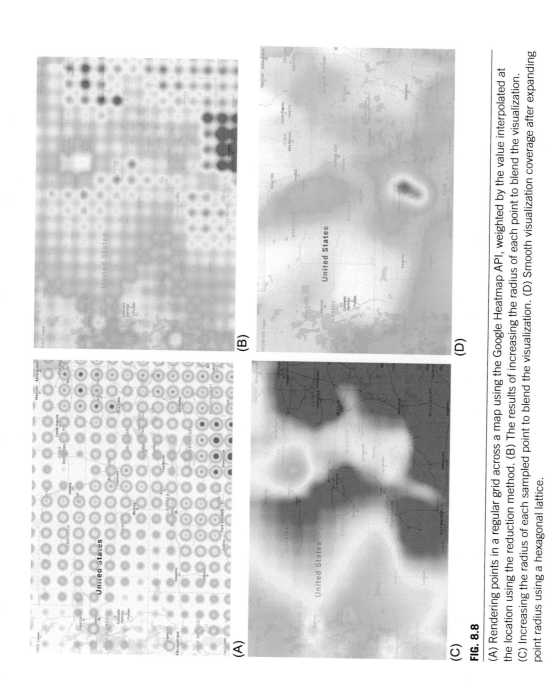

FIG. 8.8

(A) Rendering points in a regular grid across a map using the Google Heatmap API, weighted by the value interpolated at the location using the reduction method. (B) The results of increasing the radius of each point to blend the visualization. (C) Increasing the radius of each sampled point to blend the visualization. (D) Smooth visualization coverage after expanding point radius using a hexagonal lattice.

magnitude measurement values are rendered as red while lower magnitudes are rendered as green or transparent. To create the effect of a continuous heatmap visualization, Fig. 8.8B shows the results of increasing the radius of each point to blend the visualization. The visualization in Fig. 8.8B is connected, but artifacts of each individual point are still recognizable. The size of each point can be further increased to reduce this effect. Fig. 8.8C shows the radius increased further to complete the blending effect, where the individual points are significantly overlapped to reduce the artifacts present in Fig. 8.8A and B. However, the overlap of the points has the effect of magnifying the magnitude of the color gradient. To resolve this problem, as described in Section 8.4.3, we used hexagonal lattice arrangement to place the query points. The size of each heatmap point is increased to allow minimal overlap and create a smooth, accurate visualization. Fig. 8.8D shows a smooth visualization with a query points laid out in a hexagonal lattice and heatmap radius increased to create a smooth visualization.

8.5.2.3 Web application result

The developed web application supports real-time interpolation and visualization of six major pollutants (O_3, $PM_{2.5}$, PM_{10}, CO, SO_2, and NO_2) as shown in Table 8.7.

Fig. 8.9 shows the user interfacer of the real-time web application for multipollutant visualization. The left menu bar can be used to customize the map overlay. Collapsible menu options allow the user to select a pollution parameter. A legend is created to show the gradient colors which are associated with the pollution intensity on the map.

Fig. 8.10 shows the main functionalities of the web application. Users can select a date and time to explore the air quality at a particular time. The application also shows the raw data based on which interpolation was performed. Existing measurement data at monitor sites as well as interpolated data can be displayed by clicking corresponding checkboxes. The triangular mesh used to interpolate data can be also showed. When clicking markers on map, temporal trends for each air pollutant are available for viewing.

Table 8.7 The six parameters used to calculate the AQI in the United States.

Parameters	Parameter name	Parameter unites
O_3	Ozone	PPB
$PM_{2.5}{}^{a}$	$PM_{2.5}$ mass	$\mu g/m^3$
$PM_{10}{}^{b}$	PM_{10} mass	$\mu g/m^3$
CO	Carbon monoxide	PPM
SO_2	Sulfur dioxide	PPB
NO_2	Nitrogen dioxide	PPB

[a] $PM_{2.5}$ are particles with a mean aerodynamic diameter less than or equal to 2.5 μm.
[b] PM_{10} are particles with a mean aerodynamic diameter between 2.5 and 10 μm.

FIG. 8.9

The main web interface.

8.6 **Discussion and conclusion**

This chapter investigated SF-based extension and reduction spatiotemporal interpolation methods and compared them with previous IDW-based methods. We evaluated model accuracy as well as computation performance of these models. It was discovered that the choice of time scale affects the performance of both the IDW and the SF-based extension method. The SF-based extension method appears to have the best performance of the three algorithms when there is a good choice of time scale. The SF-based reduction method has the benefit of having good performance while being time-scale invariant and an order of magnitude more efficient in terms of processing time. In addition, we also visualized real-time air pollution data using the SF-based reduction method across the contiguous United States. Visualization creates a natural and intuitive interface between the raw data and the brain. A web server was deployed that collects pollution data each hour and presents it to users using the developed visualization approach.

Working with large amount of air pollution data is computationally intensive, especially for huge amount of cross-validation experiments. We used several strategies to speed up the process. First, in order to improve the runtime, we parallelized the computation process on a cloud environment. A virtual machine (VM) was provisioned using the Microsoft Azure cloud computing service. An eight-core D-series VM was provisioned to run the LOOCV experiment. Using this setup, the LOOCV method completed with one choice of time scale in just 36 h as opposed to the 25 days that was initially calculated.

Even given this speedup, it was not practical to run a large number of trials using different time scales in order to determine a best time scale using cross-validation. It is generally cheaper to initially construct a Delaunay triangulation using a full set

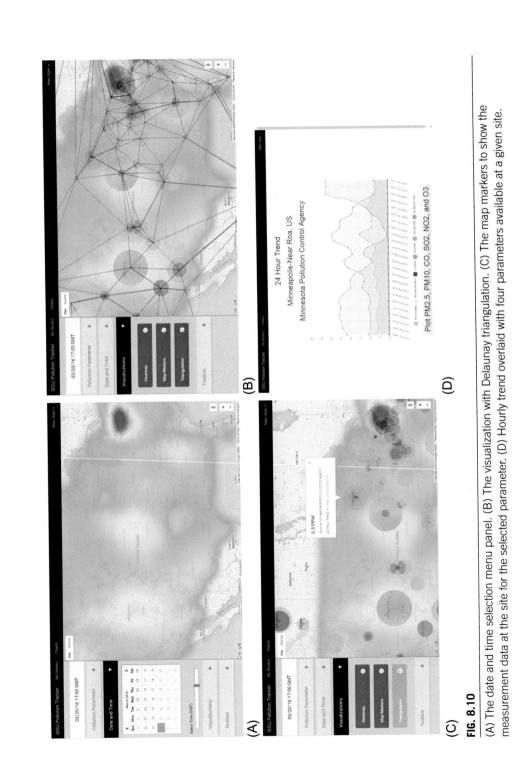

FIG. 8.10

(A) The date and time selection menu panel. (B) The visualization with Delaunay triangulation. (C) The map markers to show the measurement data at the site for the selected parameter. (D) Hourly trend overlaid with four parameters available at a given site.

of input points rather than to incrementally build it. However, for the LOOCV use case where a single point needs to be removed and added for each of the N trials, incremental construction can be much cheaper. Incremental insertion algorithms for Delaunay triangulations generally insert one point at a time by splitting the existing triangle that contains the point into three triangles, and then recursively checking the empty circle property on the new triangle and performing an edge flip if it is violated (Lawson, 1977). An alternate version of the LOOCV method was created to test the performance of the extension method using the incremental construction. Using this approach, which was single threaded in nature, the entire LOOCV method was completed with a choice of time scale in just 3.5 h. This represents a 10 times improvement from brute forcing with multiple threads.

In addition, our visualization strategy has proven to be flexible, accurate, and performant in practice. Using sample data sets, a researcher can visualize pollution data at a macrolevel across the entire country. The researcher can then pan and zoom to the state, county, city, neighborhood, or street level. In testing, the map application remains responsive across a wide sample of consumer grade and budget desktop or laptop computers. When panning or zooming, the application waits for an "idle" event to prevent flooding the system with requests to render points. When the idle event is captured, the visualization is typically rendered in less than half a second even at resolutions as high as full screen 1440p.

We have several limits in this project. Temporal gaps of hours may be accurately filed with interpolation, but longer gaps in the data set are difficult to interpolate. Likewise, some areas have few or no measurements of certain pollutants. It is hard to interpolate from stations located far away to fill in gaps in the spatial domain. Concentrations estimated from chemical diffusion models or remotely sensed data will be helpful to fill the gaps. Our future work will incorporate various approaches to better estimate air pollution concentration at multiple scales. Nevertheless, the proposed method in this study allows an efficient computation of vast amounts of data. The 2D Delaunay triangulation together with the reduction approach provides fast spatiotemporal visualization.

To summarize, this chapter investigates and compares several spatiotemporal interpolation methods with the goal to find an appropriate method that is efficient and performs well for real-time air pollution data at a large geographic scale. After being evaluated by cross-validation, we chose a suitable spatiotemporal interpolation method to visualize real-time air pollution data for the contiguous United States. The developed spatiotemporal interpolation method and visualization tools allow researchers to drill down into the vast amount of pollutant data, support policy makers to quickly respond to abnormal air pollution incidences, and provide information about pollutant exposures to the public.

Acknowledgment

This work was supported by the Office of the Vice President for Research and Economic Development at the Georgia Southern University.

References

Arthur, D., Vassilvitskii, S., 2007. *K*-means++: the advantages of careful seeding. In: SODA'07. Proceedings of the Eighteenth Annual ACM-SIAM Symposium on Discrete Algorithms, Society for Industrial and Applied Mathematics, Philadelphia, PA, pp. 1027–1035. http://dl.acm.org/citation.cfm?id=1283383.1283494.

Blond, N., Vautard, R., 2004. Three-dimensional ozone analyses and their use for short-term ozone forecasts. J. Geophys. Res. Atmos. 109 (D17). https://doi.org/10.1029/2004JD004515. D17303.

Borak, J.S., Jasinski, M.F., 2009. Effective interpolation of incomplete satellite-derived leaf-area index time series for the continental united states. Agric. Forest Meteorol. 149 (2), 320–332.

Clark, P.J., Evans, F.C., 1954. Distance to nearest neighbor as a measure of spatial relationships in populations. Ecology 35 (4), 445–453. http://www.jstor.org/stable/1931034.

Conway, J., Sloane, N.J.A., 1999. Sphere-Packings, Lattices, and Groups, third ed. Springer-Verlag, New York, NY.

de Boor, C., 2001. A Practical Guide to Splines. vol. 27 Springer, New York, NY.

Elbir, T., 2004. A GIS based decision support system for estimation, visualization and analysis of air pollution for large Turkish cities. Atmos. Environ. 38, 4509–4517. https://doi.org/10.1016/j.atmosenv.2004.05.033.

Franke, C., Schaback, R., 1998. Solving partial differential equations by collocation using radial basis functions. Appl. Math. Comput. 93, 73–82.

Geddes, A., Elston, D.A., Hodgson, M.E.A., Birnie, R.V., 2013. Stochastic model-based methods for handling uncertainty in areal interpolation. Int. J. Geograph. Inf. Sci. 27 (4), 785–803. https://doi.org/10.1080/13658816.2012.722636.

Girardot, S.P., Ryan, P.B., Smith, S.M., Davis, W.T., Hamilton, C.B., Obenour, R.A., Renfro, J.R., Tromatore, K.A., Reed, G.D., 2006. Ozone and $PM_{2.5}$ exposure and acute pulmonary health effects: a study of hikers in the great smoky mountains national park. Environ. Health Perspect. 114 (7), 1044–1052.

Gräler, B., Rehr, M., Gerharz, L., Pebesma, E., 2013. Spatio-temporal analysis and interpolation of PM_{10} measurements in Europe for 2009. In: ETC/ACM Technical Paper, pp. 1–29.

Keim, D., Qu, H., Ma, K.L., 2013. Big-data visualization. IEEE Comput. Graph. Appl. 33 (4), 20–21. https://doi.org/10.1109/MCG.2013.54.

Krige, D.G., 1966. Two dimensional weighted moving average trend surfaces for ore evaluation. J. Soc. Afr. Inst. Min. Metall. 66, 13–38.

Künzli, N., Kaiser, R., Medina, S., Studnicka, M., Chanel, O., Filliger, P., Herry, M., Horak, F.J., Puybonnieux-Texier, V., Quénel, P., Schneider, J., Seethaler, R., Vergnaud, J.C., Sommer, H., 2000. Public-health impact of outdoor and traffic-related air pollution: a European assessment. Lancet 356 (9232), 795–801. https://doi.org/10.1016/S0140-6736(00)02653-2.

Laden, F., Neas, L.M., Dockery, D.W., Schwartz, J., 2000. Association of fine particulate matter from different sources with daily mortality in six U.S. cities. Environ. Health Perspect. 108 (10), 941–947.

Lawson, C.L., 1977. Mathematical Software III; Software for C1 Surface Interpolation. In: Academic Press, New York, pp. 161–194.

Li, L., 2003. Spatiotemporal Interpolation Methods in GIS (Ph.D thesis). University of Nebraska-Lincoln, Lincoln, NE.

Li, J., Heap, A.D., 2011. A review of comparative studies of spatial interpolation methods in environmental sciences: performance and impact factors. Ecol. Inf. 6 (3–4), 228–241.

Li, L., Revesz, P., 2002. A comparison of spatio-temporal interpolation methods. In: Lecture Notes in Computer Science. Proc. of the Second International Conference on GIScience 2002, vol. 2478. Springer, pp. 145–160.

Li, L., Revesz, P., 2004. Interpolation methods for spatio-temporal geographic data. J. Comput. Environ. Urban Syst. 28 (3), 201–227.

Li, L., Zhang, X., Holt, J.B., Tian, J., Piltner, R., 2012. Estimating population exposure to fine particulate matter in the conterminous U.S. using shape function-based spatiotemporal interpolation method: a county level analysis. GSTF Int. J. Comput. 1, 24–30.

Li, L., Losser, T., Yorke, C., Piltner, R., 2014. Fast inverse distance weighting-based spatio-temporal interpolation: a web-based application of interpolating daily fine particulate matter $PM_{2.5}$ in the contiguous U.S. using parallel programming and k-d tree. Int. J. Environ. Res. Public Health 11 (9), 9101–9141.

Li, L., Zhou, X., Kalo, M., Piltner, R., 2016. Spatiotemporal interpolation methods for the application of estimating population exposure to fine particulate matter in the contiguous U.S. and a real-time web application. Int. J. Environ. Res. Public Health 13 (8), 749, 20 pp. https://doi.org/10.3390/ijerph13080749.

Liao, D., Peuquet, D.J., Duan, Y., Whitsel, E.A., Dou, J., Smith, R.L., Lin, H.M., Chen, J.C., Heiss, G., 2006. GIS approaches for the estimation of residential-level ambient PM concentrations. Environ. Health Perspect. 114 (9), 1374–1380.

Maynard, R.L., 2015. Air pollution: the last 35 years. Hum. Exp. Toxicol. 34 (12), 1253–1257. https://doi.org/10.1177/0960327115603585.   http://het.sagepub.com/content/34/12/1253. abstract.

Munkres, J., 1957. Algorithms for the assignment and transportation problems. J. Soc. Ind. Appl. Math. 5 (1), 32–38. https://doi.org/10.1137/0105003.

Pagowski, M., Grell, G.A., McKeen, S.A., Peckham, S.E., Devenyi, D., 2010. Three-dimensional variational data assimilation of ozone and fine particulate matter observations: some results using the weather research and forecasting chemistry model and grid-point statistical interpolation. Q. J. R. Meteorol. Soc. 136 (653), 2013–2024. https://doi.org/10.1002/qj.700.

Pope, C.A., Burnett, R.T., Thurston, G.D., Thun, M.J., Calle, E.E., Krewski, D., Godleski, J.J., 2004. Cardiovascular mortality and long-term exposure to particulate air pollution: epidemiological evidence of general pathophysiological pathways of disease. Circulation 109, 71–77.

Qu, H., Chan, W.Y., Xu, A., Chung, K.L., Lau, K.H., Guo, P., 2007. Visual analysis of the air pollution problem in Hong Kong. IEEE Trans. Vis. Comput. Graph. 13 (6), 1408–1415.

Robichaud, A., Ménard, R., 2014. Multi-year objective analyses of warm season ground-level ozone and $PM_{2.5}$ over North America using real-time observations and Canadian operational air quality models. Atmos. Chem. Phys. 14, 1769–1800.

Robichaud, A., Ménard, R., Zaïtseva, Y., Anselmo, D., 2016. Multi-pollutant surface objective analyses and mapping of air quality health index over North America. Air Qual. Atmos. Health, 1–17. https://doi.org/10.1007/s11869-015-0385-9.

Seaton, A., Godden, D., MacNee, W., Donaldson, K., 1995. Particulate air pollution and acute health effects. Lancet 345 (8943), 176–178.

Shepard, D., 1968. A two-dimensional interpolation function for irregularly spaced data. In: Proc. of the 23nd National Conference ACM, ACM, New York, NY, pp. 517–524.

Sibson, R., 1981. A brief description of natural neighbor interpolation. In: Barnett, V. (Ed.), Interpreting Multivariate Data. John Wiley & Sons, Inc., New York, NY, pp. 21–36.

Triche, E.W., Gent, J.F., Holford, T.R., Belanger, K., Bracken, M.B., Beckett, W.S., Naeher, L., McSharry, J.E., Leaderer, B.P., 2006. Low-level ozone exposure and respiratory symptoms in infants. Environ. Health Perspect. 114 (6), 911–916.

Zanobetti, A., Schwartz, J., 2009. The effect of fine and coarse particulate air pollution on mortality: a national analysis. Environ. Health Perspect. 117 (6), 898–903. https://doi.org/10.1289/ehp.0800108. http://www.ncbi.nlm.nih.gov/pmc/articles/PMC2702403/.

Zienkiewics, O.C., Taylor, R.L., 2000. Finite Element Method, vol. 1, The Basis. Butterworth Heinemann, London.

Zou, B., Wilson, J.G., Zhan, F.B., Zeng, Y., 2011. Air pollution exposure assessment methods utilized in epidemiological studies. J. Environ. Monit. 11, 475–490.

Zou, B., Zheng, Z., Wan, N., Qiu, Y., Wilson, J.G., 2016. An optimized spatial proximity model for fine particulate matter air pollution exposure assessment in areas of sparse monitoring. Int. J. Geograph. Inf. Sci. 30 (4), 727–747. https://doi.org/10.1080/13658816.2015.1095921.

Zurflueh, E.G., 1967. Applications of two-dimensional linear wavelength filtering. Geophysics 32, 1015–1035.

CHAPTER

Assessment methods for air pollution exposure

9

Zheng Cao

School of Geographical Sciences, Guangzhou University, Guangzhou, China

9.1 Introduction

From 1980 to 2015, world population increased at an incredible speed. According to the statistical documents from the Department of Economic and Social Affairs United Nations, nations and regions had a population increase of 0.4 billion from 1980 to 1988; whereas in 2015, the number increased to 0.8. Meanwhile, the population living in urban areas was assessed at about 39 percent of the total population in 1980. As urbanization progresses, more employment opportunities, better education, and enhanced medical services drive rural populations to urban areas. Consequently, the population living in urban areas accounted for 54 percent of the total population in 2015. Moreover, according to the World Population Prospects 2017 by the United Nations, the trend will continue to increase. Population explosion is also tied to the high consumption of fossil fuel due to an increased energy usage for air-conditioners and transportation (Dhakal, 2009; Feng et al., 2013; Chen and Chen, 2015). As a result, these activities generate many environmental problems, including air, water, and soil pollution. In recent years, fine particles with a diameter of 2.5 μm or less have attracted the most attention for their adverse health effects. Recent research estimated that the lagged PM2.5 is highly associated with the mortality increase in both developed and developing countries (Hoek et al., 2002; Kaiser et al., 2004; Kan et al., 2008; Chen et al., 2011). Therefore, it is urgent to set up and improve air pollution monitoring and health assessment systems. Improved air pollution exposure assessment methods will help determine better health protection suggestions.

Currently, three major air pollution exposure assessment methods are used: (1) proximity measures based on air pollution monitoring stations, (2) spatial analysis based on remote sensing and GIS methods, and (3) numerical simulation based on the atmospheric diffusion model. At first, air pollution exposure assessments were based on air pollution monitoring data. This method had a hypothesis that the air pollution exposure was at the same level or similar levels within a specific distance to the monitoring station. With this hypothesis, Brauer et al. (2012) investigated the global disease attribution to outdoor air pollution. The results showed that 32 percent of the global population lived in areas exceeding the WHO 1 Interim Target of 35 microg/m^3, especially in East and South Asia. Furthermore, air pollution monitoring data was also applied to the analysis of the air pollution impact from specific emission sources. Han and Naeher (2006) reviewed former investigations regarding traffic-related air pollution exposure. He reported that in the developing countries, especially some megacities, traffic-related air pollution is becoming the top potential health risk to the public. Although the fixed monitoring station data is accurate, the moving targets and distant targets from the station cannot be covered. The personal monitor device is developed and a useful tool to monitor the mobile target. Physick et al. (2011) designed an experiment to measure the nitrogen dioxide exposure in Melbourne. The result showed that air pollution data from personal monitor devices was at a high accuracy level and could be widely applied. No matter the fixed monitoring station method or the personal monitor device method, these methods cost a great deal of money and lab manpower. In addition, it may relate to personal privacy. As the remote-sensing technology develops, remote-sensing imagery becomes the essential tool to retrieve air pollutants. With the help of GIS, air pollutant distribution and vulnerability detection is much easier than before. For example, van Donkelaar et al. (2010) mapped the global PM2.5 distribution by using MODIS data. Long-term average (January 1, 2001 to December 31, 2006) PM2.5 concentrations at approximately $10 \, \text{km} \times 10 \, \text{km}$ resolution indicate a global population-weighted geometric mean of PM2.5 concentration of 20 microg/m^3. As for East China, the annual mean PM2.5 concentrations exceed 80 microg/m^3. In addition to the MODIS data for the fine-scale region, Landsat satellite imagery is a better helper. Han et al. (2014) retrieved the aerosol optical thickness based on the Landsat Enhanced Thematic Mapper Plus (ETM+) data. The results showed that the PM2.5 retrieved data is more precise than MODIS due to the high spatial resolution. Additionally, some spatial air pollution measure methods use remote sensing data to obtain a higher spatial resolution distribution of PM2.5 concentration. Land-use regression is one of the most widely used spatial analysis methods to obtain PM2.5 concentrations. This method retrieves PM2.5 value by building up empirical relationships between PM2.5 and other social economic data. For example, Eeftens et al. (2012) investigated the PM2.5 distribution in European cities. The result revealed that the median model explain variance (R2) was 71 percent. But different cities had different models and different explain variances. AOD data could also be considered as the explaining variable in the LUR model. Kloog et al. (2012) coupled the AOD data with the LUR model to obtain PM2.5 exposure in Mid-Atlantic states. He found that the LUR model with AOD data performed better than the models without AOD data. However, the LUR method is an empirical regression method. Due to the variation of time and change of study area, the explaining variable could be quite different, and a great deal of computing resources are needed. With the development of computer science and computational fluid dynamics in the 1980s, numerical simulation started to be popular. In this century, atmospheric diffusion models are more commonly used in air pollution prediction, especially with the

WRF-Chem and CMAQ stepping onto the stage. The Atmospheric diffusion model simulates air pollutants based on the atmospheric dynamics equation and atmospheric chemical processes. The terrain data, land use data, initial boundary data, and emission inventory are the basic inputs to drive this model. For the diverse parameterization scheme, high spatial resolution, and time scale continuity, numerical simulation is now widely used. Guo et al. (2016) investigated the impact of different emission schemes on regional air pollution in China during the APEC meeting by using the WRF-Chem model. He found that PM2.5 reduced dramatically in Beijing due to the restriction of emissions. Buonocore et al. (2014) studied the impact of individual power plants on public health. The result showed that power plants could increase the PM2.5 concentration, resulting in more PM2.5-related mortality. Although numerical simulation models are more and more widely used, the shortcomings cannot be ignored. Land use/cover data of most atmospheric diffusion data is updated with delays. Moreover, the emission inventory is a key factor for accuracy simulation results. Most of the emission inventory, however, is designed at a low spatial resolution. Consequently, simulation systemic errors occur. Above all, these three major air pollution exposure assessment methods have their advantages and disadvantages, and organically integrating them together is an interesting scientific challenge. The following section is related to the details of satellites used in air pollution detection and the introduction of the numerical simulation model.

9.2 Air pollution assessment based on remote sensing

Satellites can provide powerful tools to detect air pollution exposure, such as nitrogen dioxide (NO_2), sulfur dioxide (SO_2), and fine particles matter ($PM_{2.5}$) (Mikkelsen, 2002; Burgard et al., 2006; Salvador et al., 2009; Krueger et al., 2010; Tamburello et al., 2011; Dey et al., 2012; Guo et al., 2017). In this section, the characteristics of satellites used to detect different air pollutants will be introduced.

9.2.1 Monitoring of atmospheric fine particles matter using remote sensing

Currently, a great deal of satellite platforms provide imagery to analyze the general features of the global distribution of PM2.5, such as Landsat TM/ETM+ (launched in 1975); ASTER (launched in 1999); SPOT (launched in 2002); and Moderate Resolution Imaging Spectroradiometer (MODIS, launched in 1999). The key parameters to assess the fine particles with a diameter of 2.5 μm or less is the aerosol optical depth (AOD). AOD is defined as the integrated extinction coefficient over a vertical column of unit cross section. And the extinction coefficient is the fractional depletion of radiance per unit path length (Wang and Christopher, 2003; Van Donkelaar et al., 2006; Kumar et al., 2007). High AOD value relates to PM2.5 concentration, which results in high air pollution.

9.2.1.1 MODIS AOD products

MODIS is the instrument aboard the NASA's Terra and Aqua satellites. Terra MODIS and Aqua MODIS are viewing the entire Earth's surface every one to two days, obtaining data in 36 spectral bands with wavelengths ranging from 0.4 to 14.385 μm. The MODIS imagery has a spatial resolution of 250 m, 500 m, and 1 km (Kahn et al., 2009; Zhang and Reid, 2009; Shi et al., 2012; Acharya and Sreekesh, 2013). The channels with wavelengths ranging from 0.47 to 2.12 μm are used to retrieve aerosol characteristics. Daily level aerosol optical thickness data are produced at the spatial resolution of 10km×10km worldwide. MODIS AOD product of December 12, 2017 is displayed in Fig. 9.1.

9.2.1.2 Landsat 8

Landsat 8, formerly called the Landsat Data Continuity Mission, was launched in 2013. Two key instruments are carried by the satellite: the Operational Land Imager (OLI) and the Thermal Infrared Sensor (TIRS). A deep blue visible channel specially designed for water resources and coastal-zone investigation, and a new shortwave infrared channel for the detection of cirrus clouds have been recently added to the satellite. Every 16 days, the whole Earth's surface can be imaged by Landsat 8.

9.2.1.3 HJ-1A

Launched by China in 2008 HJ-1A was designed to detect environmental change and hazard monitoring. CCD camera and the hyper-spectral imager are aboard the satellite. The CCD camera has a spatial resolution of 30 m with the wavelength ranging from 0.43 to 0.90 μm. The hyper-spectral imager has a spatial resolution of 100 m with 110 to 128 multispectral.

9.2.2 Monitoring of atmospheric fine particles matter using numerical simulation model

The atmospheric diffusion model is one of the most widely used methods to study the relationship between air pollution and public health. Simple empirical models and the complex numerical models are the major branches of the mathematical models. These kinds of models are usually built on the equation governing the pollutant concentration consistent with the physical principle of mass conservation. Meteorological initial boundary conditions drive the climate or weather prediction models. The output of meteorological models and emission inventory will be input into the atmospheric diffusion modules to obtain the air pollution distribution. Currently, the most widely used numerical simulation models include WRF-Chem, MM5-CMAQ, WRF-CMAQ, and GEOS-Chem et al. (Matthias et al., 2008; Tombrou et al., 2009; Wang et al., 2009; Liu et al., 2010; Yegorova et al., 2011; Kochanski et al., 2013; Ritter et al., 2013; Wang et al., 2014). In this section, the framework of WRF-Chem and emission inventory will be covered.

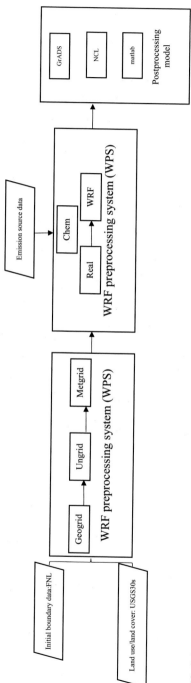

FIG. 9.1

Framework of WRF model system.

9.2.2.1 WRF-Chem model

WRF-Chem, called the next-generation atmospheric diffusion model, was developed by the National Center for Atmospheric Research (NCAR), the National Oceanic and Atmospheric Administration (represented by the National Centers for Environmental Prediction [NCEP] and the then Forecast System Laboratory [FSL]), the Air Force Weather Agency (AFWA), the Naval Research Laboratory, the University of Oklahoma, and the Federal Aviation Administration (FAA). The model is designed for the following tasks: weather or regional and local climate prediction, to simulate source and concentration of the air constituents, major air pollutant prediction (including O_3, NO_x, SO_2, PM2.5, et al.), aerosol direct and indirect forcing on global climate change investigation et al. (Ying et al., 2009; Ghude et al., 2013; Stuefer et al., 2013; Gao et al., 2014; Crippa et al., 2016). Fig. 9.1 shows the framework of the whole WRF-ARW model system.

As shown in Fig. 9.1, WRF-Chem module is a section of the WRF model. To successfully run WRF-Chem module, we need external data, including alternative observation data, conventional observation data, terrestrial data, and initial boundary data. The terrestrial data and the initial boundary data are essential, while alternative observation data and conventional observation data are essential for prediction but optional for the past condition research. All the prepared external data will be input via the WRF pre-processing system (WPS), which contains three basic sections: the geogrid, the ungrid, and the metgrid. The geogrid section is responsible for defining model domains and the interpolation of static geographical data to the grid. The ungrid section extracts the meteorological field from the initial boundary file. The metgrid section horizontally interpolates the meteorological fields extracted by the model grids as defined by the geogrid section. Output data of WPS will input the WRF model. First, the input data will be vertically interpolated based on the pressure level in the real section; then the disposed emission inventory and vertically interpolated meteorological data will be used by the WRF-Chem core module to simulate the air pollution conditions.

9.2.2.2 Emission inventory allocation method

As mentioned above, emission inventory is the indispensable input data for air pollution simulation. But the current emission inventory product is almost all at a low spatial resolution. It could not satisfy the need of air pollution exposure in urban areas. Therefore, allocating the emission to a finer spatial scale is urgent. In this section, a novel method to allocate emissions based on points of interest (POI) is introduced. This section will use Beijing as an example.

Data

The population density data is from the Beijing Statistical Bureau (http://www.bjstats.gov.cn/), and the gridded population density data of China in 2010 is from the Institute of Geographic Sciences and Natural Resources Research, CAS. The traffic flow data of major roads is obtained through field research. POI data is extracted from the Auto Navi Map, including restaurant service, places of interest, shopping malls, gasoline stations, residential areas, etc. The MEIC atmospheric emission inventory is from Tsinghua University (http://www.meicmodel.org).

Allocation method

Due to the different emission intensities, two allocation methods are designed for residential area emissions and traffic emissions. The detailed allocation method for residential areas is described below:

$$F_i = \frac{N_i}{N_{sum}} \tag{9.1}$$

$$E_i = F_i \cdot E_{sum} \tag{9.2}$$

where i represents the grid id; F_i represents the allocation weight of i grid; E_i represents the emission value in grid I; N_i represents the amount of residential building POI in grid I; N_{sum} represents the total number of residential building POI; E_{sum} represents the total emission value of the study area. The traffic emission allocation method is detailed below:

$$R_{a,b} = \frac{N_a}{N_{avg,b}} \tag{9.3}$$

$$L_i = \sum_{b=1}^{m} \sum_{a=1}^{n} L_{a,b} \cdot Q_b \cdot R_{a,b} \tag{9.4}$$

$$F_i = \frac{L_i}{L_{sum}} \tag{9.5}$$

$$E_i = F_i \cdot E_{sum} \tag{9.6}$$

where a represents the road number; b represents road level; m represents the number of road levels; n_b represents the amount of level b roads; R_{ab} represents the adjustment coefficient of road a; N_a represents the amount of parking lot POI around road a, which is a level b road; $N_{avg,b}$ represents the average amount of parking lots around level b roads; L_i represents the standard length of all roads in grid I; $L_{a,b}$ represents the total length of a road, which is a level b road. Q_b represents the conversion factor between road length and standard road length, and L_{sum} represents the standard length of all roads in the study area.

Acknowledgments

This work is supported by the National Natural Science Foundation of China (NO. 41671430). The emission inventory method is designed by Kun Wang (wkty@mail.bnu.edu.cn), and this method has been programmed with the software called ISAM.

References

Acharya, P., Sreekesh, S., 2013. Seasonal variability in aerosol optical depth over India: a spatio-temporal analysis using the MODIS aerosol product. Int. J. Remote Sens. 34 (13), 4832–4849.

Brauer, M., Amann, M., Burnett, R.T., Cohen, A., Dentener, F., Ezzati, M., Henderson, S.B., Krzyzanowski, M., Martin, R.V., Dingenen, R.V., 2012. Exposure assessment for estimation of the global burden of disease attributable to outdoor air pollution. Environ. Sci. Technol. 46 (2), 652–660.

Buonocore, J.J., Dong, X., Spengler, J.D., Fu, J.S., Levy, J.I., 2014. Using the Community Multiscale Air Quality (CMAQ) model to estimate public health impacts of PM2.5 from individual power plants. Environ. Int. 68, 200.

Burgard, D.A., Dalton, T.R., Bishop, G.A., Starkey, J.R., 2006. Nitrogen dioxide, sulfur dioxide, and ammonia detector for remote sensing of vehicle emissions. Rev. Sci. Instrum. 77 (1), 014101–014105.

Chen, S., Chen, B., 2015. Urban energy consumption: different insights from energy flow analysis, input–output analysis and ecological network analysis. Appl. Energy 138 (C), 99–107.

Chen, R., Li, Y., Ma, Y., Pan, G., Zeng, G., Xu, X., Chen, B., Kan, H., 2011. Coarse particles and mortality in three Chinese cities: the China Air Pollution and Health Effects Study (CAPES). Sci. Total Environ. 409 (23), 4934–4938.

Crippa, P., Sullivan, R.C., Thota, A., Pryor, S.C., 2016. Evaluating the skill of high-resolution WRF-Chem simulations in describing drivers of aerosol direct climate forcing on the regional scale. Atmos. Chem. Phys. 16 (1), 397–416.

Dey, S., Girolamo, L.D., Donkelaar, A.V., Tripathi, S.N., Gupta, T., Mohan, M., 2012. Variability of outdoor fine particulate (PM 2.5) concentration in the Indian Subcontinent: a remote sensing approach. Remote Sens. Environ. 127 (140), 153–161.

Dhakal, S., 2009. Urban energy use and carbon emissions from cities in China and policy implications. Energy Policy 37 (11), 4208–4219.

van Donkelaar, A., Martin, R.V., Brauer, M., Kahn, R., Levy, R., Verduzco, C., Villeneuve, P.J., 2010. Global estimates of ambient fine particulate matter concentrations from satellite-based aerosol optical depth: development and application. Environ. Health Perspect. 118 (6), 847–855.

Eeftens, M., Beelen, R., De, H.K., Bellander, T., Cesaroni, G., Cirach, M., Declercq, C., Dèdelè, A., Dons, E., De, N.A., 2012. Development of Land Use Regression models for PM(2.5), PM(2.5) absorbance, PM(10) and PM(coarse) in 20 European study areas; results of the ESCAPE project. Environ. Sci. Technol. 46 (20), 11195–11205.

Feng, Y.Y., Chen, S.Q., Zhang, L.X., 2013. System dynamics modeling for urban energy consumption and CO 2 emissions: a case study of Beijing, China. Ecol. Model. 252 (1755), 44–52.

Gao, Y., Zhao, C., Liu, X., Zhang, M., Leung, L.R., 2014. WRF-Chem simulations of aerosols and anthropogenic aerosol radiative forcing in East Asia. Atmos. Environ. 92 (92), 250–266.

Ghude, S.D., Pfister, G.G., Jena, C., van der Ronald, A., Emmons, L.K., Kumar, R., 2013. Satellite constraints of nitrogen oxide (NOx) emissions from India based on OMI observations and WRF-Chem simulations. Geophys. Res. Lett. 40 (2), 423–428.

Guo, J., He, J., Liu, H., Miao, Y., Liu, H., Zhai, P., 2016. Impact of various emission control schemes on air quality using WRF-Chem during APEC China 2014. Atmos. Environ. 140, 311–319.

Guo, J., Xia, F., Zhang, Y., Liu, H., Li, J., Lou, M., He, J., Yan, Y., Wang, F., Min, M., 2017. Impact of diurnal variability and meteorological factors on the PM2.5 - AOD relationship: implications for PM2.5 remote sensing. Environ. Pollut. 221 (94), 94.

Han, X., Naeher, L.P., 2006. A review of traffic-related air pollution exposure assessment studies in the developing world. Environ. Int. 32 (1), 106–120.

Han, W., Tong, L., Bai, J., Chen, Y., 2014. Spatial distribution of PM2.5 concentration based on aerosol optical thickness inverted by Landsat ETM+ data over Chengdu. In: Geoscience and Remote Sensing Symposium.

Hoek, G., Brunekreef, B., Goldbohm, S., Fischer, P., van den Brandt, P.A., 2002. Association between mortality and indicators of traffic-related air pollution in the Netherlands: a cohort study. Lancet (London, England) 360 (9341), 1203–1209.

Kahn, R.A., Nelson, D.L., Garay, M.J., Levy, R.C., Bull, M.A., Diner, D.J., Martonchik, J.V., Paradise, S.R., Hansen, E.G., Remer, L.A., 2009. MISR aerosol product attributes and statistical comparisons with MODIS. IEEE Trans. Geosci. Remote Sens. 47 (12), 4095–4114.

Kaiser, R., Romieu, I., Medina, S., Schwartz, J., Krzyzanowski, M., Künzli, N., 2004. Air pollution attributable postneonatal infant mortality in U.S. metropolitan areas: a risk assessment study. Environ. Health 3 (1), 4.

Kan, H., London, S.J., Chen, G., Zhang, Y., Song, G., Zhao, N., Jiang, L., Chen, B., 2008. Season, sex, age, and education as modifiers of the effects of outdoor air pollution on daily mortality in Shanghai, China: The Public Health and Air Pollution in Asia (PAPA) Study. Environ. Health Perspect. 116 (9), 1183–1188.

Kloog, I., Nordio, F., Coull, B.A., Schwartz, J., 2012. Incorporating local land use regression and satellite aerosol optical depth in a hybrid model of spatiotemporal PM2.5 exposures in the Mid-Atlantic states. Environ. Sci. Technol. 46 (21), 11913–11921.

Kochanski, A.K., Beezley, J.D., Mandel, J., Clements, C.B., 2013. Air pollution forecasting by coupled atmosphere-fire model WRF and SFIRE with WRF-Chem. Physics.

Krueger, A.J., Krotkov, N.A., Yang, K., Carn, S., Vicente, G., Schroeder, W., 2010. Applications of satellite-based sulfur dioxide monitoring. IEEE J. Sel. Top. Appl. Earth Observ. Remote Sens. 2 (4), 293–298.

Kumar, N., Chu, A., Foster, A., 2007. An empirical relationship between PM2.5 and aerosol optical depth in Delhi Metropolitan. Atmos. Environ. 41 (21), 4492–4503.

Liu, X.H., Zhang, Y., Xing, J., Zhang, Q., Wang, K., Streets, D.G., Jang, C., Wang, W.X., Hao, J.M., 2010. Understanding of regional air pollution over China using CMAQ, part II. Process analysis and sensitivity of ozone and particulate matter to precursor emissions. Atmos. Environ. 44 (30), 3719–3727.

Matthias, V., Aulinger, A., Quante, M., 2008. Adapting CMAQ to investigate air pollution in North Sea coastal regions. Environ. Model. Softw. 23 (3), 356–368.

Mikkelsen, O.A., 2002. Variation in the projected surface area of suspended particles: implications for remote sensing assessment of TSM. Remote Sens. Environ. 79 (1), 23–29.

Physick, W., Powell, J., Cope, M., Boast, K., Lee, S., 2011. Measurements of personal exposure to NO2 and modelling using ambient concentrations and activity data. Atmos. Environ. 45 (12), 2095–2102.

Ritter, M., Müller, M.D., Tsai, M.Y., Parlow, E., 2013. Air pollution modeling over very complex terrain: an evaluation of WRF-Chem over Switzerland for two 1-year periods. Atmos. Res. 132–133 (10), 209–222.

Salvador, M.Z., Resmini, R.G., Gomez, R.B., 2009. Detection of sulfur dioxide in AIRS data with the wavelet packet subspace. IEEE Geosci. Remote Sens. Lett. 6 (1), 137–141.

Shi, Y., Zhang, J., Reid, J.S., Liu, B., Deshmukh, R., 2012. Critical evaluation of cloud contamination in MISR aerosol product using collocated MODIS aerosol and cloud products. American Geophysical Union.

Stuefer, M., Freitas, S.R., Grell, G., Webley, P., Peckham, S., Mckeen, S.A., Egan, S.D., 2013. Inclusion of ash and SO_2 emissions from volcanic eruptions in WRF-Chem: development and some applications. Geosci. Model Dev. 6 (2), 457–468.

Tamburello, G., Mcgonigle, A.J.S., Kantzas, E.P., Aiuppa, A., 2011. Recent advances in ground-based ultraviolet remote sensing of volcanic SO_2 fluxes. Ann. Geophys. 54 (2), 199–208.

Tombrou, M., Bossioli, E., Protonotariou, A.P., Flocas, H., Giannakopoulos, C., Dandou, A., 2009. Coupling GEOS-CHEM with a regional air pollution model for Greece. Atmos. Environ. 43 (31), 4793–4804.

Van Donkelaar, A., Martin, R.V., Park, R.J., 2006. Estimating ground-level PM2.5 using aerosol optical depth determined from satellite remote sensing. J. Geophys. Res. Atmos. 111 (D21).

Wang, J., Christopher, S.A., 2003. Intercomparison between satellite-derived aerosol optical thickness and PM2.5 mass: implications for air quality studies. Geophys. Res. Lett. 30 (21), 267–283.

Wang, L., Wang, S., Zhang, L., Wang, Y., Zhang, Y., Nielsen, C., Mcelroy, M.B., Hao, J., 2014. Source apportionment of atmospheric mercury pollution in China using the GEOS-Chem model. Environ. Pollut. 190 (7), 166.

Wang, K., Zhang, Y., Jang, C.J., Phillips, S., Wang, B.Y., 2009. Modeling study of intercontinental air pollution transport over the trans-Pacific region in 2001 using the Community Multiscale Air Quality (CMAQ) modeling system. J. Geophys. Res. Atmos. 114 (D4).

Yegorova, E.A., Allen, D.J., Loughner, C.P., Pickering, K.E., Dickerson, R.R., 2011. Characterization of an eastern U.S. severe air pollution episode using WRF/Chem. J. Geophys. Res. Atmos. 116 (D17).

Ying, Z., Tie, X., Li, G., 2009. Sensitivity of ozone concentrations to diurnal variations of surface emissions in Mexico City: a WRF/Chem modeling study. Atmos. Environ. 43 (4), 851–859.

Zhang, J., Reid, J.S., 2009. An analysis of clear sky and contextual biases using an operational over ocean MODIS aerosol product. Geophys. Res. Lett. 36 (15), 172–173.

Applying LUR model to estimate spatial variation of PM$_{2.5}$ in the Greater Bay Area, China

10

Lingling Chen[a], Zhifeng Wu[a], Wei Tu[b], Zheng Cao[a]

[a]*School of Geographical Sciences, Guangzhou University, Guangzhou, China*
[b]*Department of Geology and Geography, Georgia Southern University, Statesboro, GA, United States*

10.1 Introduction

The Greater Bay Area (shorted from the Guangdong-Hong Kong-Macao Greater Bay Area) of China, which is considered the most rapid economic development in China during the last two decades, has been suffering severe air pollution (Chan and Yao, 2008; Matus et al., 2012). Many epidemiological studies show that exposure to PM2.5 is associated with adverse effects on cardiovascular and respiratory outcomes (Guo et al., 2009; Li et al., 2013). Accurate spatial distribution of PM2.5 is critical for air pollution control and epidemiological studies on human chronic and/or acute exposure. However, levels of PM2.5 are measured only at a limited number of ground monitoring sites, which cannot provide concentration information everywhere within study area.

Several methods have been developed to assess spatial variation and spatial-temporal dynamics of PM2.5 pollution. Land use regression (LUR) model is a widely used method in air pollution epidemiology since it can detect intra-urban variation with relatively simple and efficient computation (Levy et al., 2010).

Since PM2.5 pollution prevails over winter in the Greater Bay Area (GBA), the area is particularly unhealthy for humans, especially for sensitive groups with heart and lung diseases. This study took meteorological factors into account to develop a LUR model to predict winter mean PM2.5 concentrations in the Greater Bay Area. The accuracy of the resultant model was tested, and a surface map was created to detect the spatial characteristics of winter PM2.5 over the study area.

10.2 Material and methods

This section outlines the study area, data collection, and statistical methods used in this analysis. Ground-level winter mean PM2.5 concentrations was taken as the dependent variable in this study. Variables can represent meteorological effects, traffic conditions, land use types, and demographic and topographic characteristics that were collected as independent variables. The framework of LUR model employs a multivariate linear regression in this study. After LUR model validation, this research interpolated the PM2.5 concentration and produced winter mean PM2.5 surface maps.

10.2.1 Study area

The Greater Bay Area (GBA) is located in southeastern of China, with a population of 65 million and total area of approximately 56,000 square kilometers. The latitude and longitude of GBA are 21°27′ to 23°56′ and 111°59′ to 115°26′ (Fig. 10.1).

10.2.2 Data collection

10.2.2.1 Ground-based monitoring data

Hourly PM2.5 concentrations ($\mu g/m^3$) for winter of 2017–18 (December 1, 2017–February 28, 2018) observed at 56 monitoring sites in the mainland of GBA were obtained from the official website of China Environmental Monitoring Center (CEMC) (http://106.37.208.233:20035/). Monthly level PM2.5 of five monitoring sites within Macau and Hong Kong were obtained from direcção dos serviços meteorológicose

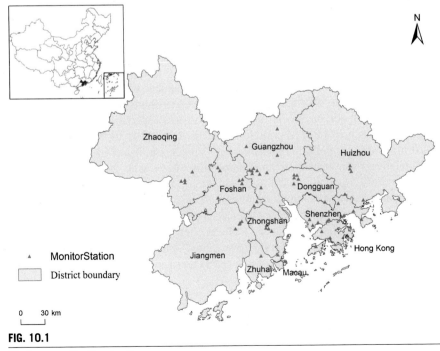

FIG. 10.1

Study area—the Greater Bay area.

geofísicos (http://www.smg.gov.mo/smg/airQuality/e_air_reports.htm). All missing or invalid data due to instrument malfunctions were removed from the original data, and only sites with at least 60% data available in each month and over 75% data available during the whole period were kept. After filtering, 58 sites were selected, and the hourly data were aggregated to winter averages, which was used as the dependent variable in the subsequent LUR model building.

10.2.2.2 Predictors of PM2.5

We created 174 potential predictor variables in five categories and 23 sub-categories (Table 10.1). The five categories were: 1) meteorological variables; 2) traffic-related indicator; 3) land use type; 4) demographic variables; 5) topographic characteristics. For each monitoring site, values of all these predictors were extracted at different buffer radii vary from 50 m to 5000 m, which was adopted from a previous study on LUR models (Zou et al., 2016). Detailed information of predictors and their assumed effect can be found in Table 10.1.

The meteorological data, containing temperature, surface pressure, wind speed, U wind speed, V wind speed, relative humidity, and boundary layer height, were obtained from the European Centre for Medium-Range Weather Forecasts (ECMWF) reanalysis datasets (ERA-Interim) at 0.125×0.125 degrees spatial resolution. Road network data, buildings and industrials area were retrieved from OpenStreetMap. The GlobalLand30 dataset of 2017 was downloaded from Tsinghua University (http://data.ess.tsinghua.edu.cn/), reclassified into impervious surface: forest, grassland, cropland, waterbody, and others. The 1 km grid-based population data in 2015 was collected from the Center for International Earth Science Information Network (CIESIN), Columbia University. DEM covering GBA with 30 m resolution generated from NASA's Shuttle Radar Topography Mission (STRM) was downloaded from 30-Meter SRTM Tile Downloader (http://dwtkns.com/srtm30m/).

10.2.3 **LUR modeling and mapping**

According to previous studies (Beelen et al., 2013; Gulliver and de Hoogh, 2015), a supervised forward regression method and a manual supervised iterative approach of variable addition and removal were used for variables selection, which was conducted in SPSS and R software. The followings are several detailed rules for selecting variables:

(1) Ensure all variables consistent with the assumed effect direction (Table 10.1);
(2) A *P*-value of <.1 for each predictor;
(3) Avoid double counting of variables as indicated by values of variance inflation factor (VIF), a variable is considered as redundant when VIF > 5 (Gulliver and de Hoogh, 2015).

The prediction model was derived; meanwhile finishing selecting predictor variables. Next, Leave-one-out-cross-validation (LOOCV) was applied to evaluate the performance of final model. Also, to evaluate the independence assumption, spatial

Table 10.1 Potential variables with units, defined buffer sizes, and assumed effect

Category (N)	Sub category	Unit	Buffer radii (m)	Assumed effect
Meteorological variables (7)	Temperature	K	NA	NA
	Surface pressure	Pa	NA	NA
	Wind speed	m/s	NA	NA
	U wind speed	m/s	NA	NA
	V wind speed	m/s	NA	NA
	Relative humidity	%	NA	NA
	Boundary layer height	m	NA	NA
Traffic-related indicator (92)	Length of all roads	m	50, 60, 70, 80, 90, 100, 150, 200, 250, 300, 350, 400, 450, 500, 1000, 1500, 2000, 2500, 3000, 3500, 4000, 4500, 5000	+
	Length of primary roads	m		+
	Length of secondary roads	m		+
	Density of all roads	km/km^2		NA
Land use type (70)	Impervious surfaces	m^2	500, 1000, 1500, 2000, 2500, 3000, 3500, 4000, 4500, 5000	+
	Forest	m^2		–
	Grassland	m^2		–
	Cropland	m^2		NA
	Waterbody	m^2		–
	Buildings	m^2		+
	Industrials	m^2		+
Demographic variables (1)	Population density	person/m^2	NA	+
Topographic characteristics (4)	DEM elevation	m	NA	NA
	Latitude	°	NA	NA
	Longitude	°	NA	NA
	Distances to the coastlines	m	NA	+

auto-correlation was tested on the residuals from the final model using Moran's I index (Bailey and Gatrell, 1995).

Based on the final LUR model, the average predicted concentration surface of PM2.5 during the winter was generated with 10×10 km spatial resolution.

10.3 Results

The final model for PM2.5 in GBA is presented in Table 10.2. Of the 174 variables generated, five variables were left in the PM2.5 LUR model: BLH, V wind speed, percentage of forest area within the 4000-m buffer, density of all roads within the

Table 10.2 Statistical results from LUR model

Predictors	Coefficients	Standard error	t score	Significance (P value)	VIF
Constant	116.111	7.369	15.756	0.000	0.000
BLH (m)	−0.183	0.023	−7.816	0.000	3.168
V wind speed (m/s)	−14.131	3.149	−4.488	0.000	2.959
Percentage of forest area with the 4000-m buffer	−18.668	3.729	−5.007	0.000	1.724
Density of all roads within the 1500-m buffer	−0.683	1.640	−4.151	0.000	3.692
Percentage of buildings area with the 5000-m buffer	107.740	34.457	3.127	0.003	3.216

1500-m buffer, which were negatively associated with PM2.5 concentration. A percentage of buildings within the 5000-m buffer were positively associated with PM2.5 concentration. The VIF values of the five variables were all less than four, which means that all predictors can be treated as uncorrelated with each other.

The model fitting R2, adjusted R2, and LOOCV R2 of the PM2.5 LUR models were 0.780, 0.756, and 0.717, respectively. The RMSE was 3.398 $\mu g/m^3$, which indicates the predicted values were well-fitted with the observed values. The histogram of residuals, as shown in Fig. 10.2A, had a bell-shaped distribution with a mean value of −0.225 and standard deviation of 3.012. The Moran's I value of the model residuals conducted in ArcGIS software was −0.009 with an insignificant P value (.973), which implied that the residuals had no significant auto-correlation. Fig. 10.2B shows the scatterplot of predicted PM2.5 concentration against observed values. The fitted linear regression line close to the 45 degrees line had a slope of 1.04 with R^2 = .808, which further suggested that the observations were estimated well by the prediction model.

After model validation, winter mean PM2.5 concentrations of surfaces in GBA at 10km resolution was mapped, as shown in Fig. 10.3.

10.4 Discussion

We have developed LUR model for predicting winter mean PM2.5 in the Greater Bay area of China using data from 58 regulatory monitoring stations. The derived model explained more than 75% of the variability in PM2.5 concentration. Two significant temporal predictors (BLH and V wind speed), and three significant spatial predictors (percentage of forest area within the 4000-m buffer, density of all roads within the 1500-m buffer, and percentage of buildings area within the 5000-m buffer), were chosen in the final LUR model.

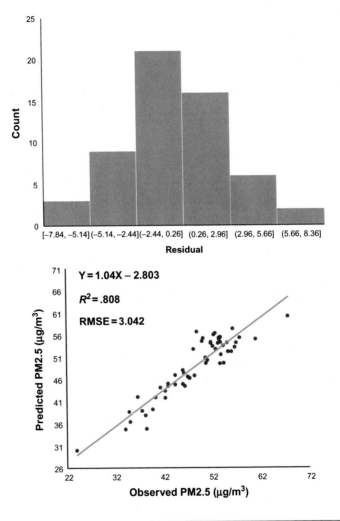

FIG. 10.2

(A) Histogram of model residuals; (B) Comparison between predicted and observed PM2.5 concentration.

BLH was found to have a negative contribution to winter mean PM2.5 concentration in the Greater Bay area. As described earlier, studies have pointed out that there exists a positive feedback between absorbing aerosols and the BLH (Ding et al., 2016; Miao et al., 2016; Petäjä et al., 2016). Su et al. (2018) found that the negative correlation between PBLH and PM2.5 is most significant during winter in China, and strong correlations between BLHs and aerosols occur in low-altitude regions. That provides a reasonable explanation of significantly negative relation between BLH and PM2.5 since the Greater Bay area locates in low-altitude regions of China, and only winter mean PM2.5 concentration was taken into consideration in this study. As another

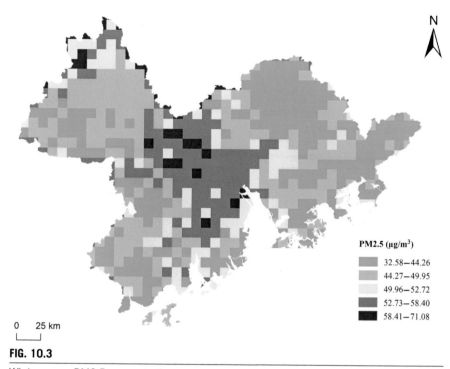

N

PM2.5 (μg/m³)

■ 32.58−44.26
■ 44.27−49.95
□ 49.96−52.72
■ 52.73−58.40
■ 58.41−71.08

0　25 km

FIG. 10.3

Winter mean PM2.5 concentrations predicted by LUR model.

temporal predictor, V wind speed value was negatively associated with winter mean PM2.5 concentrations. Wind speed is widely taken into account as the dispersion efficiency factor while modeling PM2.5 pollution specifically for short- and medium-terms using LUR (Arain et al., 2007; Reyes and Serre, 2014; Shi et al., 2017).

Additionally, the forest area turned out to be a negative predictor variable because forest can efficiently reduce the diffusion rate of PM2.5 during daytime and adsorb or deposit particulate matter at night (Liu et al., 2015). Our results suggested that density of all roads was a significant predictor variable with an unexpectedly negative effect, which is opposite to the results from many other studies that took road-related indicators as variables into LUR model building (Ross et al., 2007; Beckerman et al., 2013). That may account for the quality of road conditions and different types of roads within a certain area. Since the better the quality of roads is, the more mobile the vehicles would be, and the less pollution (Beevers and Carslaw, 2005). Moreover, it has been reported that public transport, such as subway, rail transit, et al., helps improve air quality although it would increase road density. Using daily data before and after the opening of Metro in Taipei, China, Chen and Whalley (2012) found that the opening of the Metro reduced CO emission by 5%–15%. According to Luo et al. (2017), the results showed that road density is negatively correlated with PM10. A 1% increase in the road density reduces PM10 by 0.112%–0.114%.

This study has several limitations. There remains 24.4% spatial variation unexplained by this LUR model. Satellite-derived variables, e.g., AOD and NDVI were not included in our study, which was significantly associated with PM2.5 concentrations (Kloog et al., 2015). In addition, the spatial resolution of this study was relatively coarser compared with studies based on purpose-designed monitored data. Also, only winter mean PM2.5 concentrations were aggregated as dependent variable. Thus, this LUR model is not suggested to be applied in long-term air pollution epidemiological studies.

10.5 Conclusion

In conclusion, this resultant LUR model was able to explain a majority of the variability in the ambient PM2.5 concentration in the Greater Bay area, China, with predictor variables of BLH, V wind speed, percentage of forest area, and buildings area and road density within certain buffers. The spatial autocorrelation was found insignificant, so the linear regression model was appropriate for the data. The predicted surface of PM2.5 concentration based on the LUR model was consistent with real situation over winter season in the Greater Bay area. Therefore, the model holds promises for air pollution exposure of some acute disease for sensitive groups.

References

Arain, M.A., Blair, R., Finkelstein, N., Brook, J.R., Sahsuvaroglu, T., Beckerman, B., Zhang, L., Jerrett, M., 2007. The use of wind fields in a land use regression model to predict air pollution concentrations for health exposure studies. Atmos. Environ. 41 (16), 3453–3464.

Bailey, T.C., Gatrell, A.C., 1995. Interactive spatial data analysis. vol. 413. Longman Scientific & Technical, Essex.

Beckerman, B.S., Jerrett, M., Serre, M., Martin, R.V., Lee, S.J., van Donkelaar, A., Ross, Z., Su, J., Burnett, R.T., 2013. A hybrid approach to estimating national scale spatiotemporal variability of PM2. 5 in the contiguous United States. Environ. Sci. Technol. 47 (13), 7233–7241.

Beelen, R., Hoek, G., Vienneau, D., Eeftens, M., Dimakopoulou, K., Pedeli, X., Tsai, M.-Y., Künzli, N., Schikowski, T., Marcon, A., Eriksen, K.T., Raaschou-Nielsen, O., Stephanou, E., Patelarou, E., Lanki, T., Yli-Tuomi, T., Declercq, C., Falq, G., Stempfelet, M., Birk, M., Cyrys, J., von Klot, S., Nádor, G., János Varró, M., Dédelé, A., Gražulevičienė, R., Mölter, A., Lindley, S., Madsen, C., Cesaroni, G., Ranzi, A., Badaloni, C., Hoffmann, B., Nonnemacher, M., Krämer, U., Kuhlbusch, T., Cirach, M., de Nazelle, A., Nieuwenhuijsen, M., Bellander, T., Korek, M., Olsson, D., Strömgren, M., Dons, E., Jerrett, M., Fischer, P., Wang, M., Brunekreef, B., de Hoogh, K., 2013. Development of NO$_2$ and NO$_x$ land use regression models for estimating air pollution exposure in 36 study areas in Europe—the ESCAPE project. Atmos. Environ. 72 (Suppl C), 10–23. https://doi.org/10.1016/j.atmosenv.2013.02.037.

Beevers, S.D., Carslaw, D.C., 2005. The impact of congestion charging on vehicle emissions in London. Atmos. Environ. 39 (1), 1–5.

Chan, C.K., Yao, X., 2008. Air pollution in mega cities in China. Atmos. Environ. 42 (1), 1–42. https://doi.org/10.1016/j.atmosenv.2007.09.003.

Chen, Y., Whalley, A., 2012. Green infrastructure: the effects of urban rail transit on air quality. Am. Econ. J.: Econ. Policy 4 (1), 58–97.

Ding, A.J., Huang, X., Nie, W., Sun, J.N., Kerminen, V.M., Petäjä, T., Su, H., Cheng, Y.F., Yang, X.Q., Wang, M.H., 2016. Enhanced haze pollution by black carbon in megacities in China. Geophys. Res. Lett. 43 (6), 2873–2879.

Gulliver, J., de Hoogh, K., 2015. Environmental exposure assessment: modelling air pollution concentrations. https://doi.org/10.1093/med/9780199661756.003.0135.

Guo, Y., Jia, Y., Pan, X., Liu, L., Wichmann, H.E., 2009. The association between fine particulate air pollution and hospital emergency room visits for cardiovascular diseases in Beijing, China. Sci. Total Environ. 407 (17), 4826–4830.

Kloog, I., Sorek-Hamer, M., Lyapustin, A., Coull, B., Wang, Y., Just, A.C., Schwartz, J., Broday, D.M., 2015. Estimating daily PM2.5 and PM10 across the complex geo-climate region of Israel using MAIAC satellite-based AOD data. Atmos. Environ. 122, 409–416. https://doi.org/10.1016/j.atmosenv.2015.10.004.

Levy, J.I., Clougherty, J.E., Baxter, L.K., Andres Houseman, E., Paciorek, C.J., Health Review Committee, Health Effects Institute, 2010. Evaluating heterogeneity in indoor and outdoor air pollution using land-use regression and constrained factor analysis. Research Report (Health Effects Institute) 152, 5–80. discussion 81–91.

Li, P., Xin, J., Wang, Y., Wang, S., Li, G., Pan, X., Liu, Z., Wang, L., 2013. The acute effects of fine particles on respiratory mortality and morbidity in Beijing, 2004–2009. Environ. Sci. Pollut. Res. 20 (9), 6433–6444.

Liu, X., Yu, X., Zhang, Z., 2015. PM2. 5 concentration differences between various forest types and its correlation with forest structure. Atmosphere 6 (11), 1801–1815.

Luo, Z., Wan, G., Wang, C., Zhang, X., 2017. Pollution and Road Infrastructure in Cities of the People's Republic of China. ADBI Working Paper 717. Available at https://www.adb.org/publications/pollution-and-roadinfrastructure-cities-prc.

Matus, K., Nam, K.-M., Selin, N.E., Lamsal, L.N., Reilly, J.M., Paltsev, S., 2012. Health damages from air pollution in China. Global Environ. Change 22 (1), 55–66. https://doi.org/10.1016/j.gloenvcha.2011.08.006.

Miao, Y., Liu, S., Zheng, Y., Wang, S., 2016. Modeling the feedback between aerosol and boundary layer processes: a case study in Beijing, China. Environ. Sci. Pollut. Res. 23 (4), 3342–3357.

Petäjä, T., Järvi, L., Kerminen, V.-M., Ding, A.J., Sun, J.N., Nie, W., Kujansuu, J., Virkkula, A., Yang, X., Fu, C.B., 2016. Enhanced air pollution via aerosol-boundary layer feedback in China. Sci. Rep. 6, 18998.

Reyes, J.M., Serre, M.L., 2014. An LUR/BME framework to estimate PM2.5 explained by on road mobile and stationary sources. Environ. Sci. Technol. 48 (3), 1736–1744. https://doi.org/10.1021/es4040528.

Ross, Z., Jerrett, M., Ito, K., Tempalski, B., Thurston, G.D., 2007. A land use regression for predicting fine particulate matter concentrations in the New York City region. Atmos. Environ. 41 (11), 2255–2269. https://doi.org/10.1016/j.atmosenv.2006.11.012.

Shi, Y., Ka-Lun Lau, K., Ng, E., 2017. Incorporating wind availability into land use regression modelling of air quality in mountainous high-density urban environment. Environ. Res. 157, 17–29.

Su, T., Li, Z., Kahn, R., 2018. Relationships between the planetary boundary layer height and surface pollutants derived from lidar observations over China: regional pattern and influencing factors. Atmos. Chem. Phys. 18 (21), 15921–15935. https://doi.org/10.5194/acp-18-15921-2018.

Zou, B., Chen, J., Zhai, L., Fang, X., Zheng, Z., 2016. Satellite based mapping of ground PM2. 5 concentration using generalized additive modeling. Remote Sens. 9 (1), 1.

Analysis of exposure to ambient air pollution: Case study of the link between environmental exposure and children's school performance in Memphis, TN

11

Anzhelika (Angela) Antipova

Department of Earth Sciences, University of Memphis, Memphis, TN, United States

11.1 Air pollution

Air pollution and closeness of industry to homes have been consistently rated the most severe environmental problems by the affected residents as described in the first environmental justice book called *Dumping in Dixie: Race, Class, and Environmental Quality* by Robert Bullard (Bullard, 2000). Environmental pollution impacts quality of life. The *Economist Intelligence Unit* (The EIU), the research and analysis division of The Economist Group, developed a composite quality of life index, which assesses several factors reflecting life satisfaction for 74 countries, including the United States, which ranked above the European Union-15 average on quality of life. However, the country's relatively high average masks a substantial variability among some US states offsetting the effect of high quality of life in the USA, including the historic low unemployment rate for women (4.0% in December 2017) and minorities (6.8% in December 2017, Bureau of Labor Statistics) when compared with the European countries. For example, Tennessee is faced with extant challenges of obesity and drug epidemics, lower-than-nation-wide incomes, big discrepancy in life expectancy between minority and White populations, environmental pollution (Fig. 11.1 shows current emission trends for Tennessee between 1990 and 2016 with some types of emission increasing).

The multicomponent quality of life index includes, in order of importance, health, material well-being, political stability, and security (these are the most important), followed by family relations and community life, as well as climate, job security, political freedom, and gender equality. These criteria adequately and objectively reflect actual quality of life as people with different cultural backgrounds in different countries report similar factors as vital for life satisfaction (EIU, 2005).

Air pollution represents a major environmental risk to the most important component of the given quality of life index—health (WHO, 2018). Farther, according to the *Lancet* report on pollution and health, pollution is the single largest environmental contributor to global disease and premature death today, causing an estimated 9 million premature deaths in 2015 globally, or three times as many as all the deaths due to AIDS, tuberculosis, and malaria (Landrigan et al., 2017).

11.1.1 Criteria air pollutants

The US Environmental Protection Agency (EPA) is required by the Clean Air Act, enacted by the Congress in 1970 and subsequently amended in 1977 and 1990, to set National Ambient Air Quality Standards (NAAQS) for pollutants, which can potentially damage public health and the environment. Six principal pollutants for which NAAQS have been set by the EPA are called *"criteria" air pollutants* (CAPs). These six common air pollutants include particulate matter, ground-level ozone (O_3), carbon monoxide (CO), lead (Pb), sulfur dioxide (SO_2), and nitrogen dioxide (NO_2). The World Health Organization (WHO) established air quality standards for common air pollution concentrations in outdoor air.

Fig. 11.1 shows the trends for Tier 1 categories, which distinguish pollutant emission contributions among major source types. The trends shown are for criteria air pollutants (CAPs) and precursors covered by the National Ambient Air Quality Standards (NAAQS), excluding lead. Across Tennessee, over 1990–2016, while emission due to fossil fuel combustion and chemical manufacturing has decreased (Fig. 11.1A and B), emission of criteria air pollutants, or CAPs (carbon monoxide, lead, ground-level ozone, particulate matter, nitrogen dioxide, and sulfur dioxide) due to petroleum industry and waste disposal and recycling (Fig. 11.1C and E), which harm human health and the environment, and cause property damage have been increasing by 2016 (EPA, 2017).

The two types of national ambient air quality standards identified by the Clean Air Act include *primary standards* (those protecting public health, including that of "sensitive" populations such as asthmatics, children, and the elderly), and *secondary standards* (those protecting public welfare such as against damage to crops, vegetation, buildings, reduced visibility, or haze).

Following substantial air quality improvement in the 1970s and 1980, currently studies assess the impacts of lower-level chronic exposure to various pollutants since some pollutants such as particulate matter and ozone have been found related to mortality and hospital visits at levels lower than the current NAAQS (Miranda et al., 2011). Common air pollutants and their health impacts are described here.

11.1.2 How chemical concentrations in air are measured

The standards for the monitoring of ambient air quality require pollutant measures expressed per volume of ambient air (Gaffron and Niemeier, 2015). To measure chemical concentrations in air, we standardly use the mass of chemical (such as

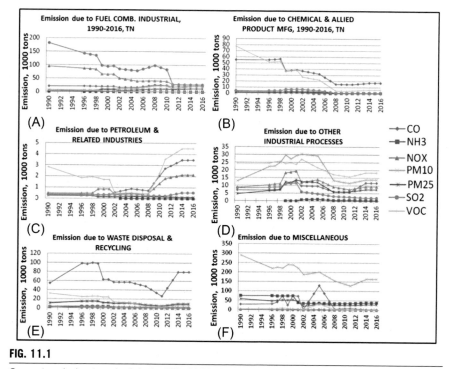

FIG. 11.1

Current emission trends data for TN, 1990–2016.

milligrams, mg, micrograms, µg, etc.) expressed per volume of air (such as cubic meter, m^3, or cubic feet, ft^3). 1 mg (mg) = 0.001 g and 1 microgram (µg) = 0.000001 g. For example, the units of micrograms per cubic meter of air ($µg/m^3$) are commonly used to express concentrations of particulate matter. Chemical concentrations in air may also be measured as parts per million (ppm) or parts per billion (ppb) by volume, e.g., to measure concentrations of ozone or oxides of nitrogen (NOx). Conversion equations can be applied to convert between measurement units such as to convert from concentration in parts per billion to concentration in micrograms per cubic meter ($µg/m^3$). For example, to convert from concentration units in ppb to µg/m3, we need a concentration in ppb multiplied by a conversion factor of 0.0409 multiplied by molecular weight of a pollutant (assuming 25°C and 1 atm pressure) (Boguski, 2006). Alternatively, WHO conversion factors may be applied for conversions between ppb and $µg/m^3$ developed for common air pollutants, for example, for ozone, 1 ppb = 1.96 $µg/m^3$ (assuming 25°C and 1 atm pressure).

In an assessment of health effects of air pollution, WHO established the Air Quality Guidelines in 2005 as well as thresholds for health-harmful pollution concentrations. However, in 2016, 91% of the global population resided in places, which did not attain the WHO air quality guidelines levels. A total of 4.2 million premature deaths globally per year was due to exposure to ambient (outdoor air pollution), specifically fine particulate matter, in both cities and rural areas in 2016 with low- and middle-income countries bearing the disproportionate burden (91% of these deaths), primarily in South-East Asia and Western Pacific (WHO, 2018). Another 3.8 million deaths occurred due to indoor air pollution from cooking, heating, and lighting their homes with polluting fuels in 2016. The $PM_{2.5}$ exposure causes cardiovascular and respiratory disease, and cancers.

In high-income countries, fossil fuel combustion is a major source (85%) of airborne particulate pollution and almost all pollution by oxides of sulfur and nitrogen. Fossil fuel combustion is responsible for the greenhouse gases, while stationary facilities, including electricity-generating plants, chemical manufacturing facilities, mining operations, emit carbon dioxide. Even within a single country, low-income, minorities, and those marginalized populations suffer disproportionately more from disease caused by pollution.

11.1.2.1 Particulate matter, PM (particle pollution)

The term PM is used for a complex mixture of various particles and droplets of organic and inorganic substances found in the air in solid and liquid form. Based on size of particles, PM is distinguished between coarse particles that are generally 10 μm (also called microns, with a symbol μm) and less (**PM_{10}**), such as dust, dirt, soot, etc., easily seen with the naked eye, and fine particles, which have diameters of 2.5 μm and less (**$PM_{2.5}$**). NO_2-based nitrate aerosols constitute an important fraction of PM2.5. Other PM2.5 pollutants that pose the greatest risks to human health include sulfate and black carbon.

Air quality is regularly measured as mean concentrations of **PM_{10}** particles (and/or of **$PM_{2.5}$** and smaller if appropriate measurement tools are available) expressed in micrograms per cubic meter of air volume daily or annually, i.e., μg/m³ (WHO, 2018). The daily metrics provide measures of *short-term peak exposures*, while *long-term exposure* is represented by annual metrics (Miranda et al., 2011).

More people are affected by particulate matter than by any other pollutant. Due to the small size (<10 μm) of airborne particles, **PM_{10}** can be inhaled and get deep into a person's lungs; however, fine particulate matter **$PM_{2.5}$** is more health damaging as it can even enter the bloodstream resulting in heart attacks, reduced lung function, deteriorating asthma symptoms, and untimely death. A recent analysis found significant associations between annual exposure to **$PM_{2.5}$** and mortality in the continental United States. To illustrate, the risk of death from all causes increases by 7.3% as annual exposure to **$PM_{2.5}$** increases by 10 μg/m³. The risk of dying increases with an increase in annual exposure to small particles; the association between PM2.5 exposure and mortality holds even at lower levels than those set by the current annual NAAQS for PM2.5 (concentrations below 12 μg/m³) (Di et al., 2017). Exposure to **$PM_{2.5}$** affects mortality even at low concentrations—even as low as about 5 μg/m³; studies find no known evidence for a no-effects threshold for fine particulate matter, i.e., there is no evidence of a threshold **$PM_{2.5}$** concentration value at which mortality is not affected (Schwartz et al., 2008; Shi et al., 2016). To provide some context, using 2002–2012 data, average **$PM_{2.5}$** concentrations across the continental United States range between 6.2 µg/m^3 (5th percentile) to 15.6 µg/m^3 (95th percentiles) yearly, with the eastern and southeastern United States and California having the greatest PM2.5 concentrations (Di et al., 2017). Further, according to the WHO Air Quality Guidelines, annual mean concentrations of **PM$_{10}$** and **PM$_{2.5}$** should not exceed 20 µg/m^3 and 10 µg/m^3, respectively, while 24-h mean concentrations should not exceed 50 µg/m^3 and 25 µg/m^3, respectively. **PM$_{2.5}$** is strongly correlated with respiratory illnesses. For example, a school site's exposure to **PM$_{2.5}$** expressed as emissions load per area was found highly significantly associated with the rate of emergency department visits for asthma among both children and adults within the school's tract (Gaffron and Niemeier, 2015). Acute and chronic health impacts occur even at the exposure to the low concentrations of fine PM.

Exposure to particles at even low levels can be lethal. A 2016 study revealed that New England's population aged 65 and older had a higher likelihood of dying prematurely from fine particle pollution even in places where current short-term particle pollution standards are being met (Shi et al., 2016), while a 2017 study observed a similar higher risk of premature death from particle pollution in Boston, which meets current standards for short-term particle pollution (Schwartz et al., 2017).

If particulate air pollution (small and fine particulates) is reduced, associated mortality will also decrease holding other factors the same (WHO, 2018). To illustrate the point, despite many urban areas in the European Union meeting the recommended limits of PM concentrations established in the 2005 WHO Air Quality Guideline, average life expectancy is 8.6 months shorter because of PM exposures from human sources. Fig. (11.2A) demonstrates a 42% decrease in national average in PM$_{2.5}$ between 2000 and 2016, while there was a further 22% decrease in PM$_{2.5}$ between 2010 and 2016 (Fig. 11.2B). Fig. 11.3 shows the location of natural coarser particles (green-colored areas), smaller particles (red areas), or a mixture of both in the atmosphere.

11.1.2.2 Ozone (O$_3$)

Ground-level ozone constitutes one of the major components of photochemical smog, so called because of the photochemical reaction of air pollutants such as nitrogen oxides (NOx) and volatile organic compounds (VOCs) with ultraviolet light (sunlight); thus, the O$_3$ levels vary strongly with season and the highest levels of ozone pollution are observed during sunny weather.

O$_3$ is a reactive gas; it causes inflammation and has large-scale temporal fluctuations (Olsson et al., 2013). There are serious risks to health from exposure to excessive ozone in the air as it is a major risk factor in asthma morbidity and mortality. Besides triggering asthma, other marked effects ozone has on human health include breathing problems, reduced lung function, and lung diseases (WHO, 2018).

A study described by Di et al. (2017) on the relationship between annual exposure to particulate matter and mortality similarly found significant associations between annual exposure to ozone and mortality in the continental United States. The risk of dying increases by 1.1% with an increase in annual exposure to ozone by 10 parts per billion (ppb). For context, average ozone concentrations vary between 36.27 and

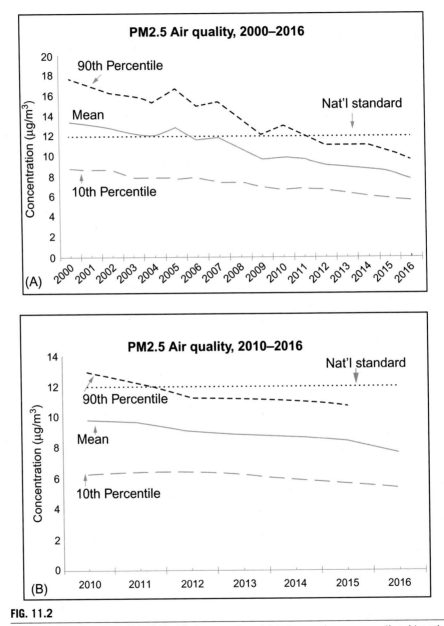

FIG. 11.2

PM$_{2.5}$ Air quality trends: (A) 2000–16. Seasonally weighted annual average, national trend based on 455 sites, (B) 2010–16. Seasonally weighted annual average, national trend based on 648 sites.

Source: Author, using data at: https://www.epa.gov/air-trends/particulate-matter-pm25-trends.

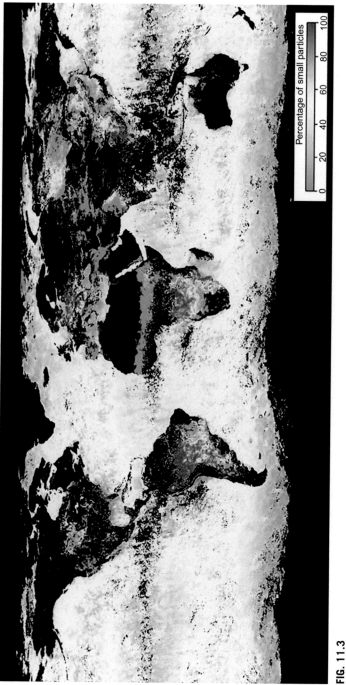

FIG. 11.3

The false-color map of Earth shows the location of the plume of natural aerosols composed of coarser particles *(green-colored areas)*, human pollution made up of smaller particles *(red areas)*, or a mixture of both large and small aerosol particles *(yellow areas)*. Aerosols are tiny solid and liquid particles suspended in the atmosphere.

Source: Aerosol Particle Radius (1 month - Aqua/MODIS, September 2016. Available at: https://neo.sci.gsfc.nasa.gov/.

55.86 ppb during the warm season with the highest amounts of ozone concentrated in the US Mountain region and California, while the US East had lower ozone concentrations during 2002–12 (Di et al., 2017). The 8-h mean level recommended for ground-level ozone by the 2005 WHO Air Quality Guidelines is $100 \mu g/m^3$ (WHO, 2018). Using a conversion equation described here, average ozone concentrations expressed in ppb can be converted to $\mu g/m^3$ resulting in average ozone values ranging between $71.1 \mu g/m^3$ to $109.5 \mu g/m^3$, which is above the WHO-recommended level.

Air pollutants play a significant role triggering asthma attacks with asthma found associated with exposure to ozone among girls below 10 years of age (Szyszkowicz, 2008).

11.1.2.3 Nitrogen dioxide (NO₂)

NOx is a generic term for the nitrogen oxides, which are a mixture of gases composed of nitrogen and oxygen. All NOx gases are harmful to human health and the environment. Two most common gases include nitric oxide (NO) and nitrogen dioxide (NO_2), which contribute to formation of smog, acid rain, and ground-level ozone and are thus most relevant for air pollution. NOx is emitted by vehicles and industry. NOx has been found to have a stronger association with SES than $PM_{2.5}$ with the former being highly dependent on proximity to major roadways with busy traffic, which may also spatially colocate with residence of communities with low SES (Hajat et al., 2013).

NO_2 is a precursor of ozone formed in the presence of sunlight. The annual mean level set to protect the public from the health effects of NO_2 by the 2005 WHO Air Quality Guidelines is $40 \mu g/m^3$. The 1-h mean level recommended not to exceed is $200 \mu g/m^3$; when short-term concentrations exceed the value, the gas is toxic gas causing significant airways inflammation (WHO, 2018). Long-term exposure to NO_2 increases symptoms of bronchitis in asthmatic children. Epidemiological studies have linked currently observed concentrations in cities of Europe and North America and reduced lung function growth.

11.1.2.4 Sulfur dioxide (SO₂)

SO_2 is an invisible sharp-smelling gas. Anthropogenic activity is the main source of sulfur dioxide in the air such as industry (e.g., the generation of electricity from sulfur-containing coal, oil, or gas; burning fossil fuels containing sulfur), motor vehicle emissions from fuel combustion. Combining with water, SO_2 forms sulfuric acid, which is the main component of acid rain.

The 10-min mean level set by the 2005 WHO Air Quality Guidelines is $500 \mu g/m^3$, and 24-h mean is $20 \mu g/m^3$. The effects are felt very quickly; averaged over a 10-min period, a SO_2 level of $500 \mu g/m^3$ should not be exceeded since most people would feel the worst symptoms after a short exposure as short as 10 min, and people with prior asthma experience or similar conditions may experience changes in pulmonary function and respiratory symptoms even after a short exposure to SO_2. On days with higher SO_2 levels, there is an increase in hospital admissions for cardiac disease and mortality.

11.1.3 **Outdoor air pollution and health**

Outdoor air pollution is a serious risk factor for public health (noncommunicable diseases) affecting urban populations in both developing and developed countries. It is a major environmental health problem. According to the American Lung Association's 2018 State of the Air report, 41% of the US population, or over four in 10 people, live in counties that have unhealthful levels of pollution emissions in terms of either ozone or particle in 2014–16.

Motor vehicle emissions are a major source of air-borne pollutants; the US urban residents who largely rely on private vehicles contribute substantially to air pollution. To illustrate, local journey-to-work flow data by Census Transportation Planning Package (CTPP) 2006–10 indicate that 93% of Memphians use motorized vehicles as a primary travel mode for commuting contributing to congestion (Fig. 11.4). Although shown for a major city in Tennessee, commuter choice is similar in other major metropolises across the US, nearly nine in 10 (86%) of Americans commute to work by car according to the latest 2016 American Community Survey (ACS) data, while according to the US Energy Information Administration's (2018) *Monthly Energy Review June 2018* report, transportation is by far the largest consumer of petroleum (e.g., motor gasoline, aviation gasoline, kerosene, hydrocarbon gas liquids, etc.), followed by industrial, residential and commercial, and electric power sectors.

Transport is a major source of outdoor air pollution (e.g., motor vehicles are the primary sources of CO emissions) with overall traffic density and diesel pollution from large trucks and buses affecting outdoor air quality in suburban and urban areas (Patel et al., 2009). To illustrate, a nearly 42% decrease in the rates of asthma-related emergency department visits and hospitalizations among Medicaid-enrolled children in Atlanta was associated with a reduction of motor vehicle traffic congestion during the 1996 Olympics (Friedman et al., 2001), while higher prevalence of asthma and

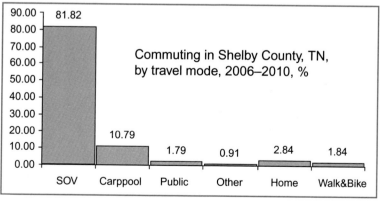

FIG. 11.4

Commuting mode in Shelby County, 2006–10.

Source: Author, using Census Transportation Planning Package (CTPP), 5-year data set (2006–10), Part 3. Available at: http://data5.ctpp.transportation.org/ctpp/Browse/browsetables.aspx

higher exacerbation rates are observed among the residents living close to interstate highways (Li et al., 2011). Besides transport, other key sources of outdoor air pollution include homes, power generation, industry, and municipal waste management (WHO, 2018). Developing and investing in cleaner transport, energy-efficiency of homes, cleaner power generation, industry and better municipal waste management could reduce ambient (outdoor) air pollution.

Research has increasingly focused on understanding the potential health effects of air pollution. Studies increasingly document that ambient air pollution negatively impacts human health. Evaluation of human health impacts of $PM_{2.5}$ and ozone is particularly needed, as $PM_{2.5}$ and ozone have been linked to adverse health outcomes through the impact on respiratory and cardiovascular systems resulting in mortality and hospital visits as well as poor birth outcomes, even at low concentrations (Miranda et al., 2011). To illustrate the latter, based on a global study of live births, including 183 individual countries in 2010, 18% of all global preterm births are estimated to be associated with anthropogenic $PM_{2.5}$, with high preterm birth rates observed in both high- and low-income countries, suggesting implementation of emission reduction strategies in order to reduce maternal $PM_{2.5}$ exposure as well as reduction of other risk factors related to preterm births (Malley et al., 2017). Similarly, mean birth weight was found lower among infants whose mothers had been exposed to ambient particulate matter (PM_{10}) from the Southern California wildfires of 2003 during any trimester (with a small for clinical relevance but significant decrease during second and third trimesters) compared with unexposed infants even after adjusting for maternal characteristics (age, educational attainment, parity, race/ethnicity), with the largest estimated effect observed in the second trimester, in a time-series study of the birth effects of acute maternal exposures to smoke from wildfires hazards (Holstius et al., 2012). Since wildfires are predicted to increase in frequency and magnitude due to climate change, the study has important infant health and development implications through the potential increase of extent of maternal air pollution exposure from wildfire hazards, especially in areas such as the western United States.

Both short-term and long-term exposures to poor air quality may increase the frequency of several health events. In 1993, the landmark Harvard Six Cities Study was published (Dockery et al., 1993). Since 1993, extensive studies provided evidence of the overall association between air pollution and human health and further discussed in the following content.

11.1.3.1 Air pollution and excessive mortality

Extensive studies document associations between air pollution and multiple adverse health outcomes. Prior studies linked exposure to air pollution with shortened life expectancy and higher morbidity (Hoek et al., 2002; Moual, et al., 2013; Brugge et al., 2007), while ambient air pollution plays a very substantial role in cardiovascular illness and death with linkages between exposure to pollution and risk of cardiovascular illnesses evidenced by many studies, including those from areas with high levels of air pollution (WHO, 2018). While all air pollutants affect health and environment, fine particulate matter ($<2.5\,\mu m$ in diameter, $PM_{2.5}$) and ozone are particularly harmful to human health, being responsible for 130,000 and 4700 excess deaths, respectively, using the 2005 air quality level (Fann et al., 2012). Older people have the highest mortality rates resulting from these pollutants. In major cities, deaths caused by $PM_{2.5}$ and ozone range from 3.5% in San Jose to 10% in Los Angeles.

I give example of a potential association between life expectancy at birth correlating visually with environmental exposure indicator in Memphis (also, here referred to as Shelby County), Tennessee (Fig. 11.5B, with darker colors indicating higher exposure to carbon monoxide, or CO). Average life expectancy at birth in Shelby County is lower than in the state (76.2 vs 76.3), while both statistics are lower than nationwide (78.88). The average statistic masks, however, big spatial variation across zip code areas in the county: Fig. 11.5A reveals that life expectancy at birth in Shelby County fluctuates substantially ranging from 70.3 in the west to 84.15 in the east. Further, recent mortality in Shelby County has been steadily increasing since 2011, as shown in Fig. 11.6, with chronic diseases, including cardiovascular, cancer, and diabetes being in the list of ten leading causes of death in Shelby County as of 2015 (Shelby County Health Department, 2015).

11.1.3.2 Industrial air pollution and health
"The average American is exposed to about ten pounds of toxic releases per year" (Bullard, 2000, p. 134). Pollution negatively impacts environment; on the other hand, polluted urban sites also represent a substantial barrier to economic urban redevelopment and remain undeveloped and undercapitalized. There is growing documented evidence to suggest that many industrial facilities are colocated with concentrated

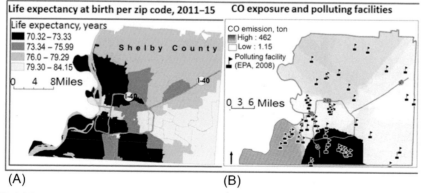

FIG. 11.5

(A) Life expectancy at birth in Memphis, Tennessee. (B) Carbon monoxide exposure and distribution of polluting facilities in Shelby County.

Source: (A) Author, using data based on Shelby County Health Department, 2015. (B) Source: Author, using data based on EPA.

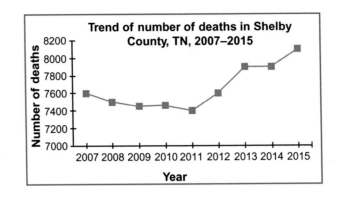

FIG. 11.6

Number of deaths in Shelby County, 2007–15.

Source: Author, using data based on Shelby County Health Department, 2015.

minority and low-income populations and households, raising important questions about environmental justice.

For example, electricity-generating power plants represent one of the major point sources of air pollution. Local-area variation in distribution and exposure to particulate-matter and nitrogen or sulfur components of air pollution is affected by proximity to power plants (Strickland et al., 2010).

Power plants emit pollution, which alone results in approximately 13,000 excess deaths annually in the United States (American Lung Association, 2011). Besides $PM_{2.5}$, which is a major component of power plant emissions (Buonocore et al., 2014), additional sources of air pollution emitted by power plants harmful to human health include toxic chemicals such as mercury, heavy metals, and acid gases, thus increasing local residents' risk of having negative health outcomes (EPA, 2012). Compared with other types of power plants, nuclear plants emit less atmospheric pollutants (e.g., $PM_{2.5}$, sulfur dioxide, and nitric oxides) (EPA, 2014). Recently, it has been suggested that dangerous emissions from power plants may be increasing (Environmental Integrity Project, 2013).

11.1.3.3 Air pollution and cardiorespiratory diseases

The range of health effects is broad; however, air pollution has been predominantly linked to specific diseases, such as cardiovascular and respiratory diseases. By lowering the levels of air pollution, the cardiovascular and respiratory health of the population will improve over long- and short term (WHO, 2018).

As Fig. 11.7 portrays, chronic exposure to outdoor air pollution is an important risk factor for deaths from cardiovascular and respiratory diseases. Ischemic heart disease and strokes accounted for 58% of outdoor air pollution-related premature deaths in 2016, chronic obstructive pulmonary disease and acute lower respiratory infections each were responsible for 18%, and lung cancer accounted for 6% of deaths (Fig. 11.7).

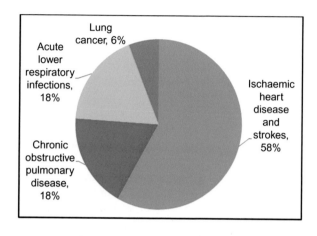

FIG. 11.7

Outdoor air pollution-related premature deaths in 2016, by disease.

Source: Author, using data based on World Health Organization (WHO), 2018. Ambient (Outdoor) Air Quality and Health. Fact sheet no. 313. Updated May 2018. World Health Organization, Geneva. Available at: http://www.who.int/news-room/fact-sheets/detail/ambient-(outdoor)-air-quality-and-health.

11.1.3.4 Air pollution and cancer

Air pollution is carcinogenic to humans. Particulate matter has been blamed for an increase in cancer incidence, specifically lung cancer. However, lung cancer-related mortality is related to both risk factors—ambient air pollution and also tobacco smoking (WHO, 2018).

11.1.3.5 Air pollution and children

Children are more sensitive to environmental exposures than the general population. It is known that children are affected by emissions in a different way than adults: children breathe in more air relative to their weight (Kleinman, 2000), and due to their organs still developing (American Lung Association, 2010; EPA, 2006; Bearer, 1995).

Miranda et al. (2011) compared communities with the most extreme air quality across the entire United States, i.e., identified the top and bottom US counties (such as 20% of the US counties with the best air quality and 20% of the US counties with the worst air quality) for daily ozone, daily $PM_{2.5}$, and design values for annual $PM_{2.5}$ and found that counties with the worst air quality are demographically younger, that is, percent of children younger than five years of age is higher compared to counties with the best air quality for each pollution metric. However, the opposite result was found at a more refined geographic scale such as that within the 5-km buffer zones constructed around air quality monitoring sites with the younger age associated with a decreased risk of being in areas around the 20% of monitors with the worst air quality in terms of $PM_{2.5}$ but increased likelihood of being in the worst ozone areas (Miranda et al., 2011). These findings may have important health implications for young children due to their higher pollution exposure to both ozone and fine particulate matter at the county level, and higher ozone exposure around air quality monitoring sites in the United States.

Children are especially vulnerable to air pollution and other environmental hazards. A recent 2017 comprehensive *Don't pollute my future!* report reviews the polluted environment's impact on children's health, stating that top 5 causes of death in children younger than 5 years of age have been linked to the environment (WHO, 2017). Among the causes listed are respiratory infections, including pneumonia, claiming over 500,000 lives of children under 5 years due to exposure to indoor and outdoor air pollution and second-hand smoke.

In the United States, one of the most prevalent chronic noncommunicable diseases is asthma, which negatively impacts quality of life and imposes a significant economic burden on healthcare services. Asthma frequently starts in the childhood. Fortunately, from 2001 to 2016, the percentage of children with asthma who had an asthma attack diminished significantly from 61.7% to 53.7%, indicating improved asthma outcomes (Asthma in US children, 2018).

Environmental conditions have the potential association with emergency treatment for asthma. In 2000, Robert Bullard wrote that the disease was "an emerging epidemic in the United States," with asthma affecting about 5 million children under 18 years of age in that year (Bullard, 2000, p. 140). According to the National Health Interview Survey, over 8% of adult Americans of at least 18 years of age (20.4 million) were affected by asthma in 2016, more Blacks than Whites in this age category (10% versus 8.1%) (CDC, 2016). For children under age 18 year, 8.4% (6.1 million) had asthma in 2016 (CDC, 2016), making it the most common chronic lung disease in childhood resulting in missed school days and hospitalizations (Asthma in US children, 2018).

Despite the aforementioned improvements in asthma outcomes, there are tremendous racial/ethnic, and socioeconomic disparities driven by social and environmental factors, including poor housing and exposure to smoking. For example, Black children are affected by asthma almost twice as much as Whites (15.5% and 6.8%, respectively) (CDC, 2016). Children from low-income households are also disproportionately affected, 11.7% of children had asthma when family income was less than $35,000, versus 7.1% of children from households with $35,000 or more (CDC, 2016).

Local context is also important with variation across neighborhoods: when children reside or attend schools close to roads with high traffic density, they are more susceptible to asthma due to exposure to higher levels of motor vehicle air pollutants. Epidemiologic research on air pollution has reported higher asthma prevalence, especially exposure to nitrogen dioxide (NO_2), nitrous oxide (N_2O), and carbon monoxide (CO) is positively related to a higher prevalence of childhood asthma, while children exposed to ozone and particulate matter (PM) have higher risk of developing wheeze (Gasana et al., 2012).

As stated earlier, asthma may be triggered by many outside and inside environmental factors, including those encountered in schools and workplaces, and it can be fatal. A Canada-based study assessed the link between the exposure to ambient air pollution (carbon monoxide (CO), nitrogen dioxide (NO_2), sulfur dioxide (SO_2), ozone (O_3)), and $PM_{2.5}$ and PM_{10} and daily emergency visits for asthma trying to quantify the short-term effects of the concentrations, i.e., what air pollution concentration and what type of pollutant result in the increased number of asthma-related visits among patients stratified by age (younger than 10 years of age and older that 10 years of age) and by gender (Szyszkowicz, 2008). The study applied the generalized linear mixed models methodology where data were grouped into clusters, which follow a hierarchical structure and constitute multilevel relations (i.e., days for which data such as asthma-related records, air pollution data, and meteorological variables are available, are nested in weeks, which in turn are nested in months, and months are nested in years). A model included the number of asthma-related visits as a response (dependent) variable and the pollutants (measured as daily mean values of the ambient air pollutant levels) and two weather parameters (temperature and relative humidity) as confounders used as independent variables. Models have been constructed for the current-day, 1-day, and 2-day lagged exposures to ambient air pollutants with the variables in the models lagged by the same number of days (0, 1, or 2). Ozone exposure was found to result in the highest percentage increase in the number of asthma-related visits (17.8%) from April to September among girls younger than 10 years old, while for the boys of the same age and during the same season the exposure to nitrogen dioxide was associated with increased asthma-related emergency visits.

11.1.3.6 Air pollution and hospitalizations

Air pollution has been related to increases in hospitalizations. Researchers use the environmental (in)justice framework in studies on the uneven burden of hospitalizations caused by environmental hazards among susceptible populations. Grineski et al. (2013) analyzed the link between to air toxics exposure and respiratory-related hospital admissions among children in El Paso, Texas.

Researchers seek to answer: (1) whether there are social inequalities in exposures to environmental hazards in El Paso, Texas, a community where the majority of residents are Hispanic; (2) whether being exposed increases children's hospitalizations for asthma and respiratory problems controlling for sociodemographic predictors; and (3) which sociodemographic variables are associated with children's hospitalizations for asthma and respiratory problems controlling for environmental exposures. Home addresses of children, who were hospitalized in 2000, have been mapped and all cases of children's asthma and respiratory infection have been used to calculate hospitalization rate variables (i.e., number of hospitalizations for children under age fifteen per 1000 children under age fifteen in the US census tract). Environmental exposure was estimated using publicly available data from the US EPA's NATA, which provide data for 177 air toxics listed in the Clean Air Act (out of total of 187), plus diesel particulate matter. *Air toxics* (also *hazardous air pollutants*) are pollutants related to cancer and other noncancer serious health effects (EPA. Assessing Outdoor Air Near Schools. Available at: https://www3.epa.gov/air/sat/about.html Accessed 8 March 2018). Under the assumption that the risks of different pollutants are additive, these pollutants can be summed to estimate an aggregate score for each census tract. The score then is used to calculate public health risks from inhalation of air toxics. Among confounders were included proportion in poverty, proportion of female-headed households as risk factors.

11.1.3.7 Air pollution and birth outcomes

A large evidence base indicates associations between exposure to air pollution, including carbon monoxide (CO), nitrogen dioxide (NO_2), particulate matter (PM), and pregnancy outcomes (Stieb et al., 2012). Poor birth outcomes may result from exposure to particulate matter and ozone (Gray et al., 2010; Ritz et al., 2007; Bell et al., 2007; Ritz et al., 2002). The following sections will define adverse birth outcomes and provide examples of studies on the effect of air pollution exposure from mobile sources (i.e., traffic-related) based in the United States and Europe and stationary sources (here, industrial facilities such as power plants) on pregnancy outcomes.

Adverse birth outcomes commonly studied include preterm birth, low birth weight (LBW), and small for gestational age (SGA). Preterm (also known as premature delivery) occurs earlier than 37 weeks of gestation. Low birth weight is defined as babies whose weight at birth is <2500 g, and SGA is defined as babies with birth weight below the 10th percentile at a specified gestational age, or weighing <2 SDs (below the mean of the population at a specified gestational age).

Preterm birth is an important risk factor of infant mortality and long-term health impact, including increased risk of death and developing various chronic physical and neurological diseases. For example, in 2013, preterm birth complications were responsible for 15% of all deaths of children under 5 globally (Malley et al., 2017).

Although the pathways linking maternal air pollution exposures and adverse birth outcomes are not well understood, women exposed to air pollution during their pregnancies may have poor pregnancy outcomes through several potential mechanisms. It is known that air pollution exposure (especially ozone) might lead to inflammation and asthma exacerbation; thus, pregnant women affected by asthma might be greater affected. PM exposure through inhalation of airborne particles may cause acute inflammation in the lungs and other organs, including the placenta increasing the likelihood of preterm labor. Increased systemic inflammation may also interfere with nutrient transport to the fetus and decrease oxygenation of maternal blood contributing to intrauterine growth restriction; air pollution exposure during pregnancy can induce oxidative stress and can trigger preterm birth (Lavigne et al., 2016). Increased respiratory symptoms and systemic inflammation due to O_3 exposure are hypothesized to increase a risk of preterm birth among asthmatic mothers more compared to nonasthmatic mothers. If the asthmatic mother controls asthma poorly, the risk of preterm delivery might increase via maternal hypoxia, thereby obstructing fetal oxygenation. Additionally, oxidative stress may lead to preeclampsia (pregnancy-related disorder with hypertension one of the symptoms), an important predictor of preterm birth (Olsson et al., 2013).

Stieb et al. (2012) reviewed sixty-two studies on the relationship between ambient air pollution, and birth weight and preterm births reporting that despite wide variation among the studies regarding pollutants and outcomes, most of the studies observed a reduction in birth weight and an increase in the odds of low birth weight associated with exposure to carbon monoxide (CO), nitrogen dioxide (NO_2), and particulate matter <10 and 2.5 μm (PM_{10} and $PM_{2.5}$). Exposure during entire pregnancy generated largest effect estimates. To illustrate, reductions in birth weight ranged from 11.4 g per 1 ppm CO to 28.1 g per 20 ppb NO_2, while pooled odds for low birth weight increased by 10% on the average per 20 μg/m3 PM_{10}. Preterm births had smaller effects estimates with mixed results with pooled odds ratios ranging from an increase by 4% per 1 ppm CO to 6% per 20 μg/m3 PM_{10}. All birth outcomes revealed less consistent results for exposure to ozone and sulfur dioxide. The observed inconsistent results with regard to the reported association between O_3 and the incidence of preterm birth may be explained by the difference between measured concentrations and actual exposure due to within-city variations and the use of air conditioning (Olsson et al., 2013).

A time-series study examined the relationship between maternal exposure to natural wildfire hazards in Southern California in 2003 and poor pregnancy outcomes, specifically measured by birth weight of infants born during the wildfire exposure window, i.e., between 21 October and 10 November 2003, and thus, potentially exposed to air pollution from the wildfire in utero, and that of those infants born before and after the wildfire event (considered unexposed in utero) (Holstius et al., 2012). Two following pathways may explain poor health outcome (here, lower birth weight) after maternal exposure to wildfire (or any other hazard): a biological category links weight decrement and exposure to air pollution from the fires, while a psychosocial pathway links lower birth weight and direct or indirect wildfires-related (or any other hazard-related) stress caused by loss of property, shelter, money, physical injury, among others. The cut point was established at average daily PM_{10} measure of 40 μg/m^3, with tracts below this value considered low exposure, while tracts with average daily levels of $PM_{10} > 40$ μg/m^3 classified as high exposure. There were slightly under 750,000 unexposed births and slightly under 140,000 exposed births within the affected counties, including Orange County, Los Angeles, San Bernardino, and Riverside. Out of births that gestated during the fires, 36% of births occurred in high-exposure census tracts. The model was developed with birth weight as a continuous outcome and an indicator of exposure for birth by trimester as an independent variable controlling for maternal socioeconomic status (proxied by education with completed high school or equivalent being reference; and race/ethnicity with non-Hispanic white as reference); and birth characteristics (Holstius et al., 2012). The estimated decrease in birth weight associated with pregnancy during the wildfires event was similar across low- and high-exposure tracts.

A recent Salt Lake Valley study by Hackmann and Sjöberg (2017) analyzed the effect of exposure to ambient air pollution ($PM_{2.5}$) during the first trimester on births. Specifically, the risk of preterm birth, an important pregnancy outcome, was studied. Logistic regression models used an outcome variable (the binary outcome term birth) and independent variables (other air pollutants such as CO, ozone, and NO_2 and individual confounders) finding strong evidence of increased probability of preterm birth resulting from $PM_{2.5}$ exposure controlling for confounding variables.

Similar results were found by Lavigne et al. (2016) in a retrospective study based in Ontario, the largest province of Canada examining the effects of air pollution on birth outcomes. Robust exposure estimates of $PM_{2.5}$, NO_2, and O_3 were assigned to a large cohort of births. Their findings for $PM_{2.5}$ and NO_2 are consistent with prior meta-analyses observing that a risk of preterm birth increases with exposure during the entire pregnancy (Stieb et al., 2012); to illustrate, the odds of preterm birth increase by 3.8% and 6.5% among women for exposure to $PM_{2.5}$ and NO_2, respectively, over the entire pregnancy, while women with preexisting diabetes exhibited much higher effects of NO_2 exposure with the odds of preterm birth increasing by 23.8%. Pollutant models were adjusted for area-level SES factors only and did not include individual risk factors such as ethnicity, income, and education. The study found that overall exposure to $PM_{2.5}$, NO_2, and O_3 during the entire pregnancy was linked to preterm birth, while exposure to O_3 appears to be positively associated with small for gestational age and term low birth weight.

A large European study using birth data from >130,000 deliveries, based in Stockholm, Sweden, examined the effect of exposure to traffic-related air pollutants during early pregnancy (i.e., the first trimester, which was defined as the first 12 weeks of gestation) on the risk of adverse birth outcomes (preterm birth and small for gestational age, SGA). SGA is more strongly associated with vehicle exhaust. Daily city-wide levels of two traffic pollution indicators, including NOx (diesel traffic is a major source of NO_2 compared to petrol cars, thus it was selected as a proxy for diesel vehicle exhaust), and ozone (O_3), were used to calculate mean pollution levels for the first trimester for each pregnancy, which were used in the logistic regression models. Although Stockholm has a relatively clean air, a higher percentage of diesel vehicles in European cities might increase both NO_2 and O_3 levels. Air pollution exposure was defined as having an air pollution concentration above the 25th percentile. It was found that preterm birth may be caused by higher levels of O_3 exposure during the first trimester, each $10\,\mu g/m3$ increase in O_3 increased the odds for being born preterm by 4%–11% adjusting for various confounders (maternal age, parity, level of education, area of origin, smoking, maternal asthma, season of conception, preeclampsia). However, in agreement with less consistent prior results with regard to the association between O_3 exposure and preterm birth (Stieb et al., 2012), no association was observed between the first trimester NOx exposure and preterm birth. Similarly, neither negative effects of exposure to higher levels of O_3 during the first trimester nor of higher first-trimester NOx levels were observed with regard to the SGA (Olsson et al., 2013). The authors note that one can find more likely an association between a variable that fluctuates strongly over time (such as the O_3 concentrations) and the studied outcome; however, when a variable exhibits a strong spatial variation rather than temporal such as the NOx level, it is less likely to find any associations. Additionally, only the first trimester exposure was studied while other exposure windows may exhibit other air pollution effects.

A prospective birth cohort study based in the Netherlands studied how residential traffic exposure impacts pregnancy-related outcomes such as birth weight, and the risks of preterm birth (defined as gestational age <37 weeks) and small size for gestational age at birth (<-2.0 SDS birth weight) (Van den Hooven et al., 2009). To proxy for traffic-related exposures, residential proximity to traffic was used accounting for within-city contrasts. Two traffic-derived indicators were used for residential proximity to traffic, including distance-weighted traffic density in a 150-m radius, and proximity to a major road defined as carrying the total traffic flows with $>10,000$ vehicles/24 h. In the study, residential proximity to traffic did not seem related to birth and pregnancy outcomes, so researchers concluded that exposure to residential traffic has no higher risk of adverse birth outcomes among pregnant women in the Netherlands.

Studies analyzing how stationary sources such as power plants might affect pregnancies, found that women residing close to power plants are more likely to suffer from adverse birth outcomes, including preterm delivery (PTD), very preterm delivery (VPTD), and term low birth weight (LBW). Ha et al. (2015) studied how residential proximity to different types of power plants may be related to negative birth outcomes. They used singleton births in Florida, which has relatively high power plant emissions (Environmental Integrity Project, 2013) using birth data for 2004–05. During the study period, there were 150 active nonrenewable-source power plants in Florida. Broken down by fuel type, most were gas-powered (66), followed by solid waste and oil (29 and 28, respectively), 17 coal plants, 7 plants with other fuel types (e.g., coke, etc.), and 3 nuclear plants. The study applied proximity-based approach as a surrogate for exposure calculating residential proximity to the nearest active nonrenewable-source power plant for each birth, with distance measured in kilometers. Several categories of buffers have been used, including <5, 5–9.9, 10–19.9, and ≥ 20 km.

Because of spatial clustering of power plants, the areas with more power plants produce higher $PM_{2.5}$ concentrations. The amount of air pollution depends on fuel type used by power plants: living closer to power plants powered by coal and solid waste exposes women to higher levels of fine particulate matter (<2.5 μm in diameter), followed by gas, oil, other, and nuclear plants.

Proximity to any power plant increases the likelihood of all negative birth outcomes with closer residents having the higher odds even after potential confounders have been adjusted for (this is known as an exposure-response relationship). As the distance to any power plant decreases by 5 km, the odds of term LBW, PTD, and VPTD increase by 1.1%, 1.8%, and 2.2%, respectively.

11.2 Socioeconomic conditions and health

Socioeconomic status (SES) is a complex construct comprising multiple domains, including income, education, and occupation. Thus, using only one indicator of SES (e.g., income or education) may not adequately reflect this relatively complex construct of SES. On the other hand, many demographic variables considered in the literature (e.g., percentage of female-headed households, etc.) may be highly correlated with key socioeconomic measures, so researchers need carefully choose which variables to select. To represent advantaged and disadvantaged communities in the United States at a national county level, Miranda et al. (2011) suggested using the following key metrics of environmental justice: race, ethnicity, age, and poverty rates.

Socioeconomic conditions have a strong influence on health outcomes. Among the general population, health outcomes may be more severe among *socially disadvantaged* groups compared to those experienced among more affluent populations groups. Bullard (2000, p. 6) includes the following characteristics associated with disadvantage: poverty, occupations below management and professional levels, low rent, and a high concentration of black residents caused by residential segregation and discriminatory housing practices and poor air quality. To illustrate, poverty is linked to higher asthma prevalence rates: between 2001 and 2003, children in families falling below the federal poverty line had 30% higher asthma prevalence rates than those in families above it (Rauh et al., 2008). Similarly, low-income children, defined as those qualified for the Medicaid coverage, are disproportionately affected by asthma, including visits to the emergency department across counties in the US South with consistently poor rankings on obesity, diabetes, and breast cancer; moreover, stress from poverty, violence, and neighborhood disadvantage, including poor housing and overcrowding, may influence the risk of asthma with tenants in poverty having little control over these factors (Malhotra et al., 2014). Adverse birth outcomes tend to occur more frequently among women whose neighborhood median family income indicator is the lowest (Lavigne et al., 2016).

Because of its links to many health outcomes (Adler and Stewart, 2010), it is hard to understand the extent to which health disparities are attributable to air pollution as observed associations of air pollution with health outcomes maybe confounded by other social factors with socioeconomic status (SES) being a major concern as a potential cause of residual confounding; in turn, making it imperative to understand how SES and air pollution exposure are related so that causality based on statistical associations between air pollution and health could be inferred and the causes of the many pollution-related health disparities could be understood (Hajat et al., 2013). Due to this confounding effect of socioeconomic factors on health, researchers apply methodologies to remove these effects of socioeconomic factors and evaluate the association between environmental risk factors and health outcomes. After many confounding variables are accounted for, including demographic characteristics such as age, sex, and socioeconomic features, environmental conditions appear to be associated with all-cause mortality rates (Pearce et al., 2010).

11.3 Socioeconomic (SES) conditions and environmental exposure

The *environmental justice* movement began with the concern that low-income and minority communities such as African American neighborhoods in the US South (Texas, Louisiana, West Virginia, and Alabama), the nation's one of the most underdeveloped regions, where they constitute the region's largest racial minority group, were disproportionally burdened by the exposure to pollution and other environmental stressors such as municipal landfills, chemical plants, paper mills, lead smelters, toxic dumping sites, and other hazardous facilities in the 1980s as described by Robert Bullard in *Dumping in Dixie: Race, Class, and Environmental Quality*, the first book on environmental justice (Bullard, 2000).

Residents who live in close proximity to industrial pollution have greater environmental and higher health risks of emphysema, chronic bronchitis, and other chronic pulmonary diseases than in the general population. To illustrate two of the five cases described by Bullard, a public housing project had been built in the mid-1950s near West Dallas' lead smelters pumping nearly 270 tons of lead particles yearly into the air and exposing the residents to this environmental poison for 50 years of its operation. The 1981 health report commissioned by the EPA revealed that young children residing in two minority neighborhoods close to smelters had high health risks associated with lead poisoning through exposure to elevated lead concentrations in the soil, air, and houses; however, no immediate results followed from the EPA or the city to improve the lead contamination problem in West Dallas. Yet another case describes how the country's largest hazardous-waste treatment, storage, and disposal facility, nicknamed the "Cadillac of dumps," has been sited in Emelle, a 90% black, small community in Sumter County, Alabama. Opened in 1978, the Emelle landfill received Superfund wastes dumped at the facility. The three most severe environmental problems in Emelle rated by community residents were hazardous waste, toxic chemical leaks, and closeness of industry to their homes, while air pollution has been consistently rated as the most severe by the residents in other communities (Bullard, 2000).

Several causes of this disproportionate environmental burden experienced by many black and low-income communities have been identified by Bullard: (1) polluting facilities follow the "path of least resistance," i.e., black communities have been disproportionately burdened with these facilities; economically poor areas where minorities tend to reside are identified and targeted for nonresidential land uses, including the siting of noxious facilities, locally unwanted land uses or LULUs, and environmental hazards such as power plants, chemical plants, hazardous waste dumps, and other polluting facilities, (2) lacking political and economic power, these communities had a few national-level environmental advocates campaigning for the environmental rights of minority and low-income populations and pushing for environmental equity and improvements in local environmental quality in the black and low-income communities. Instead, mainstream environmental organizations with an influence on the nation's environmental policy emphasized largely preservation and outdoor recreation, while leisure time and the remotely located wilderness areas and national parks were typically inaccessible to the inner-city poor, thus, finding little support among working-class persons, black community residents, and the rural poor. Additionally, national media provided only limited coverage of polluted black communities, while governmental agencies took only limited remedial actions in contaminated minority communities with the many "Black Love Canals" (such as all-black town of Triana in northern Alabama with high levels of pesticide DDT contamination) going unnoticed, (3) historically, research on low-income and minority communities did not focus on environmental quality with attention mostly being on other issues, including crime, drugs, poverty, family crisis, and unemployment, (4) minority groups have limited mobility options with racial barriers to education, employment, and housing preventing residents from moving to environmentally safer areas and resulting in limited environmental choices for black households, (5) the development of spatially differentiated metropolitan areas with segregated blacks and other "visible minorities" in central-city neighborhoods, where vehicle traffic is heaviest and environmental quality is worse than in the suburbs, was fueled and accelerated by federal policies, including government housing policies and highway construction projects causing whites decentralize to the suburbs and abandon urban cores (Bullard, 2000).

According to the *"environmental justice"* or *"environmental equity"* principle, environmental hazards or nuisances and associated negative environmental exposures should not impact disproportionately population groups of a certain socioeconomic and demographic status. *"Environmental justice"* and *"environmental inequity"* studies analyze the socioenvironmental disparities. Research in environmental inequality has important health implications, including analysis of links between environmental inequality and health (Hajat et al., 2015). Hajat et al. (2015) describe the *triple jeopardy hypothesis* arguing that communities with low SES are challenged with (1) higher environmental exposure, including air pollution and other hazards because air pollution sources (including mobile and point source emissions) are frequently spatially clustered, (2) higher vulnerability to poor health caused by greater psychosocial stress such as discrimination, unhealthy behaviors, and poorer health status, which combined lead to (3) environment-driven health disparities.

The North American literature consistently reveals that disadvantaged individuals and communities, i.e., those with lower SES factors (SES can be captured by various categories such as income, wealth, education, employment, occupation), are exposed to higher concentrations of criteria air pollutants (CAPs). In the USA, the disproportionate concentrated location of sources of pollution in some areas and not in others is promoted and perpetuated by residential segregation where members of certain class and, or race are spatially colocated and have certain health-relevant environmental attributes creating spatial inequities (Bower et al., 2014; Diez Roux, 2016). *Residential segregation* occurs when races or ethnical groups are physically separated and forced to reside in certain areas. "Racial segregation is the dominant residential pattern, and racial discrimination is the leading cause of segregated housing in America" (Bullard, 2000, p. xvi). Black residential segregation remains very high for most African Americans in the United States and is linked to historical and ongoing institutional racism (Williams and Collins, 2001). Farther, low-income and minority communities tend to experience exposure to higher concentrations of ambient pollution in areas where the air quality monitoring networks measuring ambient levels of each of the criteria pollutants are available with non-Hispanic blacks having been found consistently over-represented in communities with regard to the exposure to the poorest air quality such as $PM_{2.5}$ and ozone, both pollutants associated with adverse health outcomes even at low levels of exposure (Miranda et al., 2011). Racial differences in socioeconomic status are caused by residential segregation and in turn, contribute to racial differences in health. However, many prior studies have been conducted at a single site, while patterns of associations between SES and air pollution need to be investigated across a range of geographic areas (Hajat et al., 2013). Thus, contextual factors, such as the degree of residential segregation and the spatial location of various SES groups in regard to major air pollution sources, may alter the SES-air pollution associations (Mohai et al., 2009).

Individual and community-based studies conducted in the United States argue that racial and ethnic minorities have a greater health risk from air pollution (Ard et al., 2016). Community-level SES appear to be more strongly related to concentrations of air pollutants, while small associations have been found between individual SES measures and air pollution (Cesaroni et al., 2010; Hajat et al., 2013). The studies conducted in the USA commonly utilize proximity to measure how close various population groups reside with regard to polluting industries or major roads while finding environmental inequalities indicated by income level and ethnic origin. To illustrate, certain ethnic minorities, especially lower- income groups, are more likely to reside close to noxious land uses, including polluting industries, main roadways with high volumes of traffic, airports, incinerators, solid waste facilities, and power stations (Norton et al., 2007; Brender et al., 2011; Wilson et al., 2014; Jia et al., 2014; Pratt et al., 2015; Ha et al., 2015; Johnson et al., 2016; Ard et al., 2016). Later developments in environmental justice studies allowed modeling of ambient air concentrations and population exposures at much finer geographic scales with many studies concluding that disadvantaged populations consistently reside in areas with higher environmental contamination (WHO, 2010; Hajat et al., 2013, 2015).

However, race/ethnicity is not always an adequate sociodemographic determinant of population vulnerability associated with proximity to hazardous industrial facilities. Consider the following example. Population in the United States is being exposed to new contaminants and a wide range of environmental and public health threats stemming from a dramatic expansion of unconventional oil and gas development and use of hydraulic fracturing ("fracking") to extract gas and oil, thus making unconventional gas development an environmental justice issue (Clough and Bell, 2016). The new extraction technique involves injection of very large volumes of fluids and proppants at high pressure into extraction wells to fracture the shale and release natural gas trapped underground; after the fracking process, deep injection wells are being used as one of the strategies to dispose of fracking wastewater underground. New research increasingly indicates serious health threats from both water contamination and air pollution (Srebotnjak and Rotkin-Ellman, 2014). Examples of the latter include unhealthy levels of smog and of toxic air contaminants emitted from a dense network of hydraulically fractured extraction wells and associated infrastructure (pipelines, compressor stations, increased truck traffic, deteriorating road conditions) with emissions causing eye, nose, and throat irritation, respiratory and central nervous system diseases, cancer, premature death (Adgate et al., 2014), reproductive and infant health including lower birth weight and birth defects, mental health such as anxiety in the exposed persons. The traditional environmental justice concern is that nearby communities might potentially experience public health effects and that certain vulnerable population subgroups might be exposed to pollution from unconventional gas wells, specifically, that a disproportionate number of minority might reside in greater numbers in areas near to unconventional wells. Race was not associated with the presence of waste disposal sites wells, while block group median income was inversely associated with the presence of waste disposal sites in Ohio (Silva et al., 2018). This result was in agreement with findings from a Texas-based study observing a greater percentage of wastewater disposal wells in block groups with high poverty (defined as a mean percentage of residents living in poverty greater than the regional mean of 18.6%) (Johnston et al., 2016). A concentration of unconventional gas wells was similarly observed in higher poverty areas, while no difference was found with respect to race in a recent Pennsylvania-based study (Ogneva-Himmelberger and Huang, 2015). Another recent study found no evidence of a disproportionate proportion of minority in areas close to unconventional wells in Pennsylvania; further, researchers indicate that low-income areas are not disproportionately targeted by shale gas development industry (Clough and Bell, 2016). In their study of whether a disproportionate number of minority or low-income residents reside in areas near to unconventional wells in Pennsylvania, areas nearer and further away from gas wells were compared regarding their demographic composition (measured as the percentage of blacks and Hispanics who live on residential land at census block group level), poverty (as defined by the US government), income distribution, and educational attainment (Clough and Bell, 2016). Areas close to unconventional wells had a much lower percentage of blacks and Hispanics compared to those areas further away, as well as a slightly lower (but statistically significant) percentage of households below the poverty threshold, a lower percentage of those in the lowest income category, and a higher percentage of people who received a high school diploma or less (Clough and Bell, 2016). With research on environmental justice and unconventional gas development still being at the development stage, future research needs to better understand, for example, whether intensive shale gas development is disproportionately greater in lowest-income communities and other important questions.

Overall, SES is associated with air pollution exposure: those with higher SES can choose housing farther from busy roads or use community resources to improve air quality. Hajat et al., (2013) found negative associations between the pollutants ($PM_{2.5}$ and NOx) and neighborhood-wide SES for the overall study population, i.e., average air pollution concentrations decreased as neighborhood-wide SES increased. Similarly, increases in $PM_{2.5}$ and NO_2 exposures were found related to decreases in neighborhood median family income, and to increases in percent of visible minority in the neighborhoods in Ontario, Canada (Lavigne et al., 2016). Regarding the magnitude of effects, and using community-level educational attainment as a SES indicator, lower SES is associated with increases in $PM_{2.5}$ exposures; for example, places with 15% greater population with less than a high school education have relatively small increases in $PM_{2.5}$ concentrations ranging between $0.14\,\mu g/m^3$ to $0.9\,\mu g/m^3$ (Brochu et al., 2011; Ebisu and Bell, 2012; Gray et al., 2013; Hajat et al., 2013).

However, associations of SES with environmental exposures was found to be *context specific*, that is, results from air pollution exposure studies vary based on place, e.g., in New York (focusing on the southern Bronx and northern Manhattan), both $PM_{2.5}$ and NOx pollutants were positively associated with high-SES neighborhoods, i.e., higher concentrations of air pollutants have been observed in high-SES neighborhoods such as Upper West Side (Hajat et al., 2013). This finding may be attributable to the city development, as major roads were constructed near water bodies (rivers and lakes) providing scenic views of nature and easy access to these natural and urban amenities resulting in high-SES individuals clustering close to these facilities. Similar findings of positive relationship of high SES and air pollution exposure have been found in European cities such as Rome, Italy (Cesaroni et al., 2010). However, individuals with high SES have more resources that can protect them against increased concentrations of pollution by using private transportation rather than public, by working in the indoor environments and not outdoor, by having higher-quality housing to keep off the noxious fumes, noise, and odors and by having access to climate control and filtration for indoor environments. In contrast, lower-income households (black or white) have to adapt to a lower-quality physical environment as they cannot afford "luxury items" such as air conditioning, and bottled water for drinking (Bullard, 2000). Farther, low-SES communities have limited political power and cannot prevent undesirable land uses such as plants and roads from being constructed near their communities (Hajat et al., 2015).

Additionally, studies conducted in the USA found differences by pollutant. To illustrate, areas of low-poverty clusters and high-poverty clusters had exposure to similar concentrations of NO_2 and $PM_{2.5}$ in Los Angeles; however, high poverty areas had higher levels of exposure to other CAPs (Molitor et al., 2011). For example, with respect to ozone exposure, higher concentrations of ozone (O_3) concentrations were found in places where higher SES groups dwell. Similar associations between poverty and ozone exposure have been observed with percent of population in poverty being negatively associated with the probability of a county having the worst air quality for ozone (Miranda et al., 2011). Lavigne et al. (2016) analyzed the distribution of air pollutants across neighborhoods in Ontario, Canada, observing an increase in O_3 concentrations in neighborhoods with fewer visible minorities. Other studies found opposite results with lower SES groups being exposed to higher concentrations of O_3 in California and North Carolina (Marshall, 2008; Gray et al., 2013), and lower SES groups being exposed to higher concentrations of O_3 compared to higher SES groups in Phoenix, Arizona (Grineski et al., 2007). Across the entire United States, the percent of non-Hispanic blacks in the US counties with the worst air quality has been found more than twice the corresponding proportion in the counties with the best air quality measured by concentrations of daily ozone, daily $PM_{2.5}$, and design values for annual $PM_{2.5}$, and more children younger than 5y.o. resided in those worst air quality-counties than in the best air quality-counties, while higher rates of poverty have been found in counties with the worst air quality in terms of fine particles concentrations (both annual and daily $PM_{2.5}$) compared with counties with the best air quality (Miranda et al., 2011). In this study, positive association was found between a percent of population in poverty and the probability of a county having the worst air quality for both annual and daily $PM_{2.5}$ (Miranda et al., 2011).

As stated earlier, the place where people reside influences their health. However, social and gender inequalities influence environmental health risks, including air quality (ambient, indoor, at work), housing and residential location, waste management, climate change, and others. Unequal distribution of environmental risk factors occurs due to social inequality in income, social status, employment, and education. The existence of environmental inequalities across populations according to socioeconomic status with the concentration of sources of pollution in certain areas of a territory is referred to as *exposure differential* (Evans and Kantrowitz, 2002).

Adverse socioeconomic conditions add to environmental exposures making residents experience double effects of exposure to both adverse environment and social disadvantage. In other words, even given the same environmental exposure, the effect of exposure can be modified through some mechanism. To illustrate the effect of social aspects, particular population groups may reside more often in neighborhoods with potentially dangerous environmental conditions, such as close to municipal landfills, chemical facilities, airports, or major roads. Evidence shows that disadvantaged groups suffer the health effects of air pollution exposure more severely (WHO, 2010). This is referred to as *susceptibility differential* (Sexton et al., 1993). However, factors of the local neighborhood environment may not affect everyone equally. Further, within the same neighborhood, social individual-level factors such as education and health behavior (smoking, taking alcohol, inactive lifestyle, etc.) may exacerbate the exposure of the risk groups.

A recent analysis estimated risks of death from all causes associated with long-term exposure to fine particulate matter among racial minorities and disadvantaged population subgroups (Di et al., 2017). Exposed to $PM_{2.5}$, men, blacks, Asian, and Hispanic persons, and people of lower socioeconomic status (e.g., those with Medicaid eligibility, that is the population of persons 65 years of age or older) have a higher risk of mortality due to the exposure compared with the general population, while white and Medicaid-eligible persons have higher risks of death associated with ozone exposure. Overall, the risk of death for black persons is three times higher than for other population subgroups when exposed to $PM_{2.5}$. An increased risk of death from exposure to $PM_{2.5}$ exists among black persons even with higher socioeconomic status (not eligible for Medicaid due to higher income) (Di et al., 2017).

11.4 Other demographic factors and environmental exposure

Additionally there are noneconomic determinants of exposure to environmental risk factors such as age, gender, ethnicity, preexisting chronic diseases. In this light, specific population subgroups are considered as more vulnerable groups to environmental risks compared with the general population. Susceptible groups include children and elderly people, households with low educational attainment, unemployed people, as well as migrants and ethnic groups with the magnitude of environmental inequality varying greatly (WHO, 2010). In the United States, two-thirds of all deaths occur in people of 65 years of age or older (Di et al., 2017).

Air quality is distributed unevenly across different demographic groups in the United States: counties with the worst $PM_{2.5}$ air quality have a statistically significant larger proportion of non-Hispanic Blacks, smaller share of older people (over 64 years of age), larger percent of people in poverty, while counties with the worst ozone concentrations have a statistically significant larger percent of non-Hispanic Blacks, larger percent of young children (under 5 years of age), smaller percent in poverty, and larger populations in total compared to the US counties with the best air quality (Miranda et al., 2011). Even at a more refined geographic scale, significant links between race, age, and air quality for both $PM_{2.5}$ and ozone (Miranda et al., 2011). Since noneconomic factors, including age, racial/ethnic membership, and sex, are almost always differentially related to health outcomes, these characteristics need to be included as covariates to better inform planned epidemiological analyses (Hajat et al., 2013).

11.4.1 Extent of health impacts of air pollution

To sum up, the extent to which health disparities are attributable to air pollution is hard to understand or interpret due to several factors.

1. First, the aforementioned confounding effects of socioeconomic status (SES) make it difficult to estimate the true effect of air pollution on health outcomes.
   1.1. Similarly, many prior studies specifically focused on just a single SES domain, i.e., education, or income, while SES variables from different domains (income, wealth, education, occupation, and probably other factors) better and more broadly reflect the influence of SES.
   1.2. Further, even when a study looks at multiple neighborhood-level variables characterizing several SES domains, for example, income can be captured by the following variables: median household income, the percentage of households living under the poverty level, the percentage on public assistance, and the percentage of single-parent families; while neighborhood-level wealth can be reflected by using the following characteristics: the percentage of households owning their home; the percentage of households receiving interest, dividend, or rental income; and the median value of owner-occupied homes, etc., only some neighborhood-level characteristics are reported (e.g., median home value to capture neighborhood's wealth) in associations with air pollution while others are ignored.
   1.3. Compared with individual SES measures, community-level SES appear to be more strongly related to air pollutants concentrations making

neighborhood-wide SES an important predictor of air pollution exposures; however, interpretation of these results needs to be exercised with caution because of the difference of how individual- and neighborhood-level variables are collected and the sources of collection.

1.4. Recent studies on SES and air pollution use estimates (i.e., predicted levels of air pollution, not directly measured concentrations) to assess the association between air pollution and social factors. In turn, estimated air pollution concentrations are predicted in part based on the covariates such as population density and land use.

2. Second, the inclusion into analysis of other social factors such as residential segregation and the spatial location of various SES groups relative to key sources of air pollution may change the associations between SES and air pollution.

3. Third, a lacking heterogeneity inherent in many past studies limited to a single site instead of an analysis of patterns of SES-air pollution associations across various geographic areas similarly limit the understanding of the extent of health impacts of air pollution. Overall, the results may not be generalizable to the entire population as a whole and the associations of SES with environmental exposures appear to be context specific with negative associations in some places, while positive associations with higher SES have been found for other areas (such as New York and in Rome, Italy) indicating higher air pollution exposures with higher SES. The positive association in New York may be due to high-SES neighborhoods having high-density land use and being proximate to busy roadways and thus higher air pollutant concentrations, versus the relative isolation of lower-SES neighborhoods from roads and high-density land uses with lower levels of traffic-related air pollution.

4. Yet another reason is the difference in methodologies employed by different studies.

4.1. While both neighborhood- and individual-level SES are independently related to air pollution and are thus important to consider simultaneously in air pollution health effects research, most studies include either area-wide SES factors (such as neighborhood-level poverty and educational attainment) or individual-level SES factors (Hajat et al., 2013). Regarding the neighborhood definition, many studies use data based on census tract due to data being readily available to researchers, and confidentiality. A limitation of using census tracts is their potential considerable variation in size across metropolitan areas (e.g., census tracts at one area may be larger in size than in other sites and thus more heterogeneous) making them a less meaningful neighborhood definition in certain metropolitan areas.

4.2. Because of spatial clustering of air pollution sources (such as point source emissions, e.g., from industrial facilities such as power plants, which are often grouped together), proper modeling techniques are needed to appropriately account for observed spatial autocorrelation. Models, which are not adjusted for spatial autocorrelation, that is, which consider only the correlation within areas and do not consider the correlation of air pollution outcomes between areas, violate the independence of observations assumption and may affect inference and result in incorrect conclusion. For example, parameter estimates from models not adjusted for spatial autocorrelation (i.e., which do not account for the spatial clustering of air pollutants) tend to overestimate the magnitude of effect

4.3. Difference in statistical methods employed, such as transformation of variables: due to a high within-city variability of air pollutant, its values can be transformed to the natural log form to prevent model nonconvergence and model parameter estimates are exponentiated and presented as percentage differences from the geometric mean concentration of a pollutant (such as NOx), while other pollutant concentrations are used in models without transformation, with associations presented as differences from the mean concentration in micrograms per cubic meter (such as $PM_{2.5}$) (Hajat et al., 2013). Thus, associations measured on different scales for the two pollutants contribute to the difficulty in comparison of magnitudes of effect sizes observed in these air pollutants.

4.4. Some studies transform the ordinal variables (e.g., categories of income, education, wealth), to z-scores, and then model the z-scores as continuous variables for easier comparisons between different areas (e.g., to compare neighborhood-level SES variables with those at the individual level), however, basing z-scores on the different types of initial variables, e.g., using the original ordinal SES variables at the individual level, while using continuous SES variables for the z-scores for the neighborhood-level.

4.5. Studies of individual and area-wide SES associations with individual-level air pollutant concentrations are conducted using a cross-sectional design (Hajat et al., 2013), allowing for only associations to be made. In contrast, longitudinal approach supports stronger inferences and studies over long periods of exposures to air pollution could have important implications for population health and health disparities while examining changes in inequality over time.

11.5 **Outside air pollution and mental health**

Exposure to severe air pollution in children influences cognitive outcomes (Calderón-Garcidueñas et al., 2011) and cognitive dysfunction in healthy children (Calderón-Garcidueñas et al., 2008). Residents of the neighborhoods exposed to pollution suffer from worse cognitive outcomes (Mohai et al., 2011). Disadvantaged neighborhoods might expose their residents to multiple adverse stressors worsening depressive symptoms: Blair et al. (2014) found that certain physical and social neighborhood-level characteristics, including neighborhood disadvantage and deprivation, deterioration, disorder, residential instability, and social ties affect depression outcomes.

Scientists have provided evidence that exposure to particle pollution increases the risk of mental health disorders. For example, a 2016 study examined association between long-term exposure to fine particulate matter and major depressive disorder in Seoul, Korea (Kim et al., 2016). 27,000 residents who inhaled particle pollution over a long period of time were more likely to suffer from major depressive disorder. Particularly those persons who also had a chronic disease (say, asthma, COPD, or diabetes) had an increased risk for the disorder. A large 2017 study used data from community living groups across the United States and looked at links between ambient air pollution and depressive and anxiety symptoms in older adults (Pun et al., 2017). More symptoms of depression and anxiety were observed among older adults when concentrations of particle pollution levels increased.

Chen et al. (2017) analyzed mental health impacts of living near major roads among residents of Ontario, Canada. Adults had a higher risk of dementia if they lived closer to the road, i.e., within 300 m, while the strongest association was found among those who lived closest to the roads (i.e., <50 m), who had never changed residential address and who lived in major cities. At the same time, the study found that those who lived close to heavy traffic did not increase their risk for Parkinson's disease or multiple sclerosis.

Power et al. (2011) study examined a cohort of older men for the impacts by traffic-related air pollution on cognitive function. The result indicated older men faced increased risk of having poor cognition if they experienced long-term exposure to traffic pollution.

11.6 Indoor air pollution and health

Poor areas have a greater concentration of inadequate housing where in addition to environment outdoors, inside environment may affect health. For example, poor-quality housing may have cockroaches, dust mites, poor air filtration in ventilation systems, indoor pesticide use and mold linked to increased risk of childhood asthma (Malhotra et al., 2014), doctor visits, and missed school days (Rauh et al., 2008).

Indoor environmental conditions may be both reflective of poverty level and represent a source of indoor pollution exposure. The median year of home construction (Corburn et al., 2006) serves a proxy to exposure to indoor hazards based on the assumption that older homes in high-poverty areas in comparison with newer homes will be of poor quality containing respiratory triggers. Areas with older homes had significantly higher RHIs. Substandard housing conditions can cause stress (Stewart and Rhoden, 2006), which triggers asthma (Bloomberg and Chen, 2005). Households headed by females due to relative lack of resources might have increased difficulties coping with physical and mental health problems (Downey, 2005).

Some 3 billion people are exposed to indoor smoke, which is a serious health risk of noncommunicable diseases, including stroke, ischemic heart disease, chronic obstructive pulmonary disease (COPD), and lung cancer. Exposure comes from inefficient cooking using open fires and polluting stoves and heating their homes with biomass (such as wood, animal dung, and crop waste), kerosene fuels and coal (WHO, 2018). Breathing particulate matter (soot) from household air pollution almost doubles the risk for childhood pneumonia among children under 5 years of age. Exposure to household air pollution is attributable for 45% of all pneumonia deaths in this age category, while increasing risk for pneumonia in adults with exposure responsible for 28% of all deaths to pneumonia among adults (WHO, 2018).

11.7 Air pollution and school performance

Epidemiologic research on air pollution has reported higher poorer academic performance among children attending schools located in areas with the highest air pollution (Mohai et al., 2011).

Academic performance (here used interchangeably with *achievement*) can be measured as aggregated standardized test scores and/or rate of absenteeism (days missed from school).

The potential mechanism is through higher school absenteeism rates due to respiratory problems, e.g., asthma attacks (Pastor et al., 2004, 2006a,b) or through neurological effects on children's development (Gaffron and Niemeier, 2015; Legot et al., 2010).

We will consider the impact of the following factors upon school performance:

1. Socioeconomic factors
2. Outside environment
3. Negative health outcome such as obesity

11.7.1 Socioeconomic factors and school performance

Over 20 US states have no existing legislation on new school location with regard to manmade hazards (Gaffron and Niemeier, 2015). Schools experience the burden of traffic-related air pollution due to their location close to roads with busy traffic: higher concentrations of air pollutants at these schools are positively correlated in a highly significant way with the socioeconomic metrics of the school student bodies, including higher shares of poor students and students belonging to ethnic minorities (Black and Hispanic) (Green, et al., 2004). Students in such schools have a higher risk of developing respiratory diseases and performing poorly with regard to school achievement, even after school, family, and geographic factors are controlled for (Pastor et al., 2006a,b; Mohai et al., 2011).

Laws ban new school siting in proximity of health hazards in the 10 US states. For example, the bill was passed in 2003 in California prohibiting new school's location close to health hazards, including freeways or busy traffic corridors, that is, any new school cannot be sited closer than 500 ft from hazardous activities. However, the law does not pertain to existing schools in California. A recent study analyzed how $PM_{2.5}$ emissions from road traffic may affect the public K-12 schools in Sacramento,

California. Gaffron and Niemeier (2015) observed examples of environmental injustice by finding significant associations between higher concentrations of $PM_{2.5}$ and lower school performance as expressed in Academic Performance Indices, lower parental education levels and fewer White students. Higher $PM_{2.5}$ levels are significantly associated with students' families' weaker economic situation as expressed in a proportion of students eligible for subsidized or free school lunches (Green et al., 2004; Pastor et al., 2006a,b; Mohai et al., 2011; Gaffron and Niemeier, 2015). Schools' air pollution burden might worsen health and educational disadvantages already experienced by the affected vulnerable populations.

Studies show evidence of links between economic disadvantage indicators, including qualifying for free or reduced lunch at school (a proxy for poverty), and poorer school performance (Reardon and Galindo, 2009; US Department of Agriculture Food and Nutrition Service, 2016). Mother's education affects positively children's school performance: children whose mothers have higher levels of education tend to perform academically better (Magnuson, 2007). Race and ethnicity of mothers was linked to academic performance, mother's racial and ethnic group membership of being either Black or Hispanic was associated with poorer performance in school among their children (Duncan and Magnuson, 2005; Kao and Thompson, 2003; Bali and Alvarez, 2003).

Commonly used socioeconomic measures include the following:

1. Median household income
2. % of African-American
3. % of Poverty
4. Median house value
5. % of Females headed house
6. % of Unemployment
7. % of Households without a car
8. % of Owner-occupied housing
9. % of Population ≥ 25 yrs. high school graduate
10. % of age > 65
11. Population density
12. Crime rate

11.7.2 Outside (ambient) air pollution and school performance

Health of different communities is impacted by key health hazards, including toxic waste sites, chemical exposure, and air quality from road proximity. Poorer health may contribute to reduced educational attainment (Scharber et al., 2013). Studies provide evidence of health impacts on school performance via respiratory problems, such as asthma, which may decrease school attendance rates among students with this condition since exposure to toxins may exacerbate asthma and cause low attendance. High concentrations of carbon monoxide (CO), even at below EPA thresholds, have been shown to decrease school attendance. Missing school frequently contributes to poorer preparation for standardized tests via decreased learning opportunities. Additionally, an increase in proportions of students with disabilities may lower school performance scores, all else equal (Lucier et al., 2011). Higher traffic density has been linked to respiratory effects; children who live closer to high-volume roadways tend to have a higher prevalence of respiratory diseases, while students who live in places with higher traffic intensity, including around schools, have decreased lung function. The amount of air pollutants that downwind schools are exposed to is significantly correlated with distance to high-volume roadways and the amount of truck traffic. $PM_{2.5}$ emissions also correlate with the traffic-related health indicators such as the rate of asthma-related emergency department visits at census tract level, which may result in increased school absences leading to lost learning opportunities and thus, contributing to poorer preparation for standardized tests, all else equal (Lucier et al., 2011).

Concentrations of different air pollutants can vary widely across the areas as a function of various factors, including the amount and types of proximate industrial facilities, heavily traveled or congested roadways nearby, as well as weather patterns (EPA. Assessing Outdoor Air Near Schools. Available at: https://www3.epa.gov/air/sat/about.html Accessed 8 March 2018).

School air may be affected by the following factors: closeness to heavily trafficked roads, and exposure to diesel exhaust emitted by idling diesel engine-based vehicles, including school buses and trucks, proximate location to neighboring industrial large and small facilities ranging from chemical plants, refineries, and factories, to gasoline stations and dry cleaners, as well as toxic chemicals found in building and other materials used in school, and others.

Studies relate environmental pollution and inequalities in educational outcomes among vulnerable populations from environmental justice perspective. An idea of *"environmental ascription"* is used to explain "ascriptiveness" of places reproducing structural social inequalities, i.e., how a place's proximity to environmental pollution, especially known and suspected developmental neurotoxins, or toxins, which are especially harmful for neurological, cognitive, or social development and can thus hinder children's potential of academic success (Legot et al., 2010; Lucier et al., 2011). Environmental ascription is understood as the property for inherited discrepancies in wealth and power tending to self-perpetuate; since children have little control in choosing their place of residence and are more likely to experience economic deprivation compared to adults (Clark-Reyna et al., 2016), places and polluted environments are "inherited." Some places have highly unequal distribution of pollution, exhibiting spatially clustered patterns accompanied with high-level toxic emissions measured in terms of reported volume of air releases, thus increasing children's exposure to those specific toxins that are highly likely to limit cognitive development, and academic success. So, the residents and children also inherit educational inequalities via the limitations of learning abilities and life chances since early success in school can have lifelong impacts on life chances of a person and is reflected in future outcomes, including future success, economic attainment, and productivity, in turn, leading to the social reproduction of economic deprivation (Clark-Reyna et al., 2016). Place's proximity to toxic pollution can act to degrade human capital by limiting life chances (Lucier et al., 2011).

A prior body of research examined linkage between school-based exposure to air toxics and children's school performance (Legot et al., 2010). Research studies observed negative impacts of respiratory and neurological toxins (Lucier et al., 2011; Scharber et al., 2013; Mohai et al., 2011; Pastor et al., 2004, 2006a,b) on school performance. In one of the early studies, Pastor et al. (2004) examined standardized test scores in Los Angeles schools and air pollution risk at the census tract level and Toxic Release Inventory (TRI) data. According to the American Lung Association's 2018 *State of the Air* report, Los Angeles remains the city with the worst ozone pollution. Pastor et al.'s (2004) finding indicates a significant negative association between school-level air pollution and test scores even controlling for school demographics. Pastor et al. (2006a,b) examined relationship between air pollution and test scores for the rest of California finding similar patterns. The Toxic Release Inventory (TRI) data contain data on air pollution emissions from the largest point sources in the country, such as industrial plants across the United States. Data submitted yearly to EPA represent toxic chemicals deliberately and accidentally released into air, surface water, and the ground by industrial facilities and transferred to offsite facilities. But data are excluded on emissions from mobile sources, such as trucks, automobiles, ships, and aircraft. Similarly, small area service industries, such as gas stations, dry cleaners, and auto body shops and other facilities, are not required to report their emissions because of small size or being nonlisted industrial sectors but might potentially represent significant air polluters.

A negative relationship has been found between school site exposure to air toxics and school performance. For example, schools located near heavily trafficked roads are exposed to $PM_{2.5}$ emissions from road traffic and are negatively associated with learning outcomes. A recent study based in Sacramento, California, observed that the higher $PM_{2.5}$ emissions from road traffic around the public K-12 schools, the lower the schools' Academic Performance Index (API was used in California between 1999 and 2013 to measure academic performance and progress on statewide assessments, replaced in 2017) and the lower the average parental education levels (Gaffron and Niemeier, 2015). Educational disadvantages in terms of academic achievement, which is already experienced by the affected vulnerable populations (e.g., by students due to minority status and poverty), might be further exacerbated by a disproportionate risk from "place," i.e., school's air pollution burden. The status of "*Socioeconomically Disadvantaged*" is given to Californian school students neither of whose parents have received a high-school diploma or when a student is eligible for the free or reduced-price lunch program (the latter considered a proxy for poverty). Food insecurity may reduce diet quality leading to developmental consequences, including decreased academic achievement (Lucier et al., 2011).

Linkage between exposure to particulate matter (PM) and school performance was also studied in Texas, the greatest US consumer of energy across all sectors including residential, commercial, industrial, and transportation (Clark-Reyna et al., 2016; Grineski et al., 2016). According to the US Energy Information Administration's State Energy Data System (Available at: https://www.eia.gov/state/seds/seds-data-complete.php?sid=US), Texas consumed almost 14% of total energy consumed in the United States in 2016, it also is a major producer of pollution: according to the American Lung Association's *2018 State of the Air* report, 43% of its counties (15 out of 35) were given an F grade for ozone pollution in 2014–16, while US EPA's data show that the general public was exposed to higher average levels of fine particulate matter in Texas than across the country in 2017 (8.9 versus 8.6 µg/m^3). Exposure to particulate matter from diesel-using vehicles was found linked to reduced school achievement in schoolchildren in El Paso, where pollution level is elevated due to the developed trucking industry, and different polluting facilities, including high-volume international airport (Clark-Reyna et al., 2016; Grineski et al., 2016).

In a Louisiana-based study, Lucier et al. (2011) studied the association between environmental risk (measured by proximity to polluters emitting twelve developmental, neurological and respiratory toxicants) at school and aggregate education outcomes (e.g., test scores) in public elementary, middle, and high schools in Baton Rouge, which is according to the American Lung Association's *2018 State of the Air* report is the most polluted city in Louisiana with regard to particle pollution and high ozone concentration levels. The study found school performance scores negatively affected by exposure. Significant association was observed between all the measures of schools' proximity to toxic air pollution (described later) and school performance scores throughout East Baton Rouge Parish. School performance was measured at the school level, using standardized test results (the LEAP-21 tests in English, Mathematics, Social Studies, and Science), which are similar to the aforementioned Academic Performance Index.

Several proximity-based measures of environmental risk have been applied, including (1) proximity to Toxics Release Inventory (TRI) facilities in general, an indicator of proximity expressed as a dummy variable of the presence or absence of a TRI facility within 1 mile of the school itself, with the expectation of a negative relationship between this proximity measure and academic performance (i.e., having a TRI facility within the specified distance would decrease academic performance), (2) the total number of TRI facilities within 1, 2, and 3 miles of each school, with the expectation of a negative relationships between the School Performance Score and the measures of the numbers of polluters nearby (i.e., an increase in the number of polluters within the specified distance of each school will lower the School Performance Score), (3) proximity to high concentrations of toxic emissions, measured by the distance between each school and the geographic center of the TRI "toxic triangle" cluster described below; expecting a positive relationship, that is, as the distance from the cluster emitting the highest volume of toxins increases, academic performance will increase, and (4) proximity to high-volume polluters (HVP) of developmental neurotoxins in the Baton Rouge-based study measured as the distance between each school and the geographic center of the cluster of top five polluters with the highest toxicity scores, also expecting a positive relationship, and (5) estimates of air toxic concentration (coming from https://www.epa.gov/rsei), which represent summed up toxicity-weighted concentrations of chemicals across TRI facilities in each square kilometer cell of a grid centered on each TRI facility, and thus estimate the total exposure to industrial air toxics at the grid cell within which each school is contained.

The results from regression models indicate that the environmental risk measures, which not merely measure simple proximity to facilities (such as numbered 1 and 2 earlier), but which also measure distance from particularly toxic locations (such as numbered 3, 4, and 5 above) have higher values of regression coefficients and, thus, have more powerful effects on learning outcomes measured by school performance scores (Lucier et al., 2011).

The top three polluting facilities in Baton Rouge include an ExxonMobil Refinery, an ExxonMobil Chemical Plant, and a Honeywell facility, all three facilities comprising the "toxic triangle" accounting for most of total releases in the parish, suspected neurotoxin releases, and suspected respiratory toxin releases, and almost all developmental toxin releases and suspected developmental toxin releases, with the distance among the facilities ranging between 1 and 2 miles. The three are also among the top US emitters of developmental neurotoxins and respiratory toxins causing both neurological and respiratory symptoms that may result in school absenteeism. Other two facilities, Formosa Plastics and DSM Copolymer, rank high in terms of total volume of releases of toxins at very high local levels. All five facilities, the "Toxic 5," are located in the same zip code, while a large minority and poor population and a large number of schools are closely located to the polluters.

The importance of the environmental risk variables in impacting educational outcomes was demonstrated by including the school-related relevant independent variables as control variables associated with academic achievement. The relationships between exposure and school performance have been consistently found strong: the effect of environmental variables was significant even after controlling for the influential school-level variables such as a percentage of children receiving free school lunch, percentage of minority students, school size and teacher/student ratio (a proxy for school resources), percentage of teachers with emergency credentials (a proxy for teacher quality), attendance rates, and percentage of students with disabilities (Lucier et al., 2011).

A different Baton Rouge-based study argued that chemicals are not equally harmful to children's school performance and that negatively impact childhood development, and that do damage to the nervous and respiratory systems, short- or long-term, versus other chemicals such as known carcinogens may predict poor school performance better. The study focused on known and suspected developmental, neurological, and respiratory toxicants. By looking at the ascriptive forces of race, class, and place simultaneously, these pollutants have been found more negatively correlated with school performance scores than broader measures of pollution (Scharber et al., 2013).

In a recent 2016 study, associations were examined between residential exposure to air pollution (rather than school exposure because children spend more time at home than at school) and student academic achievement, and more specifically, between air toxics risk estimates from multiple pollutant sources and children's academic performance among 4th and 5th graders in El Paso, Texas, USA, where air quality is a serious concern due to numerous large-scale polluting facilities, developed rail freight industries, a growing military base, trucking being a major source of air pollution (Clark-Reyna et al., 2016). Rather than using standardized test scores aggregated at the level of the school, individual-level grade point averages have been employed. Risk from diesel particulate matter (PM) and respiratory risk were disaggregated by source (i.e., point, on-road mobile, nonroad mobile, and nonpoint sources) to include various risk indicators. For generation of the child-level pollution values, the US EPA's 2005 National Air Toxics Assessment (NATA) database at census block level was used to extract data on air toxics. Air toxics are known or suspected chemicals causing cancer or neurological, respiratory, immunological, and reproductive diseases and are regulated by the federal legislation such as US Clean Air Act. NATA database does not include criteria pollutants, which are an important exposure risk sources. The study indicates that the effects of exposure to pollution from airplanes, construction vehicles, and trains may be more detrimental to children than believed before. Among the eight toxics variables employed in the study, nonroad respiratory risk and nonroad diesel PM had the strongest relationship with learning outcomes as expressed in GPA. Examples of nonroad mobile air toxics are those produced by El Paso's airport, military base, and railways, which may decrease school GPAs via serious respiratory infections and school absenteeism among children.

Results showed that all except one (i.e., point respiratory risk, which includes emissions from factories, refineries, and power plants was not significant in a multivariate model, but significant in the bivariate correlations), air toxics variables used in the generalized estimating equations models remained significant even after all relevant confounding variables were controlled for (that is, parental, sociodemographic and school-level effects), suggesting that environmental exposure has an independent effect on individual student-level learning outcomes (Clark-Reyna et al., 2016).

Additionally, poor indoor environmental quality in schools decreases concentration and leads to poor test results. According to the EPA (https://www.epa.gov/iaq-schools), close to 50% of all the schools across the US have problems with indoor air quality; school air with high levels of pollution can increase absenteeism, make learning harder and aggravate behavioral disorders.

11.7.3 Negative health outcome and school performance: The impact of obesity

Previous research indicates that higher body mass indexes negatively impact school performance in children (Roberts et al., 2010; Datar and Sturm, 2006; Castelli et al., 2007; Caird et al., 2011). Grineski et al. (2016) studied the relationship between outdoor ambient air concentrations of MD chemicals and academic performance mediated by obesity. They examined the cognitive effects of obesity in childhood. They tested a hypothesis of an association between metabolic disrupting (MD) chemicals, also known as "obesogens" linked to obesity due to weight-adding property and school performance. Due to the ability of these chemical hazards to accumulate in the environment, ambient outdoor air concentrations may decrease academic achievement measured by children's grade point average (GPA) in reading, language, arts, math, social studies, and science at the individual child's level by increasing child's weight (psychosocial impacts include poor academic performance outcomes (Taras and Potts-Datema, 2005; Larsen et al., 2014; MacCann and Roberts, 2013; Kristjánsson et al., 2008; Pan et al., 2012); and biological impacts include impaired brain and memory functions contributing to worse school outcomes (Khan et al., 2014; Kamijo et al., 2012, 2014; Hillman et al., 2012).

To assess the level of the MD by tract, Grineski et al. (2016) used the summed values of "ambient concentration" (raw concentrations of toxics in outdoor air) provided by the NATA and assigned the total tract-level value of the ambient concentration of MD to each child based on the home location within the GIS environment. By using bivariate correlation analysis, Grineski et al. (2016) found a statistically significant negative association between the outdoor air concentration of known MDs and children's GPA (direct impact) and qualifying for free or reduced price school lunches, while greater weight measured by the BMI index was positively and significantly correlated with known ambient concentrations. Despite having a relatively high time lag in their data, e.g., school and controlling socioeconomic data came from 2012 survey; however, air pollution data measured by the outdoor ambient concentrations came from 2005 as the most recent available data provided by NATA, it was assumed that the concentrations between 2005 and 2011 have remained relatively constant as the infrastructure, including all major roads and freeways, as well as pollution-emitting point and mobile facilities such as refineries, factories, airports, train stations, and the like have remained in the same locations since 2005.

11.8 Identification of higher and lower levels of air pollution exposure

To analyze the links between unequal exposure to environmental hazards and negative health outcomes, areas experiencing higher and lower exposure need to be delineated. Air pollution exposure can be quantified using a proxy indicator such as proximity (distance) to polluting sites.

11.8.1 Distance-based

Exposure to air pollution may be assessed from *stationary point sources* (e.g., industrial plants). Exposure assessment to quantify air pollution exposure can be based on proximity to emission polluting facilities as a surrogate of exposure. Based on geographical distance, areas of higher and lower air pollution exposure may be identified with individual residencies, schools, or other objects of interest falling within the former treated as cases, and controls (residencies, schools, etc.) being outside.

Using a similar rationale in a study on the relationship between exposure to air toxics emitted by stationary sources and learning outcomes measured by average school-level API scores, Pastor et al. (2004) used a distance-based approach by identifying facilities emitting air toxics (e.g., TRI) located within 1 mile of the tract containing each school. For that purpose, they used a dummy variable of the presence of a TRI facility within one mile of the tract containing the school to designate schools within the specified distance (labeled "1") and schools that are not within 1 mile of a TRI facility (scored 0), then the average summary scores of school performance (e.g., Academic Performance Index, API) are compared between schools, which are within close proximity of TRI and those which are not. The limitation is that only those living within the unit of analysis (here, tracts) are assumed to exclusively experience adverse effects of an environmental hazard and thus all exposures are assumed equal throughout the unit of analysis, while the polluting facility may locate on a border between the units.

Alternatively, exposure to air pollution may be assessed from *mobile sources* using traffic-related metrics as a proxy for exposure. The method involves estimation of emissions of air pollutants from close proximity major roads rather than relying on air pollution monitor data, which incorporate sources other than road traffic, another limitation of the approach is that it utilizes the busiest network links within a specified distance of places under study (say, schools or residencies) while the cumulative emissions produced by vehicles on different road network segments near these places are not quantified (Gaffron and Niemeier, 2015). Traffic exposure studies commonly use proximity-based approaches based on the scientific evidence of the harmful effects of traffic-related pollution on human health, including adverse respiratory, cardiovascular, and pregnancy outcomes. Those persons living nearer to roads are potentially exposed to higher levels of traffic-related air pollutants, and in fact, reporting fair or poor health status compared with farther from roads, while existing evidence from epidemiological studies indicates a causal association between traffic-related exposure and worsening of asthma among children (Parker et al., 2012). To illustrate, the closer to the curbside, the greater is the concentration of traffic-related air pollutants, and vice versa, diminishing steeply with the distance from the road until it reaches background level within 300–500 m of the curbside. However, the buffer chosen is pollutant-specific, for example, ultrafine particle numbers have been found to decrease to 50% of their curbside concentration level within the buffer distance of 150 m from the roadway edge (Karner et al., 2010). Farther, when traffic moves in a stop-and-go fashion, vehicle emissions increase and degrade ambient air quality (Zhang and Batterman, 2013). Distance-based metric between roads and residences is based both on the distance to roads and traffic volume: distance from primary residence to a congested segment of a road network or the nearest high-traffic major roads can be measured, while to proxy for congested roads, one may use total traffic volume data. Proximity- and volume-based direct measures of traffic exposure are important traffic-derived measures of exposure due to the complex mixture of pollutants from mobile source emissions.

11.8.2 "Hot spot" approach

Environmental justice studies apply a "hot spot" method to identify pollution "hot spots." Maantay (2007) identified pollution "hot spots" based on proximity to noxious land uses in New York City and analyzed hospitalization rates for those people living within them and observed these to be higher compared to those living outside hot spot areas. Their poverty level, racial or ethnic status was also different. Asthma hot spots have been identified in New York City; its poverty levels were compared with those of nonhot spots (Corburn et al., 2006). Hot spot areas were also poorer.

11.8.3 Buffer-based and density-based approaches

When there is evidence of spatial clustering of industrial facilities, distance from industrial facilities is not considered the best measure of exposure (Ha et al., 2015). Instead, buffers with a predefined distance can be created around each point of interest (say, residential location) and the total number of industrial facilities (such as power plants) falling within this buffer can be determined. Residents living within the buffers with a greater amount of facilities are hypothesized to have a greater exposure to pollution. Health outcomes can then be compared for residents within this radius. In a recent retrospective study, Ha et al. (2015) used a distance of 20 km for a radius and created buffers around each birth location and counted the total number of power plants within this buffer to study associations between proximity to emissions from power plants and adverse birth outcomes, including low birth weight (LBW), preterm delivery (PTD), and very preterm delivery (VPTD). Women living in a 20-km buffer with at least 2 power plants increase odds of having adverse birth across all three categories (by 7%, 12%, and 17% for term LBW, PTD, and VPTD, respectively), while pregnant women who lived with no power plants within the radius had no such effect. Among all types of fuel, strongest associations with all adverse birth outcomes have been found for coal.

Yet other studies utilized pollution concentration- and facility density-based approach to identify pollution-impacted areas to examine asthma hospitalization rates in children at the ZIP code level controlling for the socioeconomic factors (Grineski, 2007).

A similar technique used in the density-based traffic exposure studies to understand relationships between traffic exposure and adverse health involves constructing concentric buffers of specified radii around points of interest (say, primary residential addresses) and measuring *traffic density* (e.g., counting the number of roads, or traffic vehicle miles traveled, or aggregated length of all roads, or average annual daily traffic (AADT) within the buffers around primary residence, with the radii of 100, 300, and 500 m commonly used for concentric traffic buffers due to the sharp decrease of the level of traffic-related air pollution to the background level within the distance of 300–500 m from the road) (Parker et al., 2012). Within these buffers, summary statistics (such as medians and interquartile ranges, which can be used to identify quartiles and form categorical traffic exposure variables) of traffic exposure indicators are tabulated for population subgroups (e.g., for potentially vulnerable subgroups defined by selected characteristics such as poverty status, urbanization, race/ ethnicity, education, or income below the poverty threshold). Next, to study whether living within a certain distance to one or more roads differs by several characteristics, such as whether poverty status or race/ ethnicity are associated with living within a specific distance to one or more roads, statistical testing is conducted to assess links between selected characteristics and traffic exposure variables (such as chi-square statistics) for each of the traffic buffers. To illustrate, the relationship among poverty and traffic exposure was found by linking data from the National Health and Nutrition Examination Surveys on the health and nutritional status of adults and children in the United States, and traffic indicators available from the National Highway Planning Network: those who have incomes below poverty have higher exposure to traffic-related pollution as they tend to live closer to the nearest road and closer to a larger number of roads (Parker et al., 2012). Additionally, the associations among health outcomes and traffic exposure indicators for subgroups can be examined. For example, using the aforementioned health survey data, which cover a large, nationally representative sample of the US population with about 5000 participants, persons who reside farthest from the nearest road, were the least likely to report fair or poor health status, and the proportion who report fair or poor health generally increased as another traffic density indicator increased (measured by the length of roads) (Parker et al., 2012).

11.8.4 Other approaches

Other exposure assessment techniques used to quantify pollution exposure include dispersion models, land use regression (LUR) models, or a hybrid approach combining several methods, including LUR and interpolation (e.g., inverse distance weighting, geostatistical kriging) and extrapolation (the difference between the two methods is that the former uses the same spatial extent as that of the measured locations, while in the latter, the broader spatial scale outside the original extent of the measured locations is used to generate predictions) to estimate concentrations of air pollution at unmeasured locations (Hajat et al., 2015). For example, Van den Hooven et al. (2009) employed a model based on a Gaussian distribution to predict the dispersion of motor vehicle exhaust, assuming that 96% of the emitted pollutants disperse up to 150 m from the road. Lavigne et al. (2016) developed land use regression (LUR) model to obtain LUR NO_2 yearly and monthly surfaces to estimate maternal residential exposure to ambient NO_2 over entire pregnancy and by trimester.

11.8.4.1 An example of exposure assessment method

We will illustrate how high- and low-exposure areas can be delineated using stationary sources as an example. The US Environmental Protection Agency (EPA) produces a comprehensive National Emissions Inventory (NEI) from all air emissions sources. The facility locations are expressed in latitude and longitude; the amount of the emission is provided for the major pollutants such as CO (carbon monoxide), lead, etc. This database is a good source for mapping air polluting facilities (from here onwards, "a facility") by pollution type. There were 69 stationary facilities reported for Shelby County in 2011 (refer to Fig. 11.5B illustrating the distribution of polluting facilities extracted from the NEI database for Shelby County, TN). Geographic Information System (GIS) will be used to visualize emitting facility distribution within the study area. GIS is an important component for the implementation of the project, as it allows for a comprehensive geospatial analysis. For example, it can be used for a spatial interpolation of CO emission across the study area to show areas with high values of CO exposure. By employing this technique, the project will identify the location of high- (darker areas in the map) and low-exposure (lighter areas) areas. Fig. 11.8 shows that an area to the right of the interstate I-55 and below I-240 loop stands out in terms of higher carbon monoxide (CO) emissions.

Schools falling within high-exposure areas on the map can be assigned as cases (labeled as "High exposure to CO emission: study group"), while those outside are designated controls (labeled as "Low exposure to CO emission: control group"). Statistical analysis may be performed by comparing exam results such as TCAP scores by school between both areas. For example, Fig. 11.8 used the numerical value of the school average exam score (TCAP-based) from 2014 TCAP database (obtainable at: http://www.tn.gov/education/data/tcap_2014_school.shtml). Fig. 11.8 illustrates the rates of proficient and advanced scores in English 1 from TCAP exam in 2014 for both types of exposure areas.

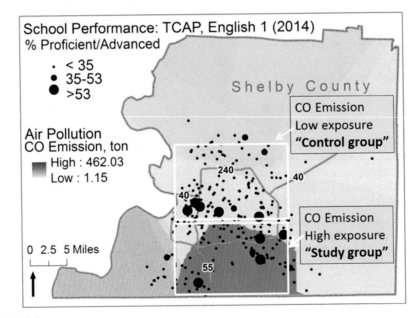

FIG. 11.8

High- and low exposure to carbon monoxide (CO) areas (study and control) and 2014 English 1 TCAP scores by school in Shelby County, TN.

Source: Author.

11.9 Case study: Analysis of exposure to ambient air pollution: The link between environmental exposure and children's school performance in Memphis, TN

Environmental justice studies make often assumptions of geographic inequalities in adverse health outcomes and academic achievement due to uneven environmental exposures. The study empirically tests this assumption. It uses estimates of air toxics exposure from stationary emission sources and models school performance rates using the average high school ACT scores in Shelby County, Tennessee.

School location matters. Schools exposed to higher air pollutants or located nearer busy roads have been documented to have higher proportions of both poor and Black students (Green et al., 2004). Students in these schools experience higher respiratory risks and have poorer academic performance, even controlling for school, family, and geographic characteristics (Pastor et al., 2006a,b; Mohai et al., 2011). However, only 14 states restrict school location at or near major sources of air pollution (Fischbach, 2006).

11.9.1 Objective

We examine the impacts of environmental exposures by stationary point facilities (e.g., industrial factories). Specifically, we focus on children, who compared with adults, are more vulnerable to air pollution exposure because the lungs continue developing during adolescence and due to their higher exposure levels as children tend to spend more time outside (Szyszkowicz, 2008). This study seeks to provide empirical evidence that would lead to a more accurate understanding of factors linked to differences in academic achievement between schools and how much of the variation between scores can be attributed to the environment (measured with proximity to pollution-emitting facilities). Specifically, this study measures education achievement scores by using the ACT scores to study how much of the variation between the scores can be attributed to exposure to environmental factors (proximity to chemical-emitting facilities). This research will provide important insight into the role played by environmental factors in academic performance and help with the future school site selection criteria.

The rationale for the study is as follows. Since air pollution can make children more susceptible to respiratory infection (Ostro et al., 2009), and daily levels of pollutants linked to hospitalizations for respiratory infections (Barnett et al., 2005; Ciencewicki and Japers, 2007; Ostro et al., 2009), higher morbidity may both impact school attendance and decrease school performance since the sick child cannot function as well as a healthy one. On the other hand, missing school may also result in underperformance in school because low attendance rates may lower preparation for standardized tests.

11.9.2 The significance of the expected results

The Tennessee Department of Education has set measurable goals for improving the education system in general. The scores of Tennessee students on standard achievement tests are below the national average, but the explanations commonly cited for this underperformance generally are not based on sound empirical evidence.

11.9.3 Background

11.9.3.1 Air pollution within the study area

Tennessee ranks high in hazardous air pollutant (HAP) emissions list nationwide. The *USA Today* (2008) in their special report ranked all schools nationwide based on modeled concentrations of chemicals causing health problems. 16,500 schools were indicated for their potential toxic exposure level that was twice those of the surrounding communities (The USA Today, 2008). The report-based data revealed that the air outside some schools in Shelby County (which is mostly African American at 62% and poor due to high rates of poverty and unemployment, low rates of health insurance and educational attainment, and high rates of health disparities) has the highest levels of dangerous toxic chemicals (*USA Today*).

Many local schools in Shelby County are dangerously close to the pollution-emitting facilities. For example, the Valero oil refinery south of downtown Memphis, producing almost all fuel used in the Mid-South, was recently found to release as much toxic gas daily as it was originally thought releasing over an entire year; at the same time a residential working neighborhood is less than a football field from the refinery, and there are educational facilities affected by the company, including Riverview Elementary and Middle Schools, Carver High School, among others. Similar to the El Paso's setting, which has been a focus of several recent studies on educational impacts of air pollution exposure (Clark-Reyna et al., 2016; Grineski et al., 2016), Shelby County has high-volume second-busiest-in-the-world cargo Memphis International airport and the developed trucking industry. However, toxic exposure may manifest many years later (Democracy on the Edge: A Discussion of Political Issues in America by Terry A. AmRhein) as cancer and other noncancer chronic health effects such as respiratory and neurological diseases resulting from exposure to air toxics (EPA. Assessing Outdoor Air Near Schools. Available at: https://www3.epa.gov/air/sat/about.html Accessed 8 March 2018).

11.9.3.2 Academic performance within Shelby County

According to the 2013–14 Tennessee Report Card, public schools perform in the bottom 5% of the state: many schools in Shelby County had substantially lower test scores in school subjects compared with the state level in 2013. Fig. 11.9 reports 2013–14 local school achievement measured by Tennessee Comprehensive Assessment Program (TCAP) scores with large percentage of students performing below grade level (shown in darker color) across many school subject.

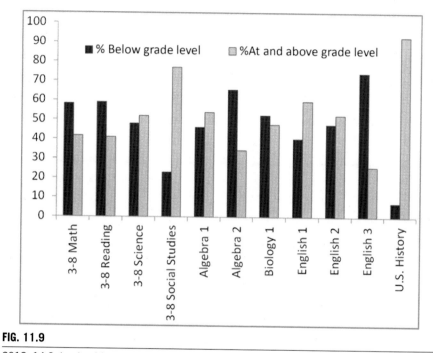

FIG. 11.9

2013–14 School achievement, Shelby County.

Source: Author.

11.9.4 **Methodology**

Many Shelby County area schools were ranked in the bottom 5% in the nation air toxic hot spots. The EPA developed mapping application titled the 2011 NATA app available on the web and on mobile devices available to the public in late 2015. Using the application we observed risks, emissions from the sources in tons per year, and monitoring data on a map noting a substantial portion of the county experiencing higher health risks and annual ambient air concentrations measured by the EPA's Cancer Risk and Respiratory Hazard Index of 50–75 with the index values ranging between 0 and >100, while suburban areas outside Memphis have lower values (EPA. 2011 National Air Toxics Assessment (NATA) App Available at: https:// gispub.epa.gov/NATA/ Accessed 8 March 2018). We used the *USA Today*'s school ranking of the school potential for poor air quality to generate the surface of areas at greatest risk of health effects due to air toxic exposures. For this purpose, the IDW interpolation was used first (Fig. 11.10) with schools superimposed to visualize spatial distribution of educational facilities ranging between the 1–11 bottom percentile among the 127,800 public, private, and parochial schools in the nation.

Next, we used pollution data from stationary point sources using the National Emissions Inventory (NEI) 2014 (the latest available) database complied by the EPA. The NEI database contains comprehensive estimate of air emissions of various air

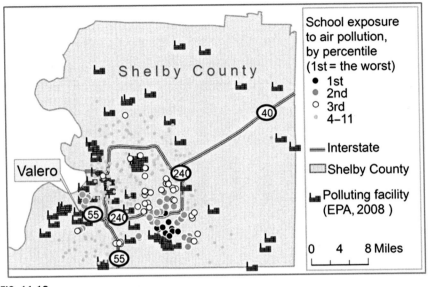

FIG. 11.10

Exposure to air pollution by school.

Source: Author.

pollutants, including criteria pollutants, criteria precursors, and hazardous air pollutants from air emissions sources. The NEI is built using the Emissions Inventory System (EIS). Specifically, we focused on NEI point sources, which include emissions estimates for larger sources located at a fixed, stationary location. NEI point sources include large industrial facilities (e.g., Cargill Corn Milling, Valero Refining Co.) and electric power plants (e.g., Allen Fossil Plant), airports, and smaller industrial, nonindustrial, and commercial facilities. In total, 29 stationary facilities have been extracted. Using latitude-longitude data, the facilities have been mapped for visualization and the subsequent analysis purposes. Next, we have created a 2-mile buffer around each polluting facility.

To measure school performance, we used 2016–17 ACT data aggregated by school within the study area of Shelby County, TN. The ACT has four sections, including English, Math, Reading, and Science. Each ACT section is scored on a scale of 1 to 36, and composite score is made up of the average of these four scores. Table 11.1 reports the average ACT scores in English, math, reading, science, and a composite score by school. We also used a percent of students scoring 21 or higher on ACT, and a percent of those scoring below 19.

Initially we identified 52 public high schools. However, 6 schools had no ACT records, so the final sample had 46 high schools. 46 schools have been added to a GIS environment and distances to the nearest polluting facility computed.

We identified high schools falling within and outside buffer areas. Those inside the buffer have been designated as "affected," or exposed to air pollution ($N=22$),

Table 11.1 Description of 2016–17 ACT scores of high schools.

Value	Average English score	Average Math score	Average Reading score	Average Science score	Average Composite score	% Scoring 21 or higher	% Scoring below 19
Min	12.2	15	13.7	14.2	13.9	5.2	8.5
Max	25.9	24.5	25.9	25	25.5	84.7	92.9
Mean	16.54	17.36	17.55	18.02	17.52	28.89	65.17
St.Dev.	3.60	2.30	3.11	2.53	2.88	22.96	24.55

while those beyond have been labeled as "not affected," or not exposed to air pollution ($N=24$). Fig. 11.11 shows the location of 2-mile buffer areas around each pollution-emitting facility. The figure also illustrates the distribution of affected (shown with a triangle symbol within the gray-colored 2-mile buffer area) and not affected high schools (shown with a triangle symbol outside buffer areas) relative to polluting facilities (shown with a star symbol) within the study area.

We applied statistical analyses using two-sample t-tests assuming unequal variances to compare school ACT performance between affected schools (i.e., high schools located within the 1-mile, 1.5-mile, 2-mile, 2.5-mile, 3-mile, and 3.5-mile buffer of the nearest polluting facility) and not affected (i.e., for schools located outside the buffers above).

We have shown the effect of the distance from the polluting facility on average ACT score performance, average percentage of students who score at least 21 on ACT, and average percentage of students who score at 19 or below on ACT. Table 11.2 reports the results of the t-test of the differences between schools exposed to air pollution (within the buffer of specified distance) and not exposed (outside the buffer area). No difference in ACT score was found for distances of <2 miles (not

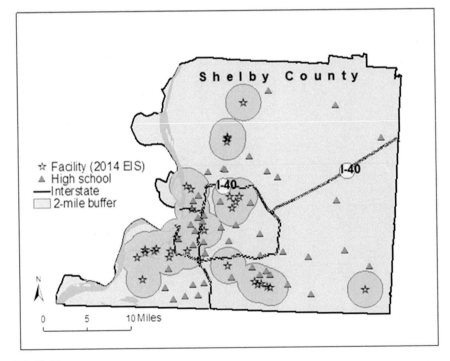

FIG. 11.11

Public high schools, and 2-mile buffers around polluting facilities (EPA's 2014 EIS) within the study area.

Source: Author.

Table 11.2 T-test results of the differences in ACT performance between schools exposed to air pollution and not exposed, by distance from the polluting facility.

Distance ACT section	2.5mile				3mile				3.5mile			
	N=27 <2.5mi	N-19 >2.5mi	t-Stat	P(T <=t)	N=32 <3mi	N=14 >3mi	t-Stat	P(T <=t)	N=35 <3.5mi	N=11 >3.5mi	t-Stat	P(T <=t)
Average English	15.91	17.61	−1.51	0.07	15.96	18.09	−1.81	**0.04 (*)**	15.92	18.79	−2.26	**0.019 (**)**
Average Math	16.90	18.07	−1.59	0.06	16.99	18.29	−1.62	0.06	16.93	18.85	−2.14	**0.025 (*)**
Average Reading	16.96	18.61	−1.68	**0.05 (*)**	17.04	19.01	−1.87	**0.037 (*)**	16.99	19.72	−2.38	**0.016(**)**
Average Science	17.50	18.90	−1.78	**0.04 (*)**	17.58	19.20	−1.91	**0.035 (*)**	17.53	19.80	−2.47	**0.013(**)**
Average Composite	16.97	18.45	−1.64	0.055	17.04	18.81	−1.83	**0.04 (*)**	16.99	19.46	−2.34	**0.017(**)**
Score 21 or Higher	25.69	34.41	−1.06	0.148	27.77	32.40	−0.56	0.291	26.74	35.40	−0.99	0.166
Score 19 or Below	67.89	58.99	1.14	0.130	67.90	55.80	1.50	0.07	68.68	50.23	2.18	**0.022(*)**

Note: bold = the result is significant. * = 0.05 significance level; ** = 0.01 significance level; *** = 0.001 significance level.

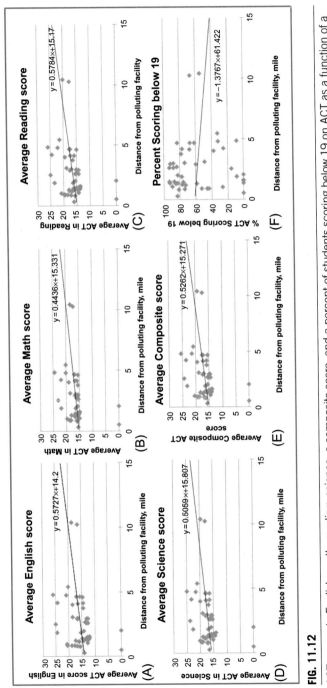

FIG. 11.12

ACT scores in English, math, reading, science, a composite score, and a percent of students scoring below 19 on ACT as a function of a distance from polluting facilities.

Source: Author.

shown in the table). The results indicate a significant difference in two ACT sections (reading and science) for schools within 2.5 miles, while all ACT sections (but math) show statistically significant difference between schools located within 3 miles and 3.5 miles compared to those located outside of these distances (English, reading, science, and a composite score). We did not find a difference in average percent of student scoring 21 or higher on ACT for any distance. However, more students perform worse in those schools located closer to the sources of pollution than in schools farther away from pollution emission: an average percent of students scoring below 19 on ACT is significantly higher in schools within 3.5 miles compared to schools located farther away. Fig. 11.12 illustrates that students' score increases significantly as the distance from pollution-emitting source increases, while the average percent of underperforming students measured as those scoring below 19 on ACT decreases (statistically significantly, at 0.05 level of significance).

While the current study did not control for socioeconomic attributes, the future study will estimate the impact of air toxics exposure from stationary emission sources on school performance rates by analyzing not only all students (the current study), but also specifically focusing on minority membership, economic disadvantage, and disability status among children in Shelby County, Tennessee.

References

Adgate, J.L., Goldstein, B.D., McKenzie, L.M., 2014. Potential public health hazards, exposures and health effects from unconventional natural gas development. Environ. Sci. Technol. 48 (15), 8307–8320.

Adler, N.E., Stewart, J., 2010. Health disparities across the lifespan: meaning, methods, and mechanisms. Ann. N. Y. Acad. Sci. 1186, 5–23.

American Lung Association, 2010. Children's Health. [Internet] The Association, Washington, DC. Available from: http://www.stateoftheair.org/2010/health-risks/health-risks-childrens.html. (Accessed January 10, 2015).

American Lung Association, 2011. Toxic Air: The Case for Cleaning up Coal-Fired Power Plants. American Lung Association, Washington, DC.

Ard, K., Colen, C., Becerra, M., Velez, T., 2016. Two mechanisms: the role of social capital and industrial pollution exposure in explaining racial disparities in self-rated health. Int. J. Environ. Res. Public Health 13 (10), 1025.

Asthma in US children, 2018. Lancet 391 (10121), 632.

Bali, V.A., Alvarez, M.R., 2003. Schools and educational outcomes: what causes the "race gap" in student test scores? Soc. Sci. Q. 84, 485–507.

Barnett, A.G., Williams, G.M., Schwartz, J., Neller, A.H., Best, T.L., Petroeschevsky, A.L., Simpson, R.W., 2005. Air pollution and child respiratory health: a case-crossover study in Australia and New Zealand. Am. J. Respir. Crit. Care Med. 171 (11), 1272–1278. https://doi.org/10.1164/rccm.200411-1586OC.

Bearer, C.F., 1995. How are children different from adults? Environ. Health Perspect. 103 (Suppl 6), 7–12.

Bell, M.L., Ebisu, K., Belanger, K., 2007. Ambient air pollution and low birth weight in Connecticut and Massachusetts. Environ. Health Perspect. 115, 1118–1124.

Blair, A., Ross, N.A., Gariepy, G., Schmitz, N., 2014. How do neighborhoods affect depression outcomes? A realist review and a call for the examination of causal pathways. Soc. Psychiatry Psychiatr. Epidemiol. 49 (6), 873–887.

Bloomberg, G.R., Chen, E., 2005. The relationship of psychological stress with childhood asthma. Immunol. Allergy Clin. North Am. 25, 83–105.

Boguski, T.K., 2006. Understanding units of measurement. In: Environmental Science and Technology Briefs for Citizens. vol. 2. Center for Hazardous Substance Research.

Bower, K.M., Thorpe Jr., R.J., Rohde, C., Gaskin, D.J., 2014. The intersection of neighborhood racial segregation, poverty, and urbanicity and its impact on food store availability in the United States. Prev. Med. 58, 33–39.

Brender, J.D., Maantay, J.A., Chakraborty, J., 2011. Residential proximity to environmental hazards and adverse health outcomes. Am. J. Public Health 101 (Suppl 1), S37–S52.

Brochu, P.J., Yanosky, J.D., Paciorek, C.J., Schwartz, J., Chen, J.T., Herrick, R.F., Suh, H.H., 2011. Particulate air pollution and socioeconomic position in rural and urban areas of the Northeastern United States. Am. J. Public Health 101 (Suppl 1), S224–S230.

Brugge, D., Durant, J.L., Rioux, C., 2007. Near-highway pollutants in motor vehicle exhaust: a review of epidemiologic evidence of cardiac and pulmonary health risks. Environ. Health 6, 23. Available at: http://www.ncbi.nlm.nih.gov/pmc/articles/PMC1971259/. (Accessed January 10, 2015).

Bullard, R., 2000. Dumping in Dixie: Race, Class, and Environmental Quality, third ed. Westview Press.

Buonocore, J.J., Dong, X., Spengler, J.D., Fu, J.S., Levy, J.I., 2014. Using the community multiscale air quality (CMAQ) model to estimate public health impacts of $PM_{2.5}$ from individual power plants. Environ. Int. 68, 200–208.

Caird, J., Kavanagh, J., Oliver, K., Oliver, S., O'Mara, A., Stansfield, C., Thomas, J., 2011. Childhood obesity and educational attainment: a systematic review. EPPI-Centre, Social Science Research Unit, Institute of Education, University of London, London.

Calderón-Garcidueñas, L., Engle, R., Mora-Tiscareño, A., Styner, M., Gómez-Garza, G., Zhu, H., et al., 2011. Exposure to severe urban air pollution influences cognitive outcomes, brain volume and systemic inflammation in clinically healthy children. Brain Cogn. 77 (3), 345–355.

Calderón-Garcidueñas, L., Mora-Tiscareño, A., Ontiveros, E., Gómez-Garza, G., Barragán-Mejía, G., et al., 2008. Air pollution, cognitive deficits and brain abnormalities: a pilot study with children and dogs. Brain Cogn. 68 (2), 117–127.

Castelli, D.M., Hillman, C.H., Buck, S.M., Erwin, H.E., 2007. Physical fitness and academic achievement in third and fifth-grade students. J. Sport Exerc. Psychol. 29, 239–252.

Centers for Disease Control and Prevention (CDC), 2016. Summary Health Statistics: National Health Interview Survey, 2016.

Cesaroni, G., Badaloni, C., Romano, V., Donato, E., Perucci, C.A., Forastiere, F., 2010. Socioeconomic position and health status of people who live near busy roads: the Rome Longitudinal Study (RoLS). Environ. Health 9, 41.

Chen, H., Kwong, J.D., Copes, R., et al., 2017. Living near major roads and the incidence of dementia, Parkinson's disease and multiple sclerosis: a population-based cohort study. Lancet 389, 718–726. https://doi.org/10.1016/S0140-6736(16)32399-6.

Ciencewicki, J., Japers, I., 2007. Air pollution and respiratory viral infection. Inhal. Toxicol. 19, 1135–1146.

Clark-Reyna, S.E., Grineski, S.E., Collins, T.W., 2016. Residential exposure to air toxics is linked to lower grade point averages among school children in El Paso, Texas, USA. Popul. Environ. 37 (3), 319–340. https://doi.org/10.1007/s11111-015-0241-8.

Clough, E., Bell, D., 2016. Just fracking: a distributive environmental justice analysis of unconventional gas development in Pennsylvania, USA. Environ. Res. Lett. 11 (2), 025001. Available at: http://iopscience.iop.org/article/10.1088/1748-9326/11/2/025001/pdf.

Corburn, J., Osleeb, J., Porter, M., 2006. Urban asthma and the neighbourhood environment in New York City. Health Place 12 (2), 167–179.

Datar, A., Sturm, R., 2006. Childhood overweight and elementary school outcomes. Int. J. Obes. (Lond) 30, 1449–1460.

Di, Q., Wang, Y., Zanobetti, A., Wang, Y., Koutrakis, P., Choirat, C., Dominici, F., Schwartz, J.D., 2017. Air pollution and mortality in the Medicare population. N. Engl. J. Med. 376 (26), 2513–2522. http://www.nejm.org/doi/full/10.1056/NEJMoa1702747.

Diez Roux, A.V., 2016. Neighborhoods and health: what do we know? What should we do? Am. J. Public Health 106 (3), 430–431.

Dockery, D.W., Pope III, C.A., Xu, X., et al., 1993. An association between air pollution and mortality in six U.S. cities. N. Engl. J. Med. 329, 1753–1759.

Downey, L., 2005. Single mother families and industrial pollution in metropolitan America. Sociol. Spectr. 25 (6), 651–675.

Duncan, G.J., Magnuson, K.A., 2005. Can family socioeconomic resources account for racial and ethnic test score gaps? Future Child. 15 (1), 35–54.

Ebisu, K., Bell, M.L., 2012. Airborne $PM_{2.5}$ chemical components and low birth weight in the northeastern and mid-Atlantic regions of the United States. Environ. Health Perspect. 120 (12), 1746–1752.

EIU, 2005. The Economist Intelligence Unit's Quality-of-Life Index. Available at: https://www.economist.com/media/pdf/QUALITY_OF_LIFE.pdf. (Accessed January 1, 2018).

Environmental Integrity Project, 2013. US Power Plant Global Warming Emissions Rising in 2013 After Years of Decline. Environmental Integrity Project, Washington, DC.

Environmental Protection Agency, 2014. Clean Energy: Air Emissions. http://www.epa.gov/cleanenergy/energy-and-you/affect/air-emissions.html. Updated May 22, 2014.

Environmental Protection Agency (EPA), 2006. A Framework for Assessing Health Risks of Environmental Exposures to Children. National Center for Environmental Assessment, Washington, DC. EPA/600/R-05/093F. Available from the National Technical Information Service, Springfield, VA, and online at http://www.epa.gov/ncea. (Accessed January 10, 2015).

Environmental Protection Agency (EPA), 2012. Reducing Air Pollution From Power Plants. http://www.epa.gov/airquality/powerplants/. Published March 27, 2012. Updated March 27, 2012.

Environmental Protection Agency (EPA), 2017. Criteria Air Pollutants. At: https://www.epa.gov/criteria-air-pollutants.

Evans, G.W., Kantrowitz, E., 2002. Socio-economic status and health: the potential role of environmental risk exposure. Annu. Rev. Public Health 23, 303–331.

Fann, N., Lamson, A.D., Anenberg, S.C., Wesson, K., Risley, D., Hubbell, B.J., 2012. Estimating the national public health burden associated with exposure to ambient $PM_{2.5}$ and ozone. Risk Anal. 32 (1), 81–95.

Fischbach, S., 2006. Not in My Schoolyard: Avoiding Environmental Hazards at School Through Improved School Site Selection Policies: A Report to the U.S. Environmental Protection Agency. Rhode Island Legal Services, Rhode Island, RI, USA.

Friedman, M.S., Powell, K.E., Hutwagner, L., Graham, L.M., Teague, W.G., 2001. Impact of changes in transportation and commuting behaviors during the 1996 Summer Olympic Games in Atlanta on air quality and childhood asthma. JAMA J. Am. Med. Assoc. 285 (7), 897–905.

Gaffron, P., Niemeier, D., 2015. School locations and traffic emissions-environmental (in) justice findings using anew screening method. Int. J. Environ. Res. Public Health 12, 2009–2025.

Gasana, J., Dillikar, D., Mendy, A., Forno, E., Vieira, E.R., 2012. Motor vehicle air pollution and asthma in children: a meta-analysis. Environ. Res. 117, 36–45.

Gray, S., Edwards, S., Miranda, M.L., 2010. Assessing exposure metrics for PM and birth weight models. J. Expo. Sci. Environ. Epidemiol. 20, 469–477.

Gray, S.C., Edwards, S.E., Miranda, M.L., 2013. Race, socioeconomic status, and air pollution exposure in North Carolina. Environ. Res. 126, 152–158.

Green, R.S., Smorodinsky, S., Kim, J.J., McLaughlin, R., Ostro, B., 2004. Proximity of California public schools to busy roads. Environ. Health Perspect. 112, 61–66.

Grineski, S.E., 2007. Incorporating health outcomes into environmental justice research: the case of children's asthma and air pollution in Phoenix, Arizona. Environ. Hazard 7 (4), 360–371.

Grineski, S., Bolin, B., Boone, C., 2007. Criteria air pollution and marginalized populations: environmental inequity in metropolitan Phoenix, Arizona. Soc. Sci. Q. 88, 535–554.

Grineski, S., Clark-Reyna, S., Collins, T., 2016. School-based exposure to hazardous air pollutants and grade point average: a multi-level study. Environ. Res. 147, 164–171. https://doi.org/10.1016/j.envres.2016.02.004.

Grineski, S.E., Collins, T.W., Chakraborty, J., McDonald, Y.J., 2013. Environmental health injustice: exposure to air toxics and children's respiratory hospital admissions in El Paso, Texas. Prof. Geogr. 65 (1), 31–46. https://doi.org/10.1080/00330124.2011.639625.

Ha, S., Hu, H., Roth, J., Kan, H., Xu, X., 2015. Associations between residential proximity to power plants and adverse birth outcomes. Am. J. Epidemiol. 182 (3), 215–224.

Hackmann, D., Sjöberg, E., 2017. Ambient air pollution and pregnancy outcomes—a study of functional form and policy implications. Air Qual. Atmos. Health 10 (2), 129–137.

Hajat, A., Diez-Roux, A.V., Adar, S.D., Auchincloss, A.H., Lovasi, G.S., O'Neill, M.S., Sheppard, L., Kaufman, J.D., 2013. Air pollution and individual and neighborhood socioeconomic status: evidence from the multi-ethnic study of atherosclerosis (MESA). Environ. Health Perspect. 121 (11−12), 1325–1333.

Hajat, A., Hsia, C., O'Neill, M.S., 2015. Socioeconomic disparities and air pollution exposure: a global review. Curr. Environ. Health Rep. 2 (4), 440–450.

Hillman, C.H., Kamijo, K., Pontifex, M.B., 2012. The relation of ERP indices of exercise to brain health and cognition. In: Boecker, H., Hillman, C.H., Scheef, L., Struder, H.K. (Eds.), Functional Neuroimaging in Exercise and Sport Sciences. Springer, New York, NY, USA, pp. 419–446.

Hoek, G., Brunekreef, B., Goldbohm, S., Fischer, P., van den Brandt, P.A., 2002. Association between mortality and indicators of traffic-related air pollution in the Netherlands: a cohort study. Lancet 360, 1203–1209.

Holstius, D.M., Reid, C.E., Jesdale, B.M., Morello-Frosch, R., 2012. Birth weight following pregnancy during the 2003 Southern California wildfires. Environ. Health Perspect. 120 (9), 1340–1345.

Jia, C., James, W., Kedia, S., 2014. Relationship of racial composition and cancer risks from air toxics exposure in Memphis, Tennessee, U.S.A. Int. J. Environ. Res. Public Health 11 (8), 7713–7724.

Johnson, R., Ramsey-White, K., Fuller, C.H., 2016. Socio-demographic differences in toxic release inventory siting and emissions in metro Atlanta. Int. J. Environ. Res. Public Health 13 (8), 747.

Johnston, J.E., Werder, E., Sebastian, D., 2016. Wastewater disposal wells, fracking, and environmental injustice in southern Texas. Am. J. Public Health 106 (3), 550–556.

Kamijo, K., Pontifex, M.B., Khan, N.A., Raine, L.B., Scudder, M.R., Drollette, E.S., Evans, E.M., Castelli, D.M., Hillman, C.H., 2012. The association of childhood obesity to neuroelectric indices of inhibition. Psychophysiology 49, 1361–1371.

Kamijo, K., Pontifex, M.B., Khan, N.A., Raine, L.B., Scudder, M.R., Drollette, E.S., Evans, E.M., Castelli, D.M., Hillman, C.H., 2014. The negative association of childhood obesity to cognitive control of action monitoring. Cereb. Cortex 24, 654–662.

Kao, G., Thompson, J.S., 2003. Racial and ethnic stratification in educational achievement and attainment. Annu. Rev. Sociol. 29, 417–442.

Karner, A.A., Eisinger, D.S., Niemeier, D., 2010. Near-roadway air quality: synthesizing the findings from real-world data. Environ. Sci. Technol. 44 (14), 5334–5344.

Khan, N.A., Raine, L.B., Donovan, S.M., Hillman, C.H., 2014. The cognitive implications of obesity and nutrition in childhood. Monogr. Soc. Res. Child Dev. 79, 51–71.

Kim, K.Y., Lim, Y.H., Bea, H.J., Kim, M., Jung, K., Hong, Y.C., 2016. Long-term fine particulate matter exposure and major depressive disorder in a community-based urban cohort. Environ. Health Perspect. 124, 1547–1553.

Kleinman, M.T., 2000. The Health Effects of Air Pollution on Children. South Coast Air Quality Management District, Diamond Bar, CA. Available at: http://www.aqmd.gov/docs/default-source/students/health-effects.pdf. (Accessed January 10, 2015).

Kristjánsson, A.L., Sigfúsdóttir, I.D., Allegrante, J.P., 2008. Health behavior and academic achievement among adolescents: the relative contribution of dietary habits, physical activity, body mass index, and self-esteem. Health Educ. Behav. 37, 51–64.

Landrigan, P.J., Fuller, R., Acosta, N.J.R., Adeyi, O., Arnold, R., et al., 2017. The Lancet Commission on pollution and health. Lancet 391, 462–512. https://doi.org/10.1016/S0140-6736(17)32345-0. Available at: http://www.thelancet.com/pdfs/journals/lancet/PIIS0140-6736(17)32345-0.pdf?code=lancet-site.

Larsen, J.K., Kleinjan, M., Engels, R.C.M.E., Fisher, J.O., Hermans, R.C.J., 2014. Higher weight, lower education: a longitudinal association between adolescents' body mass index and their subsequent educational achievement level? J. Sch. Health 84, 769–776.

Lavigne, E., Yasseen III, A.S., Stieb, D.M., Hystad, P., van Donkelaar, A., Martin, R.V., Brook, J.R., Crouse, D.L., Burnett, R.T., Chen, H., Weichenthal, S., Johnson, M., Villeneuve, P.J., Walker, M., 2016. Ambient air pollution and adverse birth outcomes: differences by maternal comorbidities. Environ. Res. 148, 457–466.

Legot, C., London, B., Shandra, J., 2010. Environmental ascription: high-volume polluters, schools, and human capital. Organ. Environ. 23, 271–290.

Li, S., Batterman, S., Wasilevich, E., Elasaad, H., Wahl, R., Mukherjee, B., 2011. Asthma exacerbation and proximity of residence to major roads: a population-based matched case-control study among the pediatric Medicaid population in Detroit, Michigan. Environ. Health 10, 34.

Lucier, C., Rosofsky, A., London, B., Scharber, H., Shandra, J., 2011. Toxic pollution and school performance scores: environmental ascription in East Baton Rouge Parish, Louisiana. Organ. Environ. 24, 1–21.

Maantay, J., 2007. Asthma and air pollution in the Bronx: methodological and data considerations in using GIS for environmental justice and health research. Health Place 13 (1), 32–56.

MacCann, C., Roberts, R.D., 2013. Just as smart but not as successful: obese students obtain lower school grades but equivalent test scores to nonobese students. Int. J. Obes. (Lond) 37, 40–46.

Magnuson, K., 2007. Maternal education and children's academic achievement during middle childhood. Dev. Psychol. 43, 1497–1512.

Malhotra, K., Baltrus, P., Zhang, S., McRoy, L., Cheng Immergluck, L., Rust, G., 2014. Geographic and racial variation in asthma prevalence and emergency department use among Medicaid-enrolled children in 14 southern states. J. Asthma 51 (9), 913–921.

Malley, C.S., Kuylenstierna, J.C.I., Vallack, H.W., Henze, D.K., Blencowe, H., Ashmore, M.R., 2017. Preterm birth associated with maternal fine particulate matter exposure: a global, regional and national assessment. Environ. Int. 101, 173–182.

Marshall, J.D., 2008. Environmental inequality: air pollution exposures in California's South Coast Air Basin. Atmos. Environ. 42, 5499–5503.

Miranda, M.L., Edwards, S.E., Keating, M.H., Paul, C.J., 2011. Making the environmental justice grade: the relative burden of air pollution exposure in the United States. Int. J. Environ. Res. Public Health 8 (6), 1755–1771.

Mohai, P., Kweon, B.-S., Lee, S., Ard, K., 2011. Air pollution around schools is linked to poorer student health and academic performance. Health Aff. 30 (5), 852–862.

Mohai, P., Pellow, D., Roberts, J.T., 2009. Environmental justice. Annu. Rev. Env. Resour. 34, 405–430.

Molitor, J., Su, J.G., Molitor, N.T., Rubio, V.G., Richardson, S., Hastie, D., Morello-Frosch, R., Jerrett, M., 2011. Identifying vulnerable populations through an examination of the association between multipollutant profiles and poverty. Environ. Sci. Technol. 45 (18), 7754–7760.

Moual, N.L., Jacquemin, B., Varraso, R., Dumas, O., Kauffmann, F., Nadif, R., 2013. Environment and asthma in adults. Presse Med. 42, e317–e333.

Norton, J.M., Wing, S., Lipscomb, H.J., Kaufman, J.S., Marshall, S.W., Cravey, A.J., 2007. Race, wealth, and solid waste facilities in North Carolina. Environ. Health Perspect. 115, 1344–1350.

Ogneva-Himmelberger, Y., Huang, L., 2015. Spatial distribution of unconventional gas wells and human populations in the Marcellus Shale in the United States: vulnerability analysis. Appl. Geogr. 60, 165–174.

Olsson, D., Mogren, I., Forsberg, B., 2013. Air pollution exposure in early pregnancy and adverse pregnancy outcomes: a register-based cohort study. BMJ Open 3, e001955.

Ostro, B., Roth, L., Malig, B., Marty, M., 2009. The effects of fine particle components on respiratory hospital admissions in children. Environ. Health Perspect. 117 (3), 475–480.

Pan, L.P., Sherry, B., Park, S., Blanck, H.M., 2012. The association of obesity and school absenteeism attributed to illness or injury among adolescents in the United States, 2009. J. Adolesc. Health 52, 64–69.

Parker, J.D., Kravets, N., Nachman, K., Sapkota, A., 2012. Linkage of the 1999–2008 National Health and Nutrition Examination Surveys to traffic indicators from the National Highway Planning Network. Natl. Health Stat. Rep. 45, 1–16.

Pastor, M., Morello-Frosch, R., Sadd, J.L., 2006a. Breathless: schools, air toxics, and environmental justice in California. Policy Stud. J. 34, 337–362.

Pastor, M., Sadd, J., Morello-Frosch, R., 2004. Reading, writing, and toxics: children's health, academic performance, and environmental justice in Los Angeles. Eviron. Plann. C. Gov. Policy 22, 271–290.

Pastor, M., Sadd, J., Morello-Frosch, R., 2006b. Breathless: schools, air toxics, and environmental justice in California. Policy Stud. J. 34, 337–362.

Patel, M.M., Chillrud, S.N., Correa, J.C., Feinberg, M., Hazi, Y., Kc, D., Prakash, S., et al., 2009. Spatial and temporal variations in traffic-related particulate matter at New York City high schools. Atmos. Environ. 43 (32), 4975–4981.

Pearce, J.R., Richardson, E.A., Mitchell, R.J., Shortt, N.K., 2010. Environmental justice and health: the implications of the socio-spatial distribution of multiple environmental deprivation for health inequalities in the United Kingdom. Trans. Inst. Br. Geogr. 35 (4), 522–539.

Power, M.C., Weisskopf, M.G., Alexeeff, S.E., et al., 2011. Traffic-related air pollution and cognitive function in a cohort of older men. Environ. Health Perspect. 119, 682–687.

Pratt, G.C., Vadali, M.L., Kvale, D.L., Ellickson, K.M., 2015. Traffic, air pollution, minority and socio-economic status: addressing inequities in exposure and risk. Int. J. Environ. Res. Public Health 12 (5), 5355–5372.

Pun, V.C., Manjourides, J., Suh, H., 2017. Association of ambient air pollution with depressive and anxiety symptoms in older adults: results from the NSHAP study. Environ. Health Perspect. 125, 342–348.

Rauh, V.A., Landrigan, P.J., Claudio, L., 2008. Housing and health: intersection of poverty and environmental exposures. Ann. N. Y. Acad. Sci. 1136 (1), 276–288.

Reardon, S.F., Galindo, C., 2009. The Hispanic-White achievement gap in math and reading in the elementary grades. Am. Educ. Res. J. 46, 853–891.

Ritz, B., Wilhelm, M., Hoggatt, K.J., Ghosh, J.K., 2007. Ambient air pollution and preterm birth in the environment and pregnancy outcomes study at the University of California, Los Angeles. Am. J. Epidemiol. 166, 1045–1052.

Ritz, B., Yu, F., Fruin, S., Chapa, G., Shaw, G.M., Harris, J.A., 2002. Ambient air pollution and risk of birth defects in Southern California. Am. J. Epidemiol. 155, 17–25.

Roberts, C.K., Freed, B., McCarthy, W.J., 2010. Low aerobic fitness and obesity are associated with lower standardized test scores in children. J. Pediatr. 156, 711–718.e1.

Scharber, H., Lucier, C., London, B., Rosofsky, A., Shandra, J., 2013. The consequences of exposure to developmental, neurological, and respiratory toxins for school performance: a closer look at environmental ascription in East Baton Rouge, Louisiana. Popul. Environ. 35 (2), 205–224.

Schwartz, J., Bind, M.A., Koutrakis, P., 2017. Estimating causal effects of local air pollution on daily deaths: effect of low levels. Environ. Health Perspect. 125, 23–29.

Schwartz, J., Coull, B., Laden, F., Ryan, L., 2008. The effect of dose and timing of dose on the association between airborne particles and survival. Environ. Health Perspect. 116, 64–69.

Sexton, K., et al., 1993. Air pollution health risks: do class and race matter? Toxicol. Ind. Health 9, 843–878.

Shelby County Health Department, Office of Epidemiology and Infectious Diseases, 2015. 2015 Annual Report. Available at: http://www.shelbytnhealth.com/DocumentCenter/View/732.

Shi, L., Zanobetti, A., Kloog, I., Coull, B.A., Koutrakis, P., Melly, S.J., Schwartz, J.D., 2016. Low-concentration $PM_{2.5}$ and mortality: estimating acute and chronic effects in a population-based study. Environ. Health Perspect. 124, 46–52.

Silva, G.S., Warren, J.L., Deziel, N.C., 2018. Spatial modeling to identify sociodemographic predictors of hydraulic fracturing: wastewater injection wells in Ohio census block groups. Environ. Health Perspect. 126 (6). 067008-1–067008-8.

Srebotnjak, T., Rotkin-Ellman, M., 2014. Fracking Fumes: Air Pollution from Hydraulic Fracturing Threatens Public Health and Communities. Natural Resources Defense Council (NRDC) Issue BRIEF. ip:14-10-a. Available at: https://www.nrdc.org/sites/default/files/fracking-air-pollution-IB.pdf.

Stewart, J., Rhoden, M., 2006. Children, housing and health. Int. J. Sociol. Soc. Policy 26 (7–8), 326.

Stieb, D.M., Chen, L., Eshoul, M., Judek, S., 2012. Ambient air pollution, birth weight and preterm birth: a systematic review and meta-analysis. Environ. Res. 117, 100–111.

Strickland, M.J., Darrow, L.A., Klein, M., Flanders, W.D., Sarnat, J.A., Waller, L.A., Sarnat, S.E., et al., 2010. Short-term associations between ambient air pollutants and pediatric asthma emergency department visits. Am. J. Respir. Crit. Care Med. 182 (3), 307–316.

Szyszkowicz, M., 2008. Ambient air pollution and daily emergency department visits for asthma in Edmonton, Canada. Int. J. Occup. Med. Environ. Health 21 (1), 25–30.

Taras, H., Potts-Datema, W., 2005. Obesity and student performance at school. J. Sch. Health 75, 291–295.

US Department of Agriculture Food and Nutrition Service, 2016. Income Eligibility Guidelines. Available online: http://www.fns.usda.gov/school-meals/income-eligibility-guidelines. (Accessed January 26, 2016).

US Energy Information Administration, 2018. Monthly Energy Review. June 2018. At: https://www.eia.gov/totalenergy/data/monthly/pdf/mer.pdf.

USA Today, 2008. The Smokestack Effect. Toxic Air and America's Schools. Special Report. Available at: www.content.usatoday.com/news/nation/environment/smokestack/school/. (Accessed January 10, 2015).

Van den Hooven, E.H., Jaddoe, V.W., de Kluizenaar, Y., Hofman, A., Mackenbach, J.P., Steegers, E.A., Miedema, H.M.E., Pierik, F.H., 2009. Residential traffic exposure and pregnancy-related outcomes: a prospective birth cohort study. Environ. Health 8, 59.

WHO, 2010. Environment and Health Risks: A Review of the Influence and Effects of Social Inequalities. Available at: http://www.euro.who.int/__data/assets/pdf_file/0003/78069/E93670.pdf. (Accessed June 10, 2018).

Williams, D.R., Collins, C., 2001. Racial residential segregation: a fundamental cause of racial disparities in health. Public Health Rep. 116, 404–416.

Wilson, S., Zhang, H., Jiang, C., Burwell, K., Rehr, R., Murray, R., Dalemarre, L., Naney, C., 2014. Being overburdened and medically underserved: assessment of this double disparity for populations in the state of Maryland. Environ. Health 13 (1), 26.

World Health Organization (WHO), 2017. Don't Pollute My Future! The Impact of the Environment on Children's Health. World Health Organization, Geneva. Licence: CC BY-NC-SA 3.0 IGO.

World Health Organization (WHO), 2018. Ambient (Outdoor) Air Quality and Health. Fact sheet no. 313. Updated May 2018 World Health Organization, Geneva. Available at: http://www.who.int/news-room/fact-sheets/detail/ambient-(outdoor)-air-quality-and-health.

Zhang, K., Batterman, S., 2013. Air pollution and health risks due to vehicle traffic. Sci. Total Environ. 450-451, 307–316. https://doi.org/10.1016/j.scitotenv.2013.01.074.

Further reading

Christine, P.J., Auchincloss, A.H., Bertoni, A.G., et al., 2015. Longitudinal associations between neighborhood physical and social environments and incident type 2 diabetes mellitus: the multi-ethnic study of atherosclerosis (MESA). JAMA Intern. Med. 175 (8), 1311–1320.

Downey, L., Hawkins, B., 2008. Single-mother families and air pollution: a national study. Soc. Sci. Q. 89 (2), 523–536.

Environmental Integrity Project, 2018. Dirty Deception: How the Wood Biomass Industry Skirts the Clean Air Act. Available at: http://www.environmentalintegrity.org/wp-content/uploads/2017/02/Biomass-Report.pdf.

Fuladlu, K., 2019. Urban sprawl negative impact: Enkomi return phase. J. Contemp. Urban Aff. 3 (1), 44–51.

Gordon-Larsen, P., 2014. Food availability/convenience and obesity. Adv. Nutr. 5 (6), 809–817.

Grasser, G., Van Dyck, D., Titze, S., Stronegger, W., 2013. Objectively measured walkability and active transport and weight-related outcomes in adults: a systematic review. Int. J. Public Health 58 (4), 615–625.

Hirsch, J.A., Moore, K.A., Clarke, P.J., et al., 2014. Changes in the built environment and changes in the amount of walking over time: longitudinal results from the multi-ethnic study of atherosclerosis. Am. J. Epidemiol. 180 (8), 799–809.

Mushtaq, A., 2018. Asthma in the USA: the good, the bad, and the disparity. Lancet Respir. Med. 6 (5), 335–336.

Concentrating risk? The geographic concentration of health risk from industrial air toxics across America

12

Kerry Ard, Clair Bullock

School of Environment & Natural Resources, The Ohio State University, Columbus, OH, United States

12.1 Introduction

Studies that have investigated longitudinal changes in air pollution exposure consistently report declines nationwide (Kahn, 1997, 1999; Maasoumi and Millimet, 2005). According to the US Environmental Protection Agency (EPA), from 1970 to 2017, emissions of the six criteria air pollutants regulated by the Clean Air Act declined by 73 percent (EPA, 2018); particulate matter declined 62 percent between 1970 and 1987 (Portney, 1990); and industrial air toxics has declined two-fold from 1995 to 2004 (Ard, 2015). While these declines are laudable, they are not consistent across the landscape. For example, Ard (2015) showed that despite large-scale declines across the US, African Americans are still experiencing twice the health risk from industrial air toxics than non-Hispanic Whites (herein referred to as Whites). Later work found evidence supporting these trends. Clark et al. (2014) examined estimated exposure to the transportation pollutant nitrogen dioxide from 2000 to 2010 across the US, finding that nonwhites were 2.5 times as likely to live in block groups with NO2 concentrations higher than World Health Organization guidelines. Smaller-scale studies report similar findings, such as a study of exposure to a handful of criteria chemicals regulated by the Clean Air Act, which demonstrated that inequalities in exposure have increased from 2003 to 2010 in Massachusetts (Rosofsky et al., 2018).

Scholars note these disparities in air pollution exposure are a part of a larger public health crisis (Juarez et al., 2014). Importantly, these gaps in exposure mirror gaps in health disparities that have been linked to exposure to air toxics. For example, exposure to air pollution has been related to infant mortality (Woodruff et al., 2008) and while infant mortality has declined overall in the United States, racial gaps in infant deaths remain. In addition, exposure to air toxics is related to cardiovascular

Spatiotemporal Analysis of Air Pollution and Its Application in Public Health. https://doi.org/10.1016/B978-0-12-815822-7.00012-1

outcomes (Chi et al., 2016; Franklin et al., 2015) and African Americans persistently have higher rates of cardiovascular disease than non-Hispanic Whites (National Academies of Sciences, 2017). Understanding how the exposure to air pollution has changed over time and across space is important for developing policy to mitigate these health disparities. In the following chapter, we will review the theories developed in the environmental justice literature to explain the unequal distribution of air toxics, as well as various ways the literature has used to measure this environmental inequality. We analyze the demographic characteristics of those block groups that are consistently hotspots for air pollution health risk over time across the six states (Illinois, Indiana, Michigan, Minnesota, Ohio, Wisconsin) that make up the EPA Region with the highest level of unequal distribution of air toxics between Whites and African Americans (Bullock et al., 2018).

12.2 Background

The effects of spatial clustering of different racial groups across the US have been a major focus of research for decades (DuBois, 1899; Park, 1915; Hawley, 1971). One such effect under study has been exposure to industrial pollution from manufacturing facilities, with much research finding that people of color have greater exposure to pollution than whites (Ash et al., 2013; Downey, 2007; Bullard et al., 2007). During the great migration in the early 1900s, southern African-Americans were disproportionately drawn to manufacturing jobs in industrializing cities (Farley et al., 2000). These migrations patterns led to large African-American communities settling around central cities to be near their primary employment locations, manufacturing factories, but are now aging industrial facilities (Wilson, 1996). Some researchers have hypothesized that the large-scale decommissioning of industrial plants associated with deindustrialization have therefore decreased African Americans' exposure to industrial pollution (Downey, 2005). African-Americans are certainly over-represented in the areas experiencing deindustrialization. In a report published by the Brookings Institute, Wial and Friedhoff (2006) found that deindustrialization was most extensive in Indiana, Wisconsin, Michigan, and southern states during the period 1980–2005. Sixty-one percent of the US African-American population, and thirty-five percent of the Hispanic, resided in these areas in 2000.[a] Economists have termed the declines in pollution exposure due to deindustrialization the "silver lining" of deindustrialization (Kahn, 1999), but the question remains if communities are experiencing these declines equally.

In order to understand the spatial distribution of air pollution, it is important to understand that industries of similar type cluster together in space. In early human ecology literature, examining the development of American cities, scholars noted that new industrial facilities were often sited near industries of similar type, largely because these areas

---

[a] These numbers were calculated from the total number of African-Americans or Hispanics living in the United States. South and the mentioned states, divided by the total number of Hispanics or Black or African-American alone population in the census.

already have the necessary infrastructure, access to markets and labor that companies need to effectively do business (Weber, 1909). This clustering of similar land use types means that industries of similar type, using similar industrial processes, and emitting similar air pollution components, cluster in space. Moreover, as regulation of air pollution is often industry, or chemical specific, we would expect that any changes in regulation would have differential effects spatially. For example, certain industries need to be near a source of water for their cooling processes, such as the steel industry. Therefore, we would expect steel industries to cluster near water sources. Moreover, because steel manufacturers use manganese and chromium to enhance durability and corrosion resistance in their industrial processes, we would expect any regulation of these two chemicals to covary across areas nearer to water. In addition, because industries require their materials and products to be transported from and away from their factories, we would expect pollution emitting from the various forms of transportation (e.g., cars, trucks, boats) to correlate with industrial air emissions. These practices and processes are generally inscribed in US zoning laws that designate areas for specific types of land use.

Despite the logic that air pollution clusters spatially, and covaries over time, the majority of studies examining the role of exposure to air toxics in explaining health disparities have focused primarily on one, or a handful of chemicals (Nadadur et al., 2007; NRC Report, 2009; Sexton, 2012). Public health scholars admittedly note, "[u]nderstanding the totality of human exposures, and how environmental exposures to multiple chemicals and chemical mixtures affect disease pathogenesis, are longstanding critical problems facing public health" (Payne-Sturges et al., 2018; p. 13). In response, this body of work is "moving away from a one-chemical one-health outcome model toward a new paradigm of monitoring the totality of exposures that individuals may experience over their lifetime" (Johnson et al., 2017; p. 1). This approach requires consideration of cumulative risk assessment, aggregate exposure assessments, and risks from chemical mixtures and co-exposures (NRC Report, 2004; p. 4). In particular, the focus on chemical mixtures versus single pollutant exposures has grown in importance. Nadadur et al. (2007; p. 326) note that as overall air quality improves with criteria pollutant regulations, there is an increasing understanding that "the importance of the mixture concept is likely to emerge as a major driver for health outcomes."

One of the ways in which scholars have examined the spatial clustering of air toxics is through measuring air pollution "hotspots." Pollution hotspots can be defined as locales where pollutant concentrations are substantially higher than concentrations located in adjacent or surrounding areas (NRC, 2004). These ambient air pollutant hotspots have been linked to higher levels of individual exposure to air toxics. For instance, Wang et al. (2009) focused their study on a known air pollution hotspot located in a predominantly low-income and African American community in Camden, New Jersey. The neighborhood studied has close proximity to several point source industrial pollution sources. Wang et al. (2009) used an individual exposure monitoring to analyze personal exposure to hotspot pollutants benzene and toluene in order to determine whether high ambient levels of hotspot pollutants translated to high personal levels as well. Their analysis found that both personal and ambient concentrations were higher in the hotspot than surrounding areas for these two chemicals. Wu et al. (2012)

conducted a similar hotspot study looking at personal and ambient exposure to ten volatile organic compounds. Their analysis looked at two different neighborhoods. While both had large populations of low-income and African American residents, only one was in a hotspot with dense industrial sources, the other a nearby reference site with no point source exposure. They found that "both disadvantaged communities were influenced by local air pollution sources, but the impact was stronger in the air pollution hotspot community" (Wu et al., 2012; p. 76). Zou et al.'s (2014) study is one of the few to examine pollution over a large spatial area. Using 1999 benzene estimates at the census tract level, the authors found that across the US, hotspots were highest in areas with higher percentage minorities, low educational attainment, and low SES.

Studies have also looked separately at how vehicle emissions could be creating localized hotspots of their own. As factors like proximity to roadways and traffic congestion increase, certain areas see corresponding spikes (or "hotspots") in traffic-related pollutants. For instance, Zhu et al. (2002) looked at how hotspots of ultrafine particles vary with proximity to roadways. They found that concentration of ultrafine particles within 100 meters of a roadway could be up to 25 times higher than concentrations 300 meters beyond the roadway. Similarly, Gately et al. (2017) estimated the impact of variations in road vehicle activity (number of vehicles, level of congestion, etc.) on certain criteria pollutants (CO, NO2, NOx, PM2.5, and CO_2). The study looked at hourly traffic patterns on individual road segments over a year, and estimated that there are likely hundreds of local hotspots where traffic pollutants become elevated due to vehicle activity throughout the year. The US EPA has taken a similar approach to understanding hotspots, by looking at specific pollutants from mobile sources. To date, the EPA's guidance on hotspot regulation has focused on addressing hotspots of singular pollutants (PM 2.5 and CO) in the context of transportation projects. Unfortunately, this policy guidance has been largely to the exclusion of hotspots created by industrial toxins and as noted earlier, the importance of having transportation hubs around industrial areas likely means multiplicative effects of both sources of pollution.

Public health researchers have increasingly become aware of the importance of measuring multiple chemicals. Indeed, studies on the cumulative impacts of pollution on health have become integral to the exposure science field, with scholars noting, "[w]hen pollutant mixtures are emitted from a myriad of sources into a mutual atmosphere over time, they may interact through chemical, physical, and biological response-based pathways to produce more complex and synergistic effects" (Krzyzanowski, 2011; p. 254). Cumulative effects research would suggest that the degree to which hotspots impact human health would depend on the totality of the pollutants interacting, and the effect of those interactions over time. However, hotspot research to date has largely omitted such a cumulative analysis. While the number of hotpot studies has expanded over recent years, those done in the US have been cross-sectional, limited number of chemicals, and limited by their focus on specific communities. The following analysis addresses these limitations by using estimated exposure to 572 air toxics at the block group level to determine which block groups are hotspots within those six states that make up EPA's region 5 regulation boundaries. By examining the characteristics of those block groups that are identified as hotpots, as well as those that lose and gain hotspot status over time, we are able to better understand which areas face the most significant risk of cumulative exposure—and thus experience the greatest health risk from these pollution hotspots.

12.3 **Data and methods**

The most far-reaching federal action to address unequal exposure to environmental toxics by race and class was President Bill Clinton's Executive Order 12898 signed in 1994. This required all federal agencies to identify and address, "disproportionately high and adverse human health or environmental effects of their actions on minority and low-income populations" (EO 12898). The years following this have been argued to be the "peak period" of EJ policy making (Abel et al., 2015) and when one would expect a perceivable change in environmental disparities by class and race. We focus our analysis on the four years immediately preceding the passing of EO 12898. In addition, we restrict our analysis to those states that make up EPA Region 5: Illinois, Indiana, Michigan, Minnesota, Ohio, and Wisconsin. Restricting our analysis to this region allows us to control for several endogenous variables. For example, EPA regions have been shown to have distinct trends in the rate of decline of their industrial air toxics (Ard, 2015); in the implementation of regulations (Konisky, 2015); and states within EPA region 5 have a distinct environmental and economic history, beginning with them comprising the Old Northwest Territory to industrial and deindustrialization (Zwickl et al., 2014). By analyzing those states within this regulatory and economic division, we do not have to worry about uncontrolled variation around these factors and are able to focus on the processes related to where health risk from exposure to air pollution is found and remains over time.

12.3.1 **Pollution exposure estimates**

Estimates of the health risk from industrial air toxics were obtained from the EPA Risk Screening Environmental Indicators Geographic Microdata (RSEI-GM). The data are thought to provide the most "realistic information on potential human health effects from air pollutants that has been available" (Ash and Fetter, 2004: 447). Because reporting requirements change every few years, the following analyses only utilize those 572 chemicals that were consistently monitored for the years 1995 to 1998. RSEI data models the emission routes of over 600 toxic chemicals from a variety of manufacturing, mining, utility operations, hazardous waste treatment, and disposal facilities, as well as from chemical distributors and federal facilities. The EPA uses information on the velocity at which chemicals are emitted into the environment, weather patterns, and properties of the chemicals to estimate where these particles exist in a 101 by 101 kilometer square around reporting facilities in the continental United States. These 101 by 101 kilometer squares are further broken down into one by one kilometer grid cells. These grid cells were overlaid with 2010 block group boundaries in ArcGIS 10.2.2 and weighted by the area of land they contributed to the block group. For example, if a one-kilometer grid cell were completely contained within the block group, it would contribute 100 percent of its pollution to the block group, if only 50 percent of the grid cell was within the block group it would contribute 50 percent of its pollution to the overall pollution of the block group. All of the weighted pollution was then aggregated up to each of the 40,474 block groups located within EPA Region 5 (see Ard, 2015 for a detailed explanation of this process). The pollution estimates used in these analyses are weighted by their toxicity to human health by the EPA, which uses epidemiological evidence to make chemicals commensurate across health outcomes (EPA, 2007). These weights take into account the severity of a chemical's health effects and its potency. This process creates a unitless score of pollution exposure for each 1 kilometer square that makes them comparable across types of polluting facilities and areas.

12.3.2 Demographic data

Demographic measures were determined for all block groups in the continental United States. We used 1990, 2000, and 2010 Census data obtained from Social Explorer, professional version (2012). These data were then linearly interpolated to obtain annual estimates for different racial and socioeconomic groups in census block groups for the following racial categories: (1) non-Hispanic African-American (herein referred to as African American/Black) (2) non-Hispanic White (herein referred to as White) and (3) Hispanic origin individuals of any race. The racial categories Asian/Pacific Islander, Native American and "Other" are not included here as the sample size of these groups was so small that slight fluctuations of pollution exposure in these groups were magnified and not generalizable. In addition to racial characteristics of a block group, household income data, broken down by race, were also obtained for census block groups from the U.S. Census via Social Explorer. Using Frey's (2003) indicators, household incomes were broken down into a dichotomous variable: (1) high income: those making $50,000 or above and (2) low income: those making less than $50,000. However, unlike data on race, these data are not available for 100 percent of the population; rather one in every six households receives a long form from the census requesting these data. This one-in-six-household statistic holds true for the 1990 and 2000 censuses; however, in 2010 the US Census Bureau moved collecting these long-form data from the decennial census to the American Community Survey (ACS). The ACS process of collecting data differs from that used by the US Census. Unlike the US Census, the ACS does not collect data at one point in time. The sample size is smaller than the one in six persons collected by the long-form decennial census, and in order to protect privacy of individuals the ACS uses values derived from several years of smaller aggregated samples. Information on which block groups were within the central city was determined by overlaying central city boundaries from the US census with 2000 census block group boundaries obtained from the US Census Bureau (Census, 2012). Those block groups located within the central city were coded as one, those that fell outside as zero.

Table 12.1 Number of block groups classified as hotspots over time in EPA Region 5

	Total hotspots	Percentage of hotpots in central city		Gained hotspot status	Lost hotspot status
1995	2194	68%	1995 to 1996	1134	555
1996	1615	62%	1996 to 1997	606	959
1997	1968	65%	1997 to 1998	715	444
1998	1697	63%			
TOTAL	7474				

A spatial weights matrix was calculated in ArcGIS for every state based on the latitude and longitude of the 2010 block group centroids. These spatial weights matrices were then used in the calculation of hotspots via the ArcGIS spatial statistics tool "Hotspot Analysis (Getis-Ord Gi*)." This tool assigns each block group a z-score, and P-value, with a high z-score and a low P-value reflecting that a block group has significantly higher exposure to air pollution than one would expect in a random distribution. We coded block groups that had a z-score above zero, and a P-value as below 0.05 with a one to indicate a hotspot, and those outside of these parameters as zero. Using this process, we identified statistically significant spatial clusters of block groups, within each state, with high values of estimated health risk from air pollution over the years. There were an average of 1869 block groups, out of 40,472, that were hotspots each year, with a high of 2194 in 1995 to a low of 1615 in 1996; 674 block groups were hotspots every year of the analysis. Table 12.1 shows descriptive statistics of how these hotspots changed over time.

12.4 **Results**

In the following analyses, we investigate what characteristics are associated with changes in hotspot status. There were an average of 1869 block groups, out of 40,472 that were hotspots each year, with a high of 2194 in 1995 to a low of 1615 in 1996; 674 block groups were hotspots every year of the analysis. Table 12.1 shows descriptive statistics of how these hotspots changed over time. Of those 555 block groups that gained their status as hotspots from 1995 to 1996, 259 remained hotspots from 1996 to 1997, and of these, 233 remained hotspots from 1997 to 1998. Of those 1134 that lost their hotspot status from 1995 to 1996, only 757 were still not considered hotspots in 1997 and of these 711 remained classified as not hotspots in 1998. While the percentage of hotspots located in central cities declined over the years examined, the majority of them were located in these areas (see Table 12.1).

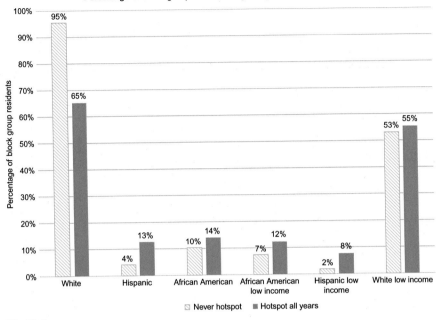

FIG. 12.1

Percentage of residents in block groups classified as hotspot across all years examined and those never classified as hotspots.

Fig. 12.1 helps us visualize the basic demographics around those 674 block groups that retained their hotspot status across all four years looked at. We can see that compared to those 36,937 block groups that were never classified as a hotpot, there were more Whites living in those block groups never identified as hotspots than those that were. The other groups, Hispanics, African Americans, African American and Hispanics of low income, as well as Whites with low income were all more represented in block groups that were hotspots continually over the four years examined. This is consistent with prior work showing non-Whites, and those of low income status have higher exposure to air toxics (Bullard et al., 2007).

To determine the characteristics of block groups that were classified as hotspots, and how this status changed over time, we ran separate logistic regression models on all block groups, coding those that became hotspots the following year as 1, the results of which are shown in Table 12.2. The first model predicts whether a block group would change from not being classified as a hotspot, to being classified as a hotspot the following year. Results in Table 12.2, Model 1, show that for every one percent increase in the percentage of African Americans in a block group, the log-odds of the block group becoming a hotspot decreased by 0.57, but this was only

Table 12.2 Results from logistic regression models predicting whether a block group gained or lost its hotspot status over time

	Model 1			Model 2		
	Odds-Ratio		95% Confidence	Odds-Ratio		95% Confidence
Percent African American	-0.57	*	-1.00 -0.13	-0.81	***	-1.22 -0.40
Percent Hispanic	1.15	*	0.22 2.09	1.46	**	0.54 2.38
Percent Low Income African American	2.59	***	2.09 3.10	3.81	***	3.35 4.28
Percent Low Income Hispanic	-0.52		-2.08 1.03	-0.19		-1.69 1.31
Percent Low Income White	2.48	***	2.21 2.75	2.60	***	2.33 2.88
Central City Location	1.00	***	0.92 1.09	0.86	***	0.78 0.95
Year	0.00		-0.04 0.03	-0.13	***	-0.16 -0.10
Constant	3.67		-62.69 70.02	252.63	***	187.02 318.24
Pseudo R^2	0.04			0.06		
N	161796			161796		

*** $P<.001$.
** $P<.01$.
* $P<.05$.

slightly significant. While every one percent increase in a block-group's percentage of Hispanic increased the log-odds of that block group becoming a hotspot by 1.15, but again this was only slightly significant. However, when considering income as well as race, we can see that for every one percent increase in a block group low-income African American and White populations, there was a 2.59 and 2.48 increase in the log-odds that block group would become a hotspot the following year. If the block group was located in the central city, the log-odds it would be a hotspot increased by 1.00.

Table 12.2, Model 2, lays out the results for characteristics associated with a block group losing status as a hotspot. Analyses were run with all block groups located in EPA's region 5 with those block groups that lost their status as hotspot the following year as 1. For every one percent increase in the African American population of a block group classified as a hotspot, the odds that block group would lose its status as a hotspot the following year decreased by 0.809. For every one percentage increase in the percentage of Hispanic residents, the odds the block group would lose its status as a hotspot increased by 1.46. Similarly, those block groups classified as a hotspot had a 3.81 increased log odds they would not be classified as hotspots the following year for every one percent increase in the percentage of low income African-American residents and 2.60 increased log odds for every one percent increase in the low income White residents. The central city location of a block group classified as a hotpot was associated with a 0.86 increased log odds it would not be classified as a hotpot the following year. The negative and significant odd ratio for the variable year reflects that over time the log odds of a block group losing their status as a hotspot decreased slightly by 0.13 and significantly.

Table 12.3 displays results for logistics regression models that predict the odds a block group will remain stable, i.e., will retain its status as a hotspot or nonhotspot. For example, if a block group was classified as a hotpot (or not a hotpot) in 1995, and it had the same status in 1996 it was coded as a one. These results help us to understand the stability of hotspots over time. Model 1 in Table 12.2 was run on all block groups in Region 5, and demonstrates that those block groups with one percentage increase in their African American population had slightly significant log odds (0.43) of retaining its status as a hotspot or nonhotspot over time. Most interestingly, those block groups that were classified as hotspots one year had a strongly significant chance of not being a hotspot the following year, with a log odds of -3.46. However, as we can see from Model 2, some of this can likely be explained by block groups going in and out of hotpot status, when the variable "Hotspot Any Year" is included the log odds for "Hotpot Prior Year" decrease. However, both the classification of a block group as a hotspot the prior year, and whether it has ever been classified as a hotspot over the time period examined, are negatively related to the log odds of it retaining its status over time. In Model 2, we can see that being classified as a hotspot one year, gives that block group −1.92 log odds it will be classified as a hotspot the following year.

Table 12.3 Results from logistic regression models predicting whether a block group retained its status as a hotspot or nonhotspot over time

	Model 1				Model 2			
	Odds-Ratio		95% Confidence		Odds-Ratio		95% Confidence	
Percent African American	0.43	*	0.10	0.75	0.35	*	0.03	0.66
Percent Hispanic	−1.47	*****	−2.18	−0.76	−1.42	***	−2.14	−0.69
Percent Low-Income African American	−2.16	***	−2.55	−1.78	−2.04	***	−2.42	−1.67
Percent Low-Income Hispanic	3.81	***	2.60	5.02	5.25	***	4.00	6.50
Percent Low-Income White	−1.46	***	−1.67	−1.25	−1.26	***	−1.47	−1.05
Central City Location	−0.52	***	−0.59	−0.46	−0.46	***	−0.53	−0.39
Year	0.06	***	0.04	0.09	0.10	***	0.07	0.13
Hotspot Prior Year	−3.46	***	−3.53	−3.40	−1.92	***	−2.04	−1.81
Hotspot Any Year					−0.60	***	−0.64	−0.56
Constant	−121.81	***	−174.47	−69.15	−197.98	***	−251.84	−144.12
Pseudo R^2	0.27				0.29			
N	161796				161796			

*** $P < .001$.
** $P < .01$.
* $P < .05$.

12.5 Discussion

In the analyses presented here, we worked to understand what demographic characteristics were associated with hotspot status over time. Using estimated air toxics from over 500 different industrial toxics weighted by health risk, we determined which block groups were classified as a hotspot for those states located within EPA Region 5 from 1995 to 1998. We focused on those four years following the signing of Executive Order 12898 in 1994 as one would expect that this would prompt identifiable mitigation in the racial and economic gap in exposure as, thus far, this has been the only federal policy act aimed at reducing racial and economic inequality in exposure to air toxics. Our results do not support that there was a perceivable mitigation of this gap over the time period examined. Rather, our results demonstrate that for every one per cent increase in a block group's low-income African-American population, and to a lesser extent low income Whites, the log odds that a block group would become a hotspot increased significantly. Results also demonstrate that those block groups identified as hotspots were slightly significantly more likely to remain hotspots if they had a greater percentage of African-American or low-income Hispanic residents. These findings support the argument that those areas where low income African-Americans and Whites, within Region 5, were more likely to have experienced becoming a hotspot over this time period and as such could have implications for health disparities in this area over this time. Future work should attempt to pull out those groups of chemicals that are specifically related to the health outcomes of concern.

The analyses presented here also examined how stable a block group was in its classification, as either a hotspot or nonhotspot. Those areas that were the most stable, and therefore had the smallest log odds of changing their status, were those with a one-unit increase in the percentage of African Americans. This likely speaks to the relatively stable relationship between race and place in the US, as has been identified in prior racial segregation research (Timberlake and Iceland, 2007). Those characteristics associated with the greatest likelihood of changing status from either a hotspot to nonhotspot or vice versa were those with low-income Hispanic population; with a one percent increase in this population being associated with a 5.25 log odds, it would change its status the following year. Although the Hispanic population was booming across the US (Flores, 2017), we do not expect that these demographic changes would have a profound impact on our results over the short time period examined here. However, future work should incorporate information about demographic turnover of areas in their relationship to becoming, or maintaining, hotspot status. Block groups that were classified as hotspots had a log odds of -3.46 of being a hotpot the next year. However, when controlling for whether that block group had ever been classified as a hotspot, the log odds decreased to -1.92. Moreover, while industrial air toxics seem to be concentrated in low-income African American and White communities, they also seem to be more likely in central cities. The log odds a block group located in the central city would be classified as a hotspot the following year was 1.00, and those block groups in central cities had significantly lower log odds of changing their status. These results suggest that the patterns of unequal exposure by race and income remained relatively stable across the time examined.

This analysis provides important insights into the demographic characteristics associated with hotspot formation and persistence over time. Our findings are largely consistent with the broader literature documenting concentration of pollution in low-income communities of color (Ard, 2015; Ash et al., 2013; Bullard et al., 2007). However looking at hotspots of a combined estimated exposure of 572 industrial toxics, weighted by their health risk, over time at a regional level, reveals the robustness of the relationship. Future work should aim to incorporate measures of nonchemical stressors, particularly psychosocial factors (e.g., low income, chronic stress, diet, unsafe neighborhoods, poor housing, lack of access to health care, etc.) that coexist in these hotspot areas (Payne-Sturges et al., 2018). Consideration of the nonchemical stressors should be particularly important in hotspot communities, which are consistently characterized by low-income populations. Hotspot communities therefore experience an exacerbated health risk because they experience more frequent and concentrated exposure to environmental hazards themselves, and they also experience greater exposure to the social factors (for instance, high stress levels associated with poverty) that increase vulnerability to pollution (Hicken et al., 2011). "Psychosocial stressors and chemical pollutants impact many of the same physiological systems (e.g., neurological, metabolic, immune, and cardiovascular), which are key regulatory systems of the body. Therefore, it is highly plausible that combined psychosocial and physical environmental exposures may interact to increase or amplify risks of adverse health. By focusing on single pollutant/chemical exposures without consideration of other risk factors acting through common biological systems, health risks may well be underestimated and vulnerable populations may be mischaracterized or overlooked altogether" (Payne-Sturges et al., 2018; p. 5).

References

Abel, T.D., Salazar, D.J., Robert, P., 2015. States of environmental justice: redistributive politics across the United States, 1993–2004. Rev. Policy Res. 32 (2), 200–225.

Ard, K., 2015. Trends in exposure to industrial air toxins for different racial and socioeconomic groups: a spatial and temporal examination of environmental inequality in the U.S. from 1995 to 2004. Soc. Sci. Res. 53, 375–390.

Ash, M., Boyce, J.K., Chang, G., Scharber, H., 2013. Is environmental justice good for white folks? industrial air toxics exposure in urban America. Soc. Sci. Quart. 94 (3), 616–636.

Ash, M., Fetter, T.R., 2004. Who lives on the wrong side of the environmental tracks? Evidence from the EPA's risk-screening environmental indicators model*. Soc. Sci. Quart. 85 (2), 441–462.

Bullard, R., Mohai, P., Saha, R., Wright, B., 2007. Toxic Wastes and Race at Twenty A Report Prepared for the United Church of Christ. Retrieved from, http://www.ucc.org/assets/pdfs/toxic20.pdf.

Bullock, C., Ard, K., Saalman, G., 2018. Measuring the relationship between state environmental justice action and air pollution inequality, 1990–2009. Rev. Policy Res. 35 (3), 466–490.

Census, 2012. Census 2000 Geographic Definitions. U.S. Census Bureau. Available at: http://www.census.gov/geo/www/geo_defn.html#MA.

Chi, G.C., Hajat, A., Bird, C.E., Cullen, M.R., Griffin, B.A., Miller, K.A., Shih, R.A., Stefanick, M.L., Vedal, S., Whitsel, E.A., Kaufman, J.D., 2016. Individual and neighborhood socioeconomic status and the association between air pollution and cardiovascular disease. Environ. Health Perspect. 124 (12), 1840–1847. Print.

Clark, L.P., Millet, D.B., Marshall, J.D., 2014. National patterns in environmental injustice and inequality: outdoor NO$_2$ air pollution in the United States. PLoS ONE 9 (4), e94431.

Downey, L., 2005. The unintended significance of race: environmental racial inequality in Detroit. Soc. Forces 83 (3), 971–1007.

Downey, L., 2007. US metropolitan-area variation in environmental inequality outcomes. Urban Stud. 44 (5–6), 953–977.

DuBois, W.E.B., 1899. The Philadelphia Negro: A Social Study. University of Pennsylvania Press, Philadelphia.

Environmental Protection Agency, 2007. EPA's Risk-Screening Environmental Indicators (RSEI) Methodology. RSEI Version 2.1.5. United States Environmental Protection Agency. Office of Pollution Prevention and Toxics, Washington, DC. Available at: http://www.epa.gov/opptintr/rsei/pubs/method_oct2007.pdf.

Executive Order 12898 of February 11, 1994. Federal actions to address environmental justice in minority populations and low-income populations. Federal Register 59 (32).

Farley, R., Danziger, S., Holzer, H., 2000. The evolution of racial segregation. In: Detroit Divided. Russell Sage Foundation, New York.

Flores, A., 2017. How the U.S. Hispanic population is changing. Pew Research Center. Retrieved from, http://www.pewresearch.org/fact-tank/2017/09/18/how-the-u-s-hispanic-population-is-changing/.

Franklin, B.A., Brook, R., Arden, P.C., 2015. Air pollution and cardiovascular disease. Curr. Probl. Cardiol. 40 (5), 207–238.

Frey, W., 2003. American demographics. Am. Demogr. 26–31.

Gately, C.K., Hutyra, L.R., Peterson, S., Wing, I.S., 2017. Urban emissions hotspots: quantifying vehicle congestion and air pollution using mobile phone GPS data. Environ. Pollut. 229, 496–504.

Hawley, A., 1971. Urban Society: An Ecological Approach, second ed. Wiley, New York.

Hicken, M., Gragg, R., Hu, H., 2011. How cumulative risks warrant a shift in our approach to racial health disparities: the case of lead, stress, and hypertension. Health Aff. (Millwood) 30 (10), 1895–1901.

Johnson, C.H., Athersuch, T.J., Collman, G.W., Dhungana, S., Grant, D.F., Jones, D.P., Patel, C.J., Vasiliou, V., 2017. Yale school of public health symposium on lifetime exposures and human health: the exposome; summary and future reflections. Hum. Genom. 11, 1.

Juarez, P.D., Matthews-Juarez, P., Hood, D.B., Im, W., Levine, R.S., Kilbourne, B.J., Langston, M.A., Al-Hamdan, M.Z., Crosson, W.L., Estes, M.G., Estes, S.M., Agboto, V.K., Robinson, P., Wilson, S., Lichtveld, M.Y., 2014. The public health exposome: a population-based, exposure science approach to health disparities research. Int. J. Environ. Res. Public Health 11, 12866–12895.

Kahn, M.E., 1997. Particulate pollution trends in the United States. Region. Sci. Urban Econ. 8 (1), 86–107.

Kahn, M.E., 1999. The silver lining of rust belt manufacturing decline. J. Urban Econ. 46 (3), 360–376.

Konisky, D., 2015. Failed Promises: Evaluating the Federal Government's Response to Environmental Justice. MIT Press, Cambridge, MA.

Krzyzanowski, J., 2011. Approaching cumulative effects through air pollution modelling. Water Air Soil Pollut. 214, 253–273.

Maasoumi, E., Millimet, D.L., 2005. Robust inference concerning recent trends in US environmental quality. J. Appl. Econometr. 20 (1), 55–77.

Nadadur, S.S., Miller, C.A., Hopke, P.K., Gordon, T., Vedal, S., Vandenberg, J.J., Costa, D.J., 2007. The complexities of air pollution regulation: the need for an integrated research and regulatory perspective. Toxicol. Sci. 100 (2), 318–327.

National Academies of Sciences, Engineering, and Medicine; Health and Medicine Division; Board on Population Health and Public Health Practice, 2017. The state of health disparities in the US. In: Baciu, A., Negussie, Y., Geller, A. (Eds.), Communities in Action: Pathways to Health Equity. National Academy of Sciences.

National Research Council, 2004. Air Quality Management in the United States. National Academies Press, Washington, D.C.

National Research Council, 2009. Science and Decisions: Advancing Risk Assessment. National Academies Press, Washington, DC, USA.

Park, R., 1915. The city: suggestions for the investigation of human behavior in the city. Am. J. Sociol. 20, 577–612.

Payne-Sturges, D., et al., 2018. Methods for evaluating the combined effects of chemical and nonchemical exposures for cumulative environmental health risk assessment. Int. J. Environ. Res. Public Health 15 (12), 2797.

Portney, P., 1990. Air pollution policy. In: Portney, P. (Ed.), Public Policies for Environmental Protection. Resources for the Future, Washington DC.

Rosofsky, A., Levy, J., Zanobetti, A., Janulewicz, P., PatriciaFabiana, M., 2018. Temporal trends in air pollution exposure inequality in Massachusetts. Environ. Res. 161, 76–86.

Sexton, K., 2012. Cumulative risk assessment: an overview of methodological approaches for evaluating combined health effects from exposure to multiple environmental stressors. Int. J. Environ. Res. Public Health 9, 370–390.

Timberlake, J.M., Iceland, J., 2007. Change in racial and ethnic residential inequality in American Cities, 1970–2000. City Commun. 6 (4), 335–365.

US EPA, 2018. Our Nation's Air: Air Quality Improves as America Grows. Retrieved, https://gispub.epa.gov/air/trendsreport/2018/documentation/AirTrends_Flyer.pdf.

Wang, S.W., Tang, X., Fan, Z.H., Wu, X., Lioy, P.J., Georgopoulos, P.G., 2009. Modeling of personal exposures to ambient air toxics in Camden, New Jersey: an evaluation study. J. Air Waste Manag. Assoc. 59 (6), 733–746.

Weber, A., 1909. Uber den Standort der Industrien (translated by Friedrich C. J. (1929)) Alfred Weber's Theory of the Location of Industries. University of Chicago Press, Chicago.

Wial, H., Friedhoff, A., 2006. Bearing the Brunt: Manufacturing Job Loss in the Great Lakes Region, 1995–2005. Brookings Institute. Metropolitan Policy Program.

Wilson, W.J., 1996. When Work Disappears: The World of the New Urban Poor. Alfred A. Knopf, New York.

Woodruff, T.J., Darrow, L.A., Parker, J.D., 2008. Air pollution and postneonatal infant mortality in the United States, 1999-2002. Environ. Health Perspect. 116 (1), 110–115.

Wu, X., Fan, Z., Zhu, X., Jung, K.H., Ohman-Strickland, P., Weisel, C.P., Lioy, P.J., 2012. Exposures to volatile organic compounds (VOCs) and associated health risks of socioeconomically disadvantaged population in a "hotpot" in Camden, New Jersey. Atmos. Environ. 57, 72e79.

Zhu, Y., Hinds, W.C., Kim, S., Sioutas, C., 2002. Concentration and size distribution of ultrafine particles near a major highway. J. Air Waste Manag. Assoc. 52, 1032–1042.

Zou, B., Peng, F., Wan, N., Mamady, K., Wilson, G.J., 2014. Spatial cluster detection of air pollution exposure inequities across the United States. PLoS ONE 9 (3), e91917. https://doi.org/10.1371/journal.pone.0091917.

Zwickl, K., Ash, M., Boyce, J.K., 2014. Regional variation in environmental inequality: industrial air toxics exposure in U.S. Cities. Ecol. Econ. 107, 494–509.

Further reading

Turaga, R.M.R., Noonan, D., Bostrom, A., 2011. Hotpots regulation and environmental justice. Ecol. Econ. 70, 1395–1405.

Travel-related exposure to air pollution and its socio-environmental inequalities: Evidence from a week-long GPS-based travel diary dataset

13

Wenbo Guo[a,b], Yanwei Chai[b], Mei-Po Kwan[c]

[a]*School of Geography and the Environment, University of Oxford, Oxford, United Kingdom*
[b]*College of Urban and Environmental Sciences, Peking University, Beijing, China*
[c]*Department of Geography and Geographic Information Science, University of Illinois at Urbana-Champaign, Urbana, IL, United States*

13.1 Introduction

Geographies of health have contributed to significant literature used in the research of health and related realms, resulting in enrichment of geography itself over the last two decades (Gatrell and Elliott, 2009; Brown et al., 2010; Gatrell, 2011). Recent developments and the widespread diffusion of geospatial data acquisition technologies enable the creation of highly accurate spatial and temporal data relevant to health research. For example, the global positioning systems (GPS) and related techniques make it possible to integrate highly accurate space-time behavior information with health data, while global information systems (GIS) provide the means to analyze and visualize integrated data (Richardson et al., 2013). Stressing time in geographical and mobility research, the origin of temporally integrated geographies seeks a dynamic travel-activity perspective for investigating social issues of health, segregation, etc. (Kwan, 2013).

The issue of health has drawn extensive attention from many fields. The newest topics of human mobility and health research include individuals' exposure to air pollution. In China, the high-speed industrialization, urbanization, and motorization for the last decades have resulted in greater air pollution, which is considered one of the major health risks in Chinese cities. According to the Beijing Municipal Committee of Transport, the number of private cars reached 93.09 million in China by the end of 2012, and the number was 4.08 million in Beijing. The proportion of car travel in Beijing reached over 30%, and the local pollution contributed 64%–72% of $PM_{2.5}$ pollution in Beijing for 2012–13, in which vehicles contributed >30% as the main source of local air pollution (Beijing Environmental Protection Bureau, 2014). The average concentrations of SO_2, NO_2, $PM_{2.5}$, and CO of Beijing were $0.028\,mg/m^3$, $0.052\,mg/m^3$, $0.109\,mg/m^3$, and $1.4\,mg/m^3$ respectively in 2012. The Chinese government has formally promulgated "Air Pollution Control Plan for Key Area" in order to reduce air pollution concentration, especially for $PM_{2.5}$. $PM_{2.5}$ is considered one of the major health risks for urban residents since $PM_{2.5}$ is very active with poisonous and harmful substances and can be easily inhaled into lungs directly. The harm of $PM_{2.5}$ on human health is mainly through the respiratory and cardiovascular systems, and the impact broadens as the concentration increases, especially for children and elder people. Based on tracking surveys in European and US cities, it shows that the mortality rate aggrandizes 0.62% and 0.46% as the average short-term concentration of PM_{10} increases every $10\,\mu g/m^3$ (Samet et al., 2000; Katsouyanni et al., 2001).

Transport-related health is usually influenced by multi-factors, including environment, neighborhood, and mobility (Saelens et al., 2003; Frank et al., 2007), whereas the relationship between travel modes preference and air pollution exposure is yet to be explored. This paper, using concentration data of $PM_{2.5}$ and CO in transport microenvironments, travel-activity, and website-based data derived from 2012. The GPS-enabled travel-activity survey in suburban Beijing, measures individuals' travel exposure to air pollution and compares the difference of $PM_{2.5}$ and CO exposure between groups through an ANOVA test. Further, a structural equation model (SEM) was used to analyze how the built environment and travel behavior is associated with an individual's daily travel-related exposure to air pollution.

13.2 Literature review

The issue of mobility has increased in prominence, not just in geography but also across other social sciences. It is discussed in highly interdisciplinary contexts, such as in sociology (Shaw and Hesse, 2010). Health is usually defined as a state of complete physical, mental, and social well-being and not merely the absence of disease or infirmity (WHO, 1947). In other words, an individual's health includes both the physical and mental aspects. Most of the existing related literature focussed on accessibility to health services (e.g., Goddard and Smith, 2001; Wang and Luo, 2005; Hawthorne and Kwan, 2012) and how environment, geographical context, and neighborhood influence health and behavior (e.g., Dahlgren and Whitehead, 1991; Jones and Moon, 1993; Kearns, 1993; Aubrey et al., 1995; Frank and Engelke, 2001). Among these studies, Dahlgren and Whitehead (1991) offered the social determinants of a health "rainbow" and explained the multilevel yet inseparable factors of individual lifestyle, social and community networks, and environmental and geographical context. Among these influential elements, the built environment was highlighted for its impacts on health, particularly of elderly and obese individuals (Berke et al., 2007; Papas et al., 2007).

However, there are very few studies discussing the relationships between mobility and health. Traditional measurement of air pollution exposure is mostly based on the outdoor environment (Frank et al., 1977; Horie and Eldon, 1979). This measurement is a substitutive measurement of potential exposure to air pollution based on residential locations, and it is considered an indirect and inaccurate measurement. On the one hand, air pollution concentration of outdoor and indoor may show significant differences because of indoor smoking (Repace and Lowrey, 1980), and the outdoor concentration may have hysteresis effect on indoor concentration (Davidson et al., 1986; Ando et al., 1996; Janssen et al., 1998). The areas near the edges and intersections of streets will significantly increase more than other places (Ott and Eliassen, 1973; Ott and Mage, 1974). On the other hand, individual exposure to air pollution is based on his/her own micro-level behavior through space over time (Cortese et al., 1976; Kwan, 2013).

The early related measurement is mainly based on direct estimates of places where activities and travels take place. Godin et al. (1972) measured the CO exposure under different behavior types (sleep, work, driving, walking, outdoor activities, etc.) in Toronto using a spectral measurement instrument. Significant difference of individual exposure was found when taking different activities, but the same results were found in similar research (Wright et al., 1975; Cortese et al., 1976). To measure an individual's exposure to air pollution more accurately, the dimension of time was taken into account. Fugas (1975) measured "Weighed weekly exposure" to SO_2, Pb, Mn at home, work place, and main streets based on an average time use structure of residents. The measurement is based on aggregated samples, and concentration and time of exposure were generalized, not considering the space/time inconsistency (different locations of streets and peak or non-peak hours) of concentration and the dynamics of individual's activity pattern (Duan, 1981).

The research on the relationship between mobility and health also stresses the diffusion of disease and mobility (Langer and Frösner, 1996), traffic congestion, traffic emission, and its impacts on human health (Colvile et al., 2004; Costabile and Allegrini, 2008; Hatzopoulou et al., 2011). However, the relationship between individual exposure to air pollution and mobility, especially for space, time, and context dimension, is yet to be studied.

Originating in environmental epidemiology, air pollution exposure was analyzed and presented using maps and GIS, including determination of pollutant sources (Bell et al., 2001), evaluation of harmful environmental factors for travel (Reynolds et al., 2003) personal pollution exposure, and pollution diffusion path assessment (Nuckols et al., 2004). The Space-Time Exposure Modeling System is a solid attempt to simulate and measure individual air pollution exposure by integrating the data of pollution sources, diffusion path, pollution measurement, and travel-activity behavior information (Gulliver and Briggs, 2005; Dhondt et al., 2013) developed an integrated modeling chain involving an activity-based transport model, emission and dispersion model, and road safety model to assess the health impact of changes in travel behavior. These authors made further suggestions for policies that would result in a win-win situation with balance between transport and health.

Beyond measurement of fixed places, including home, workplace, streets, etc., air pollution within the micro-transport environment has drawn increasing attention.

Significant differences were found within different micro-transport environments, as seen in measurements of $PM_{2.5}$ and CO that were performed in multiple cities including London (Adams et al., 2001; Kaur et al., 2007), Beijing (Huang et al., 2012a), Taipei (Tsai et al., 2008) and Mexico City (Riojas-Rodriguez et al., 2006). Setton et al. (2010) suggested that daily mobility patterns cannot be ignored when estimating individual-level air pollution exposure using paired residence- and mobility-based estimates of individual exposure to ambient nitrogen dioxide in Vancouver, British Columbia, and Southern California; this rectified the null hypothesis of epidemiological studies. In this context, health-related mobility is influenced by the environment (air quality), the neighborhood, and mobility itself (Saelens et al., 2003; Frank et al., 2007). Whereas, the interaction of mobility between demographic characteristics, geographical context, and air pollution exposure is yet to be sufficiently explored. And it is possible and necessary to study individuals' exposure to air pollution from the perspective of the dynamic activity-travel system.

13.3 **Study area and data**

The study area for this research is Shangdi-Qinghe District in Beijing, China. Beijing is a megacity with sprawling urban structure and a population of approximately 20 million at the end of 2013. Suburbanization has led to spatial expansion and urban reconstruction, which has made suburbs the forefront of socio-spatial transition in Beijing. As an outskirt town in the north-west of Beijing, Shangdi-Qinghe district is a typical suburban area for job and residence centers, which is crossed by highway and metro lines and with a close neighbor and interact with big job centers and residential areas in urban level (see Fig. 13.1). Shangdi-Qinghe district held a population of 300,000, consists of communities including various kinds of commercial housing neighborhoods, affordable housing communities, danwei residential compounds, and urban village. Shangdi-Qinghe district is a comprehensive area with multi-kinds of urban problems for research significance, especially during rapid-suburbanizing and transitional Chinese cities. The activity-travel pattern of residents living in the Shangdi-Qinghe district has good representativeness for the suburban lifestyle. And the job-house spatial mismatch has resulted in long-time exposures to transport microenvironments.

To measure individual exposure to air pollution, two main sets of data were used in this research: evaluation of CO and $PM_{2.5}$ concentrations within transport microenvironments and GPS-enabled activity-travel diary data for suburban residents in Beijing.

First, data of air pollution concentrations in transport microenvironments were derived from field-survey-based estimates of $PM_{2.5}$ and CO concentrations according to different travel modes (Huang et al., 2012a). The investigation, performed in 2010, chose two specific routes to represent average road characteristics in Beijing. The modes of buses and private cars monitored were the most commonly used vehicle types. For this study, 131 trips were monitored for concentrations of $PM_{2.5}$ and CO, which were separately estimated using a portable aerosol spectrometer and a CO monitor. This methodology offered real-time and overall trip exposures within

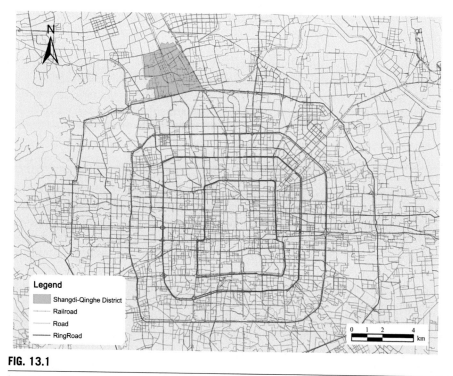

FIG. 13.1

Locations of the survey area.

different transport modes. In this research, we combined the generalized overall concentrations of PM$_{2.5}$ and CO within different travel modes with the air inhalation rate, which is normally $0.011\,m^3/min$ under no specific exercise and $0.026\,m^3/min$ when taking light exercise (US EPA, 2009). The only situation considered to be a light exercise is riding a bicycle, whilst walking, driving a car, and taking a bus and subway were measured at normal air inhalation rate. Thus, by integrating the data, we calculated how much quantities individuals were exposed to PM$_{2.5}$ and CO that occurred in each travel mode per minute.

Individuals' daily travel exposure to air pollution is measured by combining the air pollution concentration data and activity-travel dairy data. The air pollution concentration data is obtained from existing literature, which offers relatively accurate air pollution data within micro-transport environment, but it also has many limitations: (1) Though air pollution exposure within urban scale may have limited significance in the difference, the selected survey area "Dongshengxiang" may still have differences compared with other areas in Beijing; (2) The selected 18 survey days is within specific space-time context, including the season, weather, wind, etc.; these elements may be different for the activity-travel dairy survey; (3) The surveyed transport modes did not include subway. Thus, the implementation of the air pollution concentration data is based on the following hypothesis: (1) the space-time context (season, weather, temperature, wind, atmospheric characteristics, etc.) of air pollution concentration survey is consistent with the activity-travel dairy survey; (2) the surrounding environment does not affect the air pollution concentration within transport microenvironments, and the air pollution concentration within transport microenvironment stays the same; (3) microenvironment within subway is considered the same as walking, since there is substantial and frequent air exchanges between the subway microenvironment and outdoor walking environment, and electricity-powered subway do not divulge polluted air into the microenvironment.

The second type of data was obtained from the 2012 website-based and GPS-enabled Travel-Activity Survey in Shangdi-Qinghe District in Suburban Beijing. The survey was conducted from October 8 to December 20, 2012, and consisted of eight rounds, each of which lasted for a week. Based on stratified sampling through a street office and neighborhood community, 709 respondents' GPS trajectories and travel-activity diaries were collected along with individuals' socio-economic information. The data were collected using location-aware devices, a survey website and interviews with participants. A week-long travel-activity diary was collected for each participant as well as their individual and household socio-economic information from questionnaires on the specially designed survey website. Especially, the travel-activity diary was modified with GPS data derived from the location-aware devices to acquire more accurate time and space information for traveling.

The data provided for this research has many advantages. Except for the highly-accurate space-time trajectories and travel-activity diaries, the information furnished by socio-economic information is very valuable. It contains detailed attributes on individual and household socio-economic conditions, including gender, age, income, education, Hukou status, occupation and family structure information like "whether both couples are employed", "whether have child under 16 years old living together" and "whether have elder people over 60 years old living together" etc.

As for a statistical description of the participants' socio-economic status, there were more females than males (47.7% and 52.3%, respectively), and more mid-aged samples tended to live in the suburbs (people aged 30–49 comprised 66.9% of the participants). And the majority of participants have Beijing local *hukou* (89.4%). Most of the respondents were well educated (43.4% had a high school diploma, and 49.6% had a bachelor's degree or above) Mid-to-high salary earners were in the majority (58.5% earned 2000–6000 RMB per month and 21.5% earned >6000 RMB per month). The average commute distance was 8.113 km, and nearly half of participants were inclined to drive a car to commute.

13.4 Empirical analysis

Measure and mechanism of travel exposure to air pollution for individuals in Shangdi-Qinghe District was performed using three procedures: (1) Individual exposure to $PM_{2.5}$ and CO per day was estimated through a weighted sum of exposure to air pollution per minute within different transport microenvironments and travel duration by modes; (2) Comparisons of individual exposures within subgroups were examined using analysis of variance (ANOVA); (3) The mechanisms of travel mode preference and individual exposure to air pollution were explored through structural equation modeling. These procedures were implemented using a statistics package for social science (SPSS) and its add-on module for Analysis of Moment Structures (AMOS).

13.4.1 Measure and ANOVA test

To estimate individual exposure to air pollution in a time-geographical framework, this study used the weighted sum method to integrate air pollution concentrations of $PM_{2.5}$ and CO, air inhalation rate and travel information with high space-time accuracy as well as inhalation rates (Fig. 13.2).

Based on a 2012 activity dairy data, Beijing activity-travel GPS survey and air pollution density data within a transport microenvironment obtained from existing literature (Huang et al., 2012a), the measurement of individual's daily travel air pollution exposure is measured through the equation:

$$E = \sum di \cdot ti \cdot hi$$

where E represents individual daily travel air pollution exposure to $PM_{2.5}$ and CO, d_i is the air pollution density within transport mode i, t represents the travel duration of choosing transport mode i, and hi is the inhalation rates within different transport microenvironments.

Using this measurement implementation, the average individual travel exposure to $PM_{2.5}$ and CO for participants were calculated 64.575 μg and 6.147 mg per day, respectively.

Significant variation was found between different days within a week both for exposure to $PM_{2.5}$ and CO (Table 13.1). More precisely, an individual's exposure to $PM_{2.5}$ on Sunday was significantly lower than any weekdays (sig. < .01), while less significance was found on Saturday compared with weekdays (sig. < .05 for Tues., Wed., Thur., Fri.; sig. < .1 for Mon.); individual's exposure to CO can only find a significant difference between Monday and Saturday (Sig. = .016).

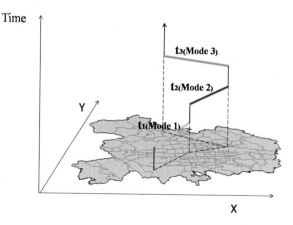

FIG. 13.2

Measurement of an individual's travel-related exposure to air pollution. *Notes:* Different colors represent air pollution density within different transport microenvironments.

Table 13.1 ANOVA test of differences in travel exposure to $PM_{2.5}$ and CO between days

	$PM_{2.5}$ (μg)	F-statistic (Significance)	CO (mg)	F-statistic (Significance)
Mon.	65.912		5.593	
Tues.	66.857		6.032	
Wed.	66.649	3.041[a] (0.006)	6.015	2.121[b] (0.048)
Thu.	66.402		6.235	
Fri.	67.054		6.296	
Sat.	58.973		7.010	
Sun.	55.548		6.067	

Notes: [a,b] indicates that the significance level is lower than 1% and 5% (P<.01, .05).

To examine whether there were significant differences between subgroups, the participants were grouped by gender, age, marital status, Beijing *hukou* (i.e., they are registered residents of Beijing, and thus are local residents), educational background, occupation, income, driver's license, and family structure (whether both couples are employed, whether they have a child under 16 years old, and whether they people over 60 years old living together.). First, the whole samples were divided into four different age subgroups for even distribution (under 29, 30–39, 40–49, and over 49 years old). There were five categories for personal income and four groups for education level (less than high school, high school, undergraduate, and graduate) and occupation (full-time, part-time, students, and retired). Differences within these subsamples were examined using an analysis of variance.

As shown in Table 13.2, results of the ANOVA tests indicated that there were significant differences based on *Hukou* status, income, and driving license for individual travel exposure to both $PM_{2.5}$ and CO. Residents holding *Beijing Hukou* suffered significantly higher levels of travel exposure to both $PM_{2.5}$ and CO than those who didn't. Low-income and mid-income people (monthly income under 6000 RMB)

Table 13.2 ANOVA test of differences in travel exposure to $PM_{2.5}$ and CO within subgroups

Group characteristics	$PM_{2.5}$ (μg)	F-statistic (Significance)	CO (mg)	F-statistic (Significance)
Gender				
Male	64.436	0.022 (0.883)	6.772	34.125[a] (0.000)
Female	64.710		5.554	
Age				
Under 29	62.063		6.285	
30–39	60.490	10,078[a] (0.000)	6.256	1.236 (0.295)
40–49	71.590		6.078	
Over 49	71.048		5.634	

Table 13.2 ANOVA test of differences in travel exposure to PM$_{2.5}$ and CO within subgroups–*Continued*

Group characteristics	PM$_{2.5}$ (µg)	F-statistic (Significance)	CO (mg)	F-statistic (Significance)
Marital status				
Unmarried	65.699	1.882	6.055	0.050
Married	64.545	(0.152)	6.158	(0.951)
Hukou status				
Non-Beijing *hukou*	56.640	10.604[a]	5.397	7.282[a]
Beijing *hukou*	65.700	(0.001)	6.254	(0.007)
Education				
Less than high school	69.649		5.851	
High school	66.861	3.625[b]	6.286	1.723
Undergraduate	60.709	(0.013)	5.882	(0.160)
Graduate	64.747		6.467	
Occupation				
Full-time	63.468		6.165	
Part-time	73.075	3.536[b]	6.138	0.783
Students	75.783	(0.014)	4.598	(0.503)
Retired	71.537		6.162	
Income per month				
Below 2000 RMB	81.237		6.237	
2001–4000 RMB	62.454	20.743[a]	5.882	3.250[b]
4001–6000 RMB	63.479	(0.000)	6.521	(0.011)
6001–10,000 RMB	54.391		5.766	
Above 10,001 RMB	60.433		6.985	
Driver's license				
Yes	71.711	66.961[a]	5.206	90.238[a]
No	56.847	(0.000)	7.167	(0.000)
Both couples are employed				
Yes	62.911	6.458[b]	6.224	1.045
No	67.848	(0.011)	5.997	(0.307)
Child under 16 years old living together				
Yes	60.214	25.357[a]	6.286	1.940
No	69.441	(0.000)	5.993	(0.164)
Old people over 60 years old living together.				
Yes	62.185	3.397[c]	5.974	1.373
No	65.775	(0.065)	6.234	(0.241)

Notes: [a,b,c] *indicates that the significance level is lower than 1%, 5%, and 10% (P <.01, .05, .1).*

suffered significantly higher levels of travel exposure to $PM_{2.5}$, especially for those who earn the least income (monthly income under 2000 RMB) and high income (monthly income ranges from 6001 to 10,000 RMB) separately suffered significantly highest levels and the lowest level of travel exposure to $PM_{2.5}$ compare with other income groups. For individuals' travel exposure to CO, high income (monthly income ranges from 6001 to 10,000 RMB) also experienced the lowest level of exposure, especially compared with low-income groups (monthly income under 2000 RMB and monthly income ranges from 2001 to 4000 RMB). There is also a significant difference between those who have or do not have a driving license and those who have a driving license exposed to a lower level of CO while a higher level of $PM_{2.5}$ than those who don't own a driving license. There is a significant difference in subgroups of age, education, occupation, and family structure for individual travel exposure only to $PM_{2.5}$. Mid-aged and older-aged persons were exposed to significant higher levels of $PM_{2.5}$, especially for groups of over 39 years old compared with those 19 to 39 years old; while only the group of aged over 49 years old has significant difference with groups of under 29 years old and 29 to 39 years old for CO exposure. And better educated people seem to experience significantly lower levels of $PM_{2.5}$ air pollution exposure, there is significant difference within groups of educational backgrounds of high school and undergraduate. Full-time workers were exposed to significantly lower levels of $PM_{2.5}$ than part-time workers and those who do not work. And students' exposure to CO level was significantly higher than full-time workers with a little difference but much lower than part-time workers and the retired. Family structure has significant impacts on $PM_{2.5}$ air pollution exposure. For those both couples in family are employed, the $PM_{2.5}$ air pollution exposure is significant lower. And if there is a child under 16 years old or old people over 60 years old, the individual will be exposed to significant lower level of $PM_{2.5}$ air pollution. The gender difference is significant in an individual's travel exposure to CO, males suffered a significant higher level of CO than female, while the exposure to $PM_{2.5}$ is almost the same for male and female.

13.4.2 Structural equation modeling

A structural equation model was used to dig into how built environments and travel behavior is associated with individual exposure to air pollution when controlling individual socio-economic status. SEM facilitates exploration of the relationships within and among groups of endogenous and exogenous variables (Bollen and Long, 1993), and can be separately defined as a measurement model or a structural model in which latent variable indicators can be dichotomous, continuous, or ordered categorical (Muthén, 1984).

   In this model, we used endogenous variables of travel times by different modes per day, built environment of transport and facility services, and individual exposure to $PM_{2.5}$ and CO per day and individual/household socio-economic status as the exogenous variables. Built environment variables include whether: they work in the inner city, the number of bus stations near home (1 km buffer), number of stores near home (1 km buffer), number of bus stations near workplace (1 km buffer), and number of stores near workplace (1 km buffer), and variables of travel times by different modes basically include bus, car, subway, bicycle, and walk. All the endogenous variables were continuous. The exogenous variables of individual/household socio-economic status were separately set as dichotomous and ordered categorical variables. Dichotomous variables consisted of gender, marital status, *hukou* status, housing source, occupation, driver's license, and whether both couples were employed, whether there were children under 18 years of age or people over 60 years of age in the household. Ordered categorical variables included education level, personal income, age, etc.

The same structured models were conducted, respectively, for individual exposure to $PM_{2.5}$ and CO per day. The individual exposure to $PM_{2.5}$ or CO per day is influenced directly by travel times by modes, which shows the preference of mode choices. And the mode choices are associated with the built environment, especially the bus stations near home, and workplace will affect the preference of travel by bus, and the stores around home and workplace will influence the walking rates. And the work-home distance is also a key factor for modal preferences. All these factors endogenous and exogenous variables result in individual exposure to $PM_{2.5}$ and CO. The results show relatively desirable goodness-of-fit for a model of individual exposure to $PM_{2.5}$ and CO, Chi-square is separately 7024.14 and 6765.75.

As shown in Table 13.3, the effects of the built environment on travel modal preference and thus on an individual's daily travel exposure to air pollution were

Table 13.3 Effect between built environment, travel times and individual travel exposure

Dependent variable	Independent variable	$PM_{2.5}$ Estimate	P	CO Estimate	P
Travel times by bus	Number of bus stations near home	0.001[a]	.021	0.008[a]	.001
Travel times by bus	Number of bus stations near the workplace	0.002[a]	.000	0.001[a]	.023
Travel times by walk	Number of stores near home	0.010[b]	.000	0.007[a]	.000
Travel times by walk	Number of stores near the workplace	0.126	.797	1.881	.919
Travel times by bicycle	Number of stores near home	0.022[b]	.001	0.019[b]	.001
Travel times by car	Work inner cities	0.037	.154	0.050[c]	.049
Daily exposure	Travel times by car	236.399	.170	24.753[c]	.060
	Travel times by subway	−2.863	.341	0.707[b]	.011
	Travel times by bus	61.910[a]	.002	−1.033	.129
	Travel times by bicycle	−102.858[c]	.078	2.797[a]	.001
	Travel times by walk	223.872[c]	.099	−6.071[a]	.001

Notes: [a,b,c] *indicates that the significance level is lower than 1%, 5%, and 10% (P<0.01, 0.05, 0.1).*

checked. For individuals' daily travel exposure to both $PM_{2.5}$ and CO, the number of bus stations near home and number of bus stations near the workplace are contributing positively to travel times by bus per day, number of stores near home, and workplace affect positively on travel times by walk per day, since residents prefer walking for shopping in a short distance. While working in inner city only has a significant effect on an individual's daily travel exposure to CO, working in an inner city will increase significantly an individual's daily travel exposure to CO but not to $PM_{2.5}$. Travel times per day by bus, bicycle, and walk have a significant effect on an individual's daily travel exposure to $PM_{2.5}$ in which travel times by bicycle has a negative effect. And travel times per day by car, subway, bicycle, and walk have a significant effect on an individual's daily travel exposure to $PM_{2.5}$ in which travel times by walk has a negative effect.

The exogenous variable effects show how an individual's socio-economic characteristics affect individual's daily exposure to air pollution of $PM_{2.5}$ and CO (Table 13.4). Males' travel exposure to both $PM_{2.5}$ and CO are significantly higher than females, since they usually spent more time than females on daily commute. Age plays a conspicuous role in individuals' exposure to $PM_{2.5}$ and CO. The elder tends to make use more often than the mid-aged and young persons of the public transport modes, which was more concentrated with PM2.5 and less intensive in CO. This mechanism also applied to the influence of driving licenses on daily travel exposure, as the old have lower possibilities to own a driving license. Car ownership

Table 13.4 Effect of demographic variable on individual travel exposure to $PM_{2.5}$ and CO

Socio-economic information	$PM_{2.5}$		CO	
	t	Sig.	t	Sig.
Gender	-2.362^b	0.018	-4.377^a	0.000
Age	2.555^b	0.011	-2.723^a	0.007
Hukou status	-2.250^b	0.025	-2.505^b	0.012
Education	1.872^c	0.061	0.109	0.913
Marital status	-0.588	0.557	0.294	0.769
Driving license	-7.045^a	0.000	7.938^a	0.000
Occupation	0.731	0.465	1.267	0.205
Personal income	-3.910^a	0.000	-0.937	0.349
Whether have children under 18 years of age in the household	-2.407^b	0.016	0.553	0.580
Whether both couples are employed	0.897	0.370	0.525	0.600
Whether have old people over 60 years of age in the household	0.045	0.964	-2.229^b	0.026

Notes: [a,b,c] *indicates that the significance level is lower than 1%, 5%, and 10% (P<0.01, 0.05, 0.1).*

and license card possession were highly correlated with motorized travel rates and further influenced travel-related air pollution exposure.

A positive answer to the question of "whether there are children under 18 years of age in the household" significantly reduced individuals' exposure to $PM_{2.5}$, while the positive answer to "Whether have old people over 60 years of age in the household" would significantly lower individuals' exposure to CO. This would be due to the differentiation in duration and share of the daily travel modes for the different family structure and corresponding family accompanying duties. Income and education level contributed positively to $PM_{2.5}$ exposure and brought higher $PM_{2.5}$-related health risks. Marital status and occupation had no significant effect on $PM_{2.5}$ or CO exposure.

13.5 Discussion and conclusion

This study explored mobility and health research by integrating air pollution concentrations of $PM_{2.5}$ and CO within different transport microenvironments: the air inhalation rates, and accurate travel pattern. The difference between subgroups of gender, education, income, marital status, occupation, accommodation type, housing source, driver's license, and Beijing *hukou* were examined through an ANOVA test. The associations of travel mode preference and individual exposure to air pollution were investigated using a structural equation model. Endogenous variables of motorized travel rates, travel duration, times, commuting distance, individual exposure to $PM_{2.5}$ and CO per day and exogenous variables regarding participants' socioeconomic status were used in the model.

Results indicate that mid-income suburbanites suffer the highest $PM_{2.5}$ and CO exposure per day, while high-income earners suffer the lowest. Rental tenants experience higher exposure to both $PM_{2.5}$ and CO than do house owners. Unmarried people share higher levels of $PM_{2.5}$ exposure per day than married people, and driver's license holders suffer higher levels of CO exposure per day than those who do not have licenses. This offers an insight into health inequalities and social justice. Mid-income people and rental tenants living in the suburbs, who commute long distances and have poorer living conditions, are also exposed to higher levels of air pollution. Home-work spatial mismatch resulted from suburbanization seems to not benefit mid-income people or rental tenants, placing them at risk of being socially disadvantaged because of exposure to air pollution from daily travel.

Acknowledgment

This research was made possible by the support of China's 12th five-year National Key Technology R&D Project "Space-time behavior and smart travel service platform".

References

Adams, H.S., Nieuwenhuijsen, M.J., Colvile, R.N., et al., 2001. Fine particle ($PM_{2.5}$) personal exposure levels in transport microenvironments, London, UK. Sci. Total Environ. 279, 29–44.

Ando, M., Katagiri, K., Tamura, K., et al., 1996. Indoor and outdoor air pollution in Tokyo and Beijing supercities. Atmos. Environ. 30 (5), 695–702.

Aubrey, T., Tefft, B., Currie, R., 1995. Public attitudes and intentions regarding tenants of community mental health residences who are neighbors. Commun. Mental Health J. 31 (1), 39–52.

Beijing Environmental Protection Bureau, 2014. Beijing Environmental Status Bulletin in 2013.

Bell, E.M., Hertz-Picciotto, I., Beaumont, J., 2001. A case-control study of pesticides and fetal death due to congenital anomalies. Epidemiology 12, 148–156.

Berke, E.M., Koepsell, T.D., Moudon, A.V., et al., 2007. Association of the built environment with physical activity and obesity in older persons. Am. J. Public Health 97 (3), 486–492.

Bollen, K.A., Long, J.S., 1993. Testing Structural Equation Models. SAGE, California.

Brown, T., McLafferty, S., Moon, G. (Eds.), 2010. A Companion to Health and Medical Geography. Wiley-Blackwell, Chichester.

Colvile, R.N., Kaur, S., Britter, R., et al., 2004. Sustainable development of urban transport systems and human exposure to air pollution. Sci. Total Environ. 334, 481–487.

Cortese, D.A., Rodarte, J.R., Rehder, K., et al., 1976. Effect of posture on the single-breath oxygen test in normal subjects. J. Appl. Phys. 41 (4), 474–479.

Costabile, F., Allegrini, I., 2008. A new approach to link transport emissions and air quality: An intelligent transport system based on the control of traffic air pollution. Environ. Modell. Softw. 23 (3), 258–267.

Dahlgren, G., Whitehead, M., 1991. The main determinants of health" model, version accessible. European strategies for tackling social inequities in health: Levelling up Part 2. World Health Organization.

Davidson, C.I., Lin, S.F., Osborn, J.F., et al., 1986. Indoor and outdoor air pollution in the Himalayas. Environ. Sci. Technol. 20 (6), 561–567.

Dhondt, S., Kochan, B., Beckx, C., et al., 2013. Integrated health impact assessment of travel behaviour: model exploration and application to a fuel price increase. Environ. Int. 51, 45–58.

Duan, N., 1981. Micro-Environment Types: A Model for Human Exposure to Air Pollution. Department of Statistics, Stanford University.

Frank, L.D., Engelke, P.O., 2001. The built environment and human activity patterns: exploring the impacts of urban form on public health. J. Plan. Lit. 16 (2), 202–218.

Frank, N.H., Hunt Jr., W.F., Cox, W.M., 1977. Population exposure: an indicator or air quality improvement. Paper No. 77-44.2. In: Presented at the 70th Annual Meeting of the Air Pollution Control Association, Toronto, Ontario, Canada.

Frank, L.D., Sallis, J.F., Conway, T.L., et al., 2007. Many pathways from land use to health: associations between neighborhood walkability and active transportation, body mass index, and air quality. J. Am. Plan. Assoc. 72 (1), 75–87.

Fugas, M., 1975. Assessment of total exposure to an air pollutant. In: Proc. Int. Conf. Environ. monitoring. Las Vegas, Nevada. vol. 2. Institute of Electrical and Electronics Engineers, Inc, New York, pp. 38–45.

Gatrell, A.C., 2011. Mobilities and Health. Ashgate, Surrey, UK.

Gatrell, A.C., Elliott, S.J., 2009. Geographies of Health—An Introduction, second ed. Blackwell Publishing, Oxford.

Goddard, M., Smith, P., 2001. Equity of access to health care services: theory and evidence from the UK. Soc. Sci. Med. 53 (9), 1149–1162.

Godin, G., Wright, G., Shephard, R.J., 1972. Urban exposure to carbon monoxide. Arch. Environ. Health: Int. J. 25 (5), 305–313.

Gulliver, J., Briggs, D.J., 2005. Time-space modeling of journey-time exposure to traffic-related air pollution using GIS. Environ. Res. 97 (1), 10–25.

Hatzopoulou, M., Hao, J.Y., Miller, E.J., 2011. Simulating the impacts of household travel on greenhouse gas emissions, urban air quality, and population exposure. Transportation 38 (6), 871.

Horie, Y., Eldon, J.A., 1979. Utility of fixed-station air monitoring data for assessing human exposure to air pollution/specialty conference on quality assurance in air pollution measurement, air pollution control association. . New Orleans, LA.

Huang, J., Deng, F.R., Wu, S.W., et al., 2012a. Comparisons of personal exposure to PM $_{2.5}$ and CO by different commuting modes in Beijing, China. Sci. Total Environ. 425, 52–59.

Janssen, N.A.H., Hoek, G., Brunekreef, B., et al., 1998. Personal sampling of particles in adults: relation among personal, indoor, and outdoor air concentrations. Am. J. Epidemiol. 147 (6), 537–547.

Jones, K., Moon, G., 1993. Medical geography: taking space seriously. Prog. Hum. Geogr. 17 (4), 515–524.

Katsouyanni, K., Touloumi, G., Samoli, E., et al., 2001. Confounding and effect modification in the short-term effects of ambient particles on total mortality: results from 29 European cities within the APHEA2 project. Epidemiology 12 (5), 521–531.

Kaur, S., Nieuwenhuijsen, M.J., Colvile, R.N., 2007. Fine particle matter and carbon monoxide exposure concentrations in urban street transport microenvironments. Atmos. Environ. 41, 4781–4790.

Kearns, R.A., 1993. Place and health: towards a reformed medical geography. Prof. Geogr. 45 (2), 139–147.

Kwan, M.P., 2012. The uncertain geographic context problem. Ann. Assoc. Am. Geogr. 102 (5), 958–968.

Kwan, M.P., 2013. Beyond space (as we knew it): toward temporally integrated geographies of segregation, health, and accessibility: space–time integration in geography and GIScience. Ann. Assoc. Am. Geogr. 103 (5), 1078–1086.

Langer, B.C.A., Frösner, G.G., 1996. Relative Importance of the Enterically Transmitted Human Hepatitis Viruses Type A and E as a Cause of Foreign Travel Associated Hepatitis/Imported Virus Infections. Springer, Vienna, pp. 171–179.

Muthén, B., 1984. A general structural equation model with dichotomous, ordered categorical, and continuous latent variable indicators. Psychometrika 9 (1), 115–132.

Nuckols, J.R., Ward, M.H., Jarup, L., 2004. Using geographic information systems for exposure assessment in environmental epidemiology studies. Environ. Health Perspect. 112 (9), 1007–1015.

Ott, W.R., Eliassen, R., 1973. A survey technique for determining the representativeness of urban air monitoring stations with respect to carbon monoxide. J. Air Pollut. Control Assoc. 23, 685–690.

Ott, W.R., Mage, D.T., 1974. Method for simulating the true human exposure of critical population groups to air pollutants. In: Proceedings of the International Symposium: Recent Advances in Assessing the Health Effects of Environmental Pollution, Paris, France.

Papas, M.A., Alberg, A.J., Ewing, R., et al., 2007. The built environment and obesity. Epidemiol. Rev. 29 (1), 129–143.

Repace, J.L., Lowrey, A.H., 1980. Indoor air pollution, tobacco smoke, and public health. Science 208 (4443), 464–472.

Reynolds, P., Von Behren, J., Gunier, R.B., et al., 2003. Childhood cancer incidence rates and hazardous air pollutants in California: an exploratory analysis. Environ. Health Perspect. 111, 663–668.

Richardson, D.B., Volkow, N.D., Kwan, M.-P., et al., 2013a. Spatial turn in health research. Science 339 (6126), 1390–1392.

Riojas-Rodriguez, H., Escamilla-Cejudo, J.A., Gonzalez-Hermosillo, J.A., et al., 2006. Personal $PM_{2.5}$ and CO exposure and heart rate variability in subjects with known ischemic heart disease in Mexico City. J. Expo. Sci. Environ. Epidemiol. 16, 131–137.

Saelens, B.E., Sallis, J.F., Frank, L.D., 2003. Environmental correlates of walking and cycling: findings from the transportation, urban design, and planning literatures. Ann. Behav. Med. 25 (2), 80–91.

Samet, J.M., Dominici, F., Curriero, F.C., et al., 2000. Fine particulate air pollution and mortality in 20 US cities, 1987–1994. N. Eng. J. Med. 343 (24), 1742–1749.

Setton, E., Marshall, J.D., Brauer, M., et al., 2010. The impact of daily mobility on exposure to traffic-related air pollution and health effect estimates. J. Expo. Sci. Environ. Epidemiol. 21, 42–48.

Shaw, J., Hesse, M., 2010. Transport, geography and the 'new' mobilities. Trans. Inst. Br. Geograph. 35 (3), 305–312.

Tsai, D.H., Wu, Y.H., Chan, C.C., 2008. Comparisons of commuter's exposure to particulate matters while using different transportation modes. Sci. Total Environ. 405, 71–77.

United States Environmental Protection Agency (US EPA), 2009. U.S. EPA Exposure Factors Handbook.

Wang, F., Luo, W., 2005. Assessing spatial and nonspatial factors for healthcare access: towards an integrated approach to defining health professional shortage areas. Health Place 11 (2), 131–146.

World Health Organization (WHO), 1947. Chronicle of the WHO. WHO, Geneva.

Wright, G.R., Jewczyk, S., Onrot, J., et al., 1975. Carbon monoxide in the urban atmosphere: hazards to the pedestrian and the street-worker. Arch. Environ. Health Int. J. 30 (3), 123–129.

Further reading

Finkelstein, M.M., Jerrett, M., DeLuca, P., et al., 2003. Relation between income, air pollution and mortality: a cohort study. Can. Med. Assoc. J. 169 (5), 397–402.

Frank, L.D., Schmid, T.L., Sallis, J.F., et al., 2005. Linking objectively measured physical activity with objectively measured urban form: findings from SMARTRAQ. Am. J. Prev. Med. 28 (2), 117–125.

Fugaš, M., 1986. Assessment of true human exposure to air pollution. Environ. Int. 12 (1), 363–367.

Gualtieri, G., Tartaglia, M., 1998. Predicting urban traffic air pollution: a GIS framework. Transp. Res. Part D: Transp. Environ. 3 (5), 329–336.

Hoek, G., Brunekreef, B., Goldbohm, S., et al., 2002. Association between mortality and indicators of traffic-related air pollution in the Netherlands: a cohort study. Lancet 360 (9341), 1203–1209.

Horie, Y., Stern, A.C., 1976. Analysis of population exposure to air pollution in New York-New Jersey-Connecticut Tri-State Region. Environmental Protection Agency, Office of Air and Waste Management, Office of Air Quality Planning and Standards.

Huang, J., Deng, F.R., Wu, S.W., et al., 2012b. Comparisons of personal exposure to PM 2.5 and CO by different commuting modes in Beijing, China. Sci. Total Environ. 425, 52–59.

Kearns, R., Moewaka-Barnes, H., McCreanor, T., 2009. Placing racism in public health: a perspective from Aotearoa/New Zealand. GeoJournal 74 (2), 123–129.

Kearns, R., Moon, G., 2002. From medical to health geography: novelty, place and theory after a decade of change. Prog. Hum. Geogr. 26 (5), 605–625.

Kwan, M.P., 2004. GIS methods in time-geographic research: geocomputation and geovisualization of human activity patterns. Geografiska Annaler: Ser B: Hum. Geogr. 86 (4), 267–280.

Kwan, M.P., 2009. From place-based to people-based exposure measures. Soc. Sci. Med. 69 (9), 1311–1313.

Learmonth, A., 1988. Disease Ecology: An Introduction Continued. Basil Blackwell Ltd.

Lebret, E., Briggs, D., Van Reeuwijk, H., et al., 2000. Small area variations in ambient NO$_2$ concentrations in four European areas. Atmos. Environ. 34 (2), 177–185.

Lee, S.C., Chang, M., 2000. Indoor and outdoor air quality investigation at schools in Hong Kong. Chemosphere 41 (1), 109–113.

Lo, C.P., Quattrochi, D.A., 2003. Land-use and land-cover change, urban heat island phenomenon, and health implications: a remote sensing approach. Photogramm. Eng. Remote. Sens. 69 (9), 1053.

Monn, C., 2001. Exposure assessment of air pollutants: a review on spatial heterogeneity and indoor/outdoor/personal exposure to suspended particulate matter, nitrogen dioxide and ozone. Atmos. Enviorn. 35, 1–32.

Morrison, D.S., Petticrew, M., Thomson, H., 2003. What are the most effective ways of improving population health through transport interventions? Evidence from systematic reviews. J. Epidemiol. Community Health 57 (5), 327–333.

Ott, W.R., 1982. Concepts of human exposure to air pollution. Environ. Int. 7 (3), 179–196.

Patterson, K.D., Pyle, G.F., 1983. The diffusion of influenza in sub-Saharan Africa during the 1918–1919 pandemic. Soc. Sci. Med. 17 (17), 1299–1307.

Patz, J.A., Daszak, P., Tabor, G.M., et al., 2004. Unhealthy landscapes: policy recommendations on land use change and infectious disease emergence. Environ. Health Perspect. 112 (10), 1092.

Pope, C.A., Burnett, R.T., Thun, M.J., et al., 2002. Lung cancer, cardiopulmonary mortality, and long-term exposure to fine particulate air pollution. JAMA 287 (9), 1132–1141.

Pope, C.A., Burnett, R.T., Thurston, G.D., et al., 2004. Cardiovascular mortality and long-term exposure to particulate air pollution epidemiological evidence of general pathophysiological pathways of disease. Circulation 109 (1), 71–77.

Ritz, B., Wilhelm, M., Zhao, Y., 2006. Air pollution and infant death in southern California, 1989–2000. Pediatrics 118 (2), 493–502.

Xiang, H., Nuckols, J.R., Stallones, L., 2000. A geographic information assessment of birth weight and crop production patterns around mother's residence. Environ. Res. 82, 160–167.

Index

Note: Page numbers followed by *f* indicate figures, *t* indicate tables, and *b* indicate boxes.

Multilayer perceptions (MLP), 124
Multivariate linear regression, 207–208
Multivariate probability distribution function, 95
Multivariate regression method, 140–141
Municipal waste management, 225–226

N

NARR. *See* North American Regional Reanalysis (NARR)
National Air Toxics Assessment (NATA), 252–253
National Ambient Air Quality Standards (NAAQS), 1, 218
National Center for Atmospheric Prediction (NCEP), 112
National Center for Atmospheric Research (NCAR), 202
National Climatic Data Center, 112
National Elevation Dataset (NED), 112
National Emissions Inventory (NEI), 112, 257–258, 261–262
National Health and Nutrition Examination Surveys (NHANES), 6–7
National Health Interview Survey, 230
National Highway Planning Network, 256–257
National Land Cover Database (NLCD), 112
National Map, 112
National Oceanic and Atmospheric Administration, 202
National Program of Cancer Registries (NPCR), 70
Natural aerosols, false-color map, 221, 223f
Neighborhood air pollution exposure impact, 92–93
Neighborhood-level SES variable, 245
Neural network (NN), 109–110, 121–124
Neurocognitive impairment, 4
Neurological disease, 232, 260
New-generation nighttime lights image product, 138
Nitric oxide (NO), 224
Nitrogen dioxide (NO_2), 224
  exposure, 197–199
  ozone production, 15–16
  public health impacts, 16–17
  spatiotemporal variation, 16
Nitrogen oxides (NOx), 7–8, 11–12, 15–16, 221, 241
NLCD. *See* National Land Cover Database (NLCD)
Noncancer chronic health effects, 231–232, 260
Nonlinear support vector machine, 120–121
Nonpoint source, 252–253
Nonroad mobile air toxics, 252–253

Nonroad respiratory risk, 252–253
Normalized difference vegetation index (NDVI), 112, 138
North American Land Data Assimilation System (NLDAS), 112
North American Regional Reanalysis (NARR), 112
NPCR. *See* National Program of Cancer Registries (NPCR)
NTL, 140, 145–147
Nucleation, 12
Numerical simulation model, 197–203

O

Obesity, school performance, 247, 253–254
Obesogens. *See* Metabolic disrupting (MD) chemicals
Oceanic CO emission, 2–3
One-chemical one-health outcome model, 279
Operational Land Imager (OLI), 200
Ordered categorical variable, 302–303
Outdoor air pollution, 169, 219–220, 245–246
  and birth outcomes, 232–235
  and cancer, 229
  and cardiorespiratory diseases, 228, 229f
  commuting mode, 225, 225f
  and excessive mortality, 226–227, 227–228f
  industrial air pollution, 227–228
  and mental health, 245–246
  motor vehicle emissions, 225
  $PM_{2.5}$ and ozone, 226
  and school performance, 248–253
  short-term and long-term exposures, 226
  transport, 225–226
Outdoor air quality, 225–226
Ozone (O_3), 92, 100, 104, 218, 221–224, 241–242
  ground-level ozone, 7–8
  public health impacts, 9–11, 169
  spatiotemporal variations, 8–9
  stratospheric ozone, 7–8
  tropospheric ozone, 7–8
Ozone Monitoring Instrument (OMI), 18–19

P

Parallel programming techniques, 154, 171–172
Parameter-elevation Regressions on Independent Slopes Model (PRISM), 137–138
Parameter model, 96–98, 100, 102–103
Partial autocorrelation function (PACF), 53–66, 75, 76f
Partial dependence plot, 141, 145–147, 146f